Contemporary Bamboo Architecture in China

K. W. Liu · Q. F. Xu · G. Wang · F. M. Chen ·
Y. B. Leng · J. Yang · K. A. Harries

Contemporary Bamboo Architecture in China

清華大學出版社
TSINGHUA UNIVERSITY PRESS

Springer

K. W. Liu
International Bamboo and Rattan Organisation (INBAR)
Beijing, China

Q. F. Xu
Shanghai Research Institute of Building Sciences Co., Ltd. (SRIBS)
Shanghai, China

G. Wang
International Center for Bamboo and Rattan (ICBR)
Beijing, China

F. M. Chen
International Center for Bamboo and Rattan (ICBR)
Beijing, China

Y. B. Leng
Shanghai Research Institute of Building Sciences Co., Ltd. (SRIBS)
Shanghai, China

J. Yang
Department of Civil Engineering
Tsinghua University
Beijing, China

K. A. Harries
University of Pittsburgh
Pittsburgh, PA, USA

ISBN 978-981-16-8308-4 ISBN 978-981-16-8309-1 (eBook)
https://doi.org/10.1007/978-981-16-8309-1

Jointly published with Tsinghua University Press
The print edition is not for sale in China (Mainland). Customers from China (Mainland) please order the print book from: Tsinghua University Press.

This Springer imprint is published by the registered company Springer Nature Singapore Pte Ltd.
The registered company address is: 152 Beach Road, #21-01/04 Gateway East, Singapore 189721, Singapore

Foreword by Ali Mchumo

Bamboo is one of the fastest growing species on earth, and woody bamboos have been used in buildings for thousands of years. For example, in the Sidama Region of southern Ethiopia, the majority of local farmers still live in traditional bamboo houses, commonly known as "Sidama" which account for 1% of the total housing in Ethiopia. Bamboo's relatively lightweight and high natural strength makes it an excellent construction material, earning it the name "vegetal steel" among architects around the world. In Ecuador, an assessment of the effects of the 2016 earthquake on local buildings emphasized that there was little or no impact on structures built with bamboo and wood. With the development of modern building technology, the application of bamboo in modern construction is gradually increasing, from small dwellings to medium-sized public buildings, as well as large outdoor landscape projects, even involving urban infrastructure construction. The Guadua bamboo pavilion constructed for the 2000 EXPO in Hannover, Germany, opened the eyes of many in the Western world. In 2004, the use of fire-resistant bamboo material for the 230,000 m^2 ceiling of the Madrid-Barajas Airport in Spain was the starting point of a new chapter in the application of bamboo as an interior decorative material. Meanwhile, in Bali, Indonesia, the "Green School" perfectly combined this environmentally friendly material with education for young generations. There is now a Green Village of majestic bamboo buildings nearby. Bamboo is now used throughout Europe for interior design, including the completed City Life shopping complex in Milan, which was designed by the late Zahad Hadid.

Under the background of global bamboo construction development, China, which possesses the largest bamboo forest area in the world, has made significant contributions to promote the development of bamboo construction materials and bamboo constructions. China produces a large amount of bamboo construction materials every year and exports them to many countries. For example, the annual export volume of China's bamboo flooring, for which statistics are available, accounts for more than 90% of the total global export trade. In the past 10 years, China has begun to use bamboo in large quantities in many public buildings and urban environments. For instance, the indoor and outdoor decorative materials in art galleries, theaters, hotels and high-end office buildings, as well as some small public transport facilities, and even all the outdoor flooring of the Civic Service Center, newly built in Xiong'an New Area, Hebei Province, all adopt high-durability bamboo materials. In addition, bamboo, as a structural material, has begun to develop in some small-sized buildings, such as 1–3-story bamboo-structured villas and public spaces, more than 90% of which are made of bamboo materials.

China has also adopted bamboo construction in sustainable rural construction. The First International Bamboo Construction Biennale was launched in 2016, at which 18 characteristic bamboo constructions were built, making a small remote village with rich bamboo resources a cultural tourism destination featuring bamboo constructions, thus greatly improving the income level of locals. Moreover, more and more young people are willing to return from the big cities to their hometowns for employment and entrepreneurship which provides vitality and a sustainable exploration path for the future development of the countryside.

Contemporary Bamboo Architecture in China was published in Chinese in May 2019 with a free full-text e-book issued simultaneously on INBAR's official Web site. Through the INBAR platform, these successful stories of bamboo architecture in China have been spread to more than 50 countries around the world. Through a large number of exquisite pictures, non-Chinese-language readers can understand stories of classic bamboo architecture. INBAR received many positive comments from our partners that encouraged us to publish an updated English version for our international audience of those interested in bamboo architecture, including researchers, designers, policy and standard makers and developers.

In the process of updating and translating the Chinese-language version, on behalf of INBAR, I would like to express my gratitude to Shanghai Research Institute of Building Sciences Co., Ltd., International Centre for Bamboo and Rattan, Tsinghua University, the University of Pittsburgh, Tsinghua University Press and Springer Nature for their support.

Last but not least, I hope that everyone can get useful information from this book and be inspired to continue exploring the potential of bamboo construction development around the world.

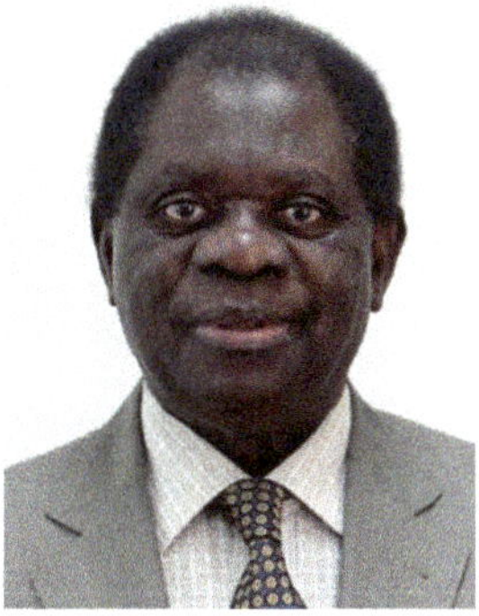

Dar es Salaam, Tanzania
March, 2021

Ali Mchumo
Director-General of International
Bamboo and Rattan Organisation
(INBAR)

Foreword by Simón Veléz

Since the creation of INBAR in 1997, 24 years has passed. It has been very impressive to see the evolution and acceptance of these materials. Rattan only exists in Asia, and it has been believed that bamboo was also limited. However, in Central and South America, starting from Mexico and all the way to Argentina, with an epicenter in Colombia and Ecuador, native bamboo has had enormous importance not only in the landscape, but also in popular culture. Unfortunately, it also has a stigma attached to it that it is the "wood of the poor people" and the poor are the ones who most reject bamboo. For someone in Asia, one does not understand life without bamboo: It is in the food, textiles, in the fibers used for basket weaving, in the artisanal crafts, mat making and as temporary supports and scaffolding in the construction industry.

Environmental concerns and global warming are now integral parts of any political, economic and social agenda. The construction industry is one of the biggest contributors to pollution of the earth—through the transport, mining and transformation of construction materials.

Bamboo in construction and rattan in furniture are clean and natural alternatives so that the construction industry no longer abuses minerals in a disproportionate manner. We are exaggerating in the use of concrete, steel, brick, glass, and what I consider even worse, whatever construction materials are derived from petroleum. We must become more vegetarian.

This book shows us a very complex evolution of how a humble vegetal material with vernacular use and an association with poverty has already been converted into an alternative for construction that competes equally with, and in some cases supersedes, wood, steel and concrete. The fibers of bamboo are much stronger than steel and are equivalent to other high-tech materials. When institutions such as INBAR, the academic world, the technology industry and the finance industry become interested in materials such as bamboo and rattan, revolutions are produced such as those that are shown in this book and tell us that the future is green.

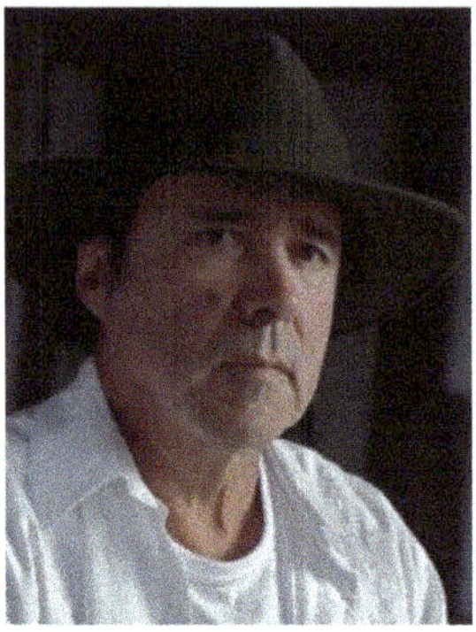

Bogotá, Colombia
March 2021

Simón Veléz
Chief Architect of Simón Veléz Architects

Foreword by Prof. Ying Hei Chui

The world is facing a serious challenge in terms of continued greenhouse gas (GHG) emission through various human actions. It is known that globally the construction sector is the third largest emitter of GHG into the atmosphere, accounting for 12% of the total emission. Increased use of low carbon footprint and renewable materials in construction will be an effective way to significantly reduce GHG emission. It is for this reason that wood has attracted worldwide attention in construction over the last decade because it is the only renewable material among the common construction materials. Wood also requires significantly less energy and emits substantially less pollutants during the manufacturing process, compared with steel, concrete and masonry.

Recently, bamboo is also attracting the attention of designers and architects as a viable construction material. This is not surprising because from a material property standpoint, bamboo has many of the same characteristics as wood. Bamboo offers even more environmental benefits than wood, in that it can produce more oxygen and biomass than conventional tree species. Obviously, bamboo has a long way to go before it reaches the same status as wood in terms of its acceptance as a common construction material. The wood products industry is more mature with well-established manufacturing processes and a regulatory framework that ensures the quality of end products and that the performance of structures built with wood products meet the requirements of building regulations and design standards. By comparison, use of bamboo in construction is still in its infancy.

An important step in the process of gaining acceptance and recognition of bamboo as a new construction material is the need to demonstrate to designers, developers, consumers and code authorities that bamboo can be used to construct safe, economical and aesthetically pleasing structures. A good example for bamboo to follow is the rapid development of cross laminated timber (CLT) as a product for constructing large and tall buildings. The interest in CLT increased by leaps and bounds about a decade ago after a few CLT buildings were designed and constructed by early adopters who believed in the product and persisted with their construction projects despite limited design guidelines for the product.

This book, *Contemporary Bamboo Architecture in China*, is an excellent resource for anyone interested in bamboo products and structures. The bamboo structures presented in Chap. 6 of this book, enhanced through the use of professionally taken color photographs, is a testimony to the creative ingenuity of their designers. These bamboo structures demonstrate the range of possibilities in architectural and structural designs with bamboo products. As a structural engineer myself, it is obvious that some of these impressive bamboo structures presented serious challenges to structural engineers in terms of performing rational structural analysis. One of the main features of the book is that the information presented is recent and up-to-date. The structures shown in Chap. 6 were constructed over the last 8 years.

I believe this book will have an impact on promoting the use of bamboo beyond the use of low-cost housing construction, not only in China but also internationally. Despite what the title implies, the scope of the book extends beyond China with structures from around the world illustrated in Chap. 6, and the discussion in various chapters provides an international context, most importantly for codes and standards in Chap. 4. In addition to showcasing various structures, this book also presents a wide range of technical topics. These topics include an introduction to bamboo properties, bamboo resource distribution around the world, the range of bamboo construction products commercially produced, the major types of bamboo structural systems, codes and standards development and organizations involved in bamboo promotion and research from around the world. The final chapter of the book discusses results from a strength–weakness–opportunity–threat (SWOT) analysis conducted to evaluate the potential of using bamboo in construction of residential, commercial and institutional structures.

The authors have done an excellent job in formatting and packaging the information in the book. Without a doubt, this book is the most comprehensive publication to date that addresses the use of bamboo in modern construction. Given the wide range of topics it covers, it is suitable as a textbook for university programs and can serve as an excellent reference for researchers, consultant structural engineers, practicing architects and government departments responsible for infrastructures.

I congratulate INBAR and its collaborating partners in not only taking the initiative to producing the first edition of the book in Chinese, but also the vision to translate the book into English, so that a wider audience interested in bamboo structures can benefit from this excellent publication.

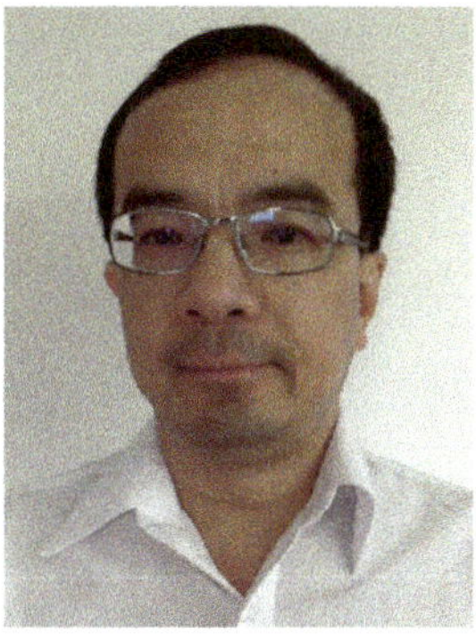

Edmonton, Canada
March 2021

Prof. Ying Hei Chui, Ph.D., P.Eng., FCAE, FIAWS
Chair, ISO Technical Committee 165 Timber Structures; NSERC Industrial Research Chair in Engineered Wood and Building Systems, University of Alberta, Canada

Preface

With the largest population in the world, China needs a large number of buildings to provide necessary support for people's work and life. However, large-scale urban construction, dominated by concrete and steel, results in increased carbon emissions and places a burden on energy resources and the environment. Sustainable urban development faces enormous challenges. In recent years, people have begun to explore more environmentally friendly building materials and methods, and the use of natural building materials is popular among urban planners and builders. However, due to the lack of forest resources in China and the need to import large amounts of wood from other countries, bamboo has garnered attention as a natural native alternative to wood.

China not only has the richest bamboo resources in the world, but also has a deep-rooted bamboo cultural tradition. How to make rational use of this natural material so that bamboo architecture can play its unique role in sustainable urban and rural construction has become a new direction of exploration.

Against the background of global bamboo architecture development, this book describes the distribution of bamboo forest and bamboo species for construction, the types and characteristics of bamboo materials for construction, the development history and research status of different forms of bamboo architecture. We go on to describe standards, relevant international organizations, research institutions and production and processing enterprises and typical cases. Starting from six aspects, this book systematically describes bamboo building development in China, analyzes the opportunities and challenges faced by the bamboo construction industry in China and provides guidance for the development of the bamboo construction industry in China.

In the first chapter, the main species of bamboo used for construction and the distribution of resources are presented. Chapter 2 focuses on the introduction of two types of bamboo materials which are widely used in the construction sector: full-culm bamboo and engineered bamboo. The chapter introduces each type from the perspective of three potential uses: structural load-bearing materials, enclosure and decorative materials and other functional materials. Chapter 2 also introduces the physical and mechanical properties of these bamboo materials with relevant test methods, as well as the research status of long-term performance, durability and fire resistance. At the same time, the application and research status of bamboo in other construction realms are briefly introduced, including bamboo scaffolding, bamboo-reinforced concrete and masonry, bamboo soil reinforcement, bamboo reinforcement for existing structures, etc. Chapter 3 introduces the development history and research status of full-culm and engineered bamboo structures. Chapter 4 systematically analyzes international standards related to bamboo structures, as well as the development of China's current national and provincial standards and industry and association standards. Chapter 5 briefly introduces the international organizations, research institutions and production and processing enterprises which are engaged in research, application and promotion of bamboo architecture in China. In Chap. 6, more than 70 exemplary (commercial) cases constructed mostly since 2014 are selected to provide a detailed overview of the use of bamboo as decorative and structural materials. In order to fully explore the potential of bamboo in engineering applications, the authors introduce the use of bamboo construction for transportation facilities (bridges, highway landscape fences and bus stations), landscape, water pipelines and urban municipal tunnels. The authors hope readers are inspired by these most vivid cases and experience the charm of modern Chinese bamboo architecture. In Chap. 7, the authors apply strength–weakness–opportunity–threat (SWOT) analysis to enumerate opportunities and challenges faced by China's modern bamboo construction industry. Approaches to leveraging opportunities and overcoming challenges are presented.

This book seeks to objectively describe commercial cases to showcase the status of the bamboo construction sector in China. No commercial endorsement is intended or should be implied in any instance. Please enjoy this book as it was intended: as a unique source of reference.

Beijing, China	K. W. Liu
Shanghai, China	Q. F. Xu
Beijing, China	G. Wang
Beijing, China	F. M. Chen
Shanghai, China	Y. B. Leng
Beijing, China	J. Yang
Pittsburgh, USA	K. A. Harries

Acknowledgements

In the process of writing and publishing this book, the authors want to express our sincerely gratitude to the leaders and colleagues of the authors' organizations; it is thanks to their advice and support that this book was successfully published. Thanks go to Ali Mchumo, Director-General of INBAR; Lu Wenming, Deputy Director-General of INBAR; Durai Jayaraman, Director of Global Programme of INBAR; Dr. Wu Junqi, Director of Communications and Outreach of INBAR; Zhu Lei, President of Shanghai Research Institute of Building Sciences Group Co., Ltd.; Li Xiangmin, Vice President of Shanghai Research Institute of Building Sciences Group Co., Ltd.; Fei Benhua, Executive Vice Director of International Centre for Bamboo and Rattan (ICBR) and President of China Bamboo Industry Association.

We also want to thank the following individuals and their universities, research institutes, design institutes, as well as other institutions for their support in helping to collect information (presented in no particular order): Dr. Fu Jinhe, Director of East Africa Region of INBAR; Chen Zhaoyuan, Professor of Tsinghua University and Fellow of Chinese Academy of Engineering; Li Zhiyong, Chief Expert of Green Economy of ICBR and Vice President of China Bamboo Industry Association; Liu Xianmiao, Associate Professor of ICBR; Hu Tao, Associate Director of Institute of Ornamental Plants and Landscape of ICBR; Huang Biao, Planning Designer of Institute of Ornamental Plants and Landscape of ICBR; Wang Zheng, Professor of Chinese Academy of Forestry (CAF); Yu Wenji, Professor of CAF; Yu Yanglun, Associate Professor of CAF; Gao Li, Associate Professor of CAF; Wang Fusheng, Professor of Nanjing Forestry University; Tan Tianfang, Chairman of the Bamboo Industry Committee of Federation of Hong Kong Industries; Dong Jianning, Head of the Architecture Division of Tongji Architectural Design (Group) Co., Ltd; Xiang Linfei, Director of the Collection and Editing Center of World Architecture; Zhang Yun, Director of PR Department of Urbanus Architecture & Design Inc; Chen Wenyun and Deng Wei, Officer of PR Department of Urbanus Architecture & Design Inc; Xie Xiaozhang, PR Manager of MAD Architects; Shang Jingjing, Public Relations of Vector Architects; Fu Xiaoming, Architect and Media Director of Land-based Rationalism D.R.C of China Architecture Design & Research Group; Lai Linli, General Manager of Shanghai Office of PES-Architects Ltd; Xu Minmin, Marketing

Coordinator of Shanghai Office of PES-Architects Ltd; Peng Baoning, Director of Technology Centre of Xi'an Construction Engineering Green Construction Group Co., Ltd; Du Hang, Business Manager of Design Consulting Division of UDS Co., Ltd; Han Lu, Partner of SUP Atelier of Architectural Design and Research Institute of Tsinghua University; Tian Xiaoxiao, Architect of Shanghai Archi-Scientific Creation Center of East China Architectural Design & Research Institute Co., Ltd; Wang Tao, Architect of Nicolas Godelet Architects; Li Xinyu, Project Manager of China Jingye Engineering Technology Company.

We thank the following manufacturing enterprises for their support (presented in no particular order): Wang Zhong, CEO of Hongyazhuyuan Science and Technology Company (Bamboo Era); Xin Junxian, Board Trustee of Bamboo Era; Li Yongjing, CEO of Bamboo Era; Wu Hui, Manager of Bamboo Era; Luo Hui, Manager of Bamboo Era; Lin Hai, CEO of Hangzhou Dasso Technology Co., Ltd. (Dasso); Liu Hongzheng, Vice General Manager of Dasso; Tang Gangyi, Manager of Dasso; Xiong Zhenhua, General Manager of Ganzhou Sentai Bamboo & Wood Co., Ltd.; Yan Ziwen, Ganzhou Sentai Bamboo & Wood Co., Ltd.; Cai Wei, CEO of Anji Zhujing Bamboo Industry Technology Co., Ltd.; Mo Yujin, Manager of Anji Zhujing Bamboo Industry Technology Co., Ltd; Xue Zhicheng, CEO of Hunan Taohuajiang Bamboo Technology Co., Ltd.; Ye Ling, CEO of Zhejiang Xinzhou Bamboo-Based Composite Technology Co., Ltd.; Zhu Qinghua, CEO of Hangzhou Fangqiao Transportation Facilities Co., Ltd.; Jin Peng, Vice General Manager of Hangzhou Fangqiao Transportation Facilities Co., Ltd.; Chen Yongjie, CEO of Jiangsu Jianzhu Green Bamboo Construction Technology Co., Ltd.

Thanks to the following colleagues for providing support for the translation of the original text from Chinese: Zhang Mingming and Cui Heqiong, Intern of Global Bamboo Construction Programme of INBAR; Charlotte King, Communications and Press Specialist of INBAR; Sang Wei, Officer of Member Country Affairs of INBAR.

The authors gratefully acknowledge the support of the China Academy of Engineering Consulting Project *Research on Development Strategy and Key Technologies of Bamboo Construction Sector in China towards 2035* (No. 2018-ZCQ-06). And the authors also thanks to the support of the China National Key R&D Program during the 13th Five-Year Plan Period (2017YFD0600800). The last author acknowledges the continued support of the Center for Nonconventional Materials and Alternative Technologies at the University of Pittsburgh (PITT-NOCMAT).

The authors acknowledge the contributions of Professor Chen Zhaoyuan and Professor Li Zhiyong to this work. Sadly, Professor Chen and Professor Li have since passed away; this work honors their memory.

Authors' Note for the Second Edition

This book was originally published in Chinese by INBAR, SRIBS and ICBR in early 2019. This edition is a translation and substantial revision of the original. Some new information has been added, especially in Chap. 2. Chapter 5 has been rewritten. Twenty-one new case studies have been added to Chap. 6. The book's entire content has been rewritten, and in some cases reorganized, with an international readership in mind. This book is necessarily a snapshot in time of a very rapidly developing industry and practice. Revisions were made in from late 2020 to mid-2021.

About the Authors' Organizations

INBAR
INTERNATIONAL BAMBOO
AND RATTAN ORGANISATION

INBAR: Established in 1997, the International Bamboo and Rattan Organisation (INBAR) is an intergovernmental development organization that promotes environmentally sustainable development using bamboo and rattan. It is currently made up of 48 member states. In addition to its Secretariat Headquarters in China, INBAR has five regional offices in Cameroon, Ecuador, Ethiopia, Ghana and India. Bamboo, the fast-growing grass plant, and rattan, the spiky climbing palm, can be important nature-based solutions to a number of pressing global challenges, for poverty alleviation, green trade, climate change mitigation and adaptation, resilient construction and environmental protection. INBAR's mission is to improve the well-being of producers and users of bamboo and rattan within the context of a sustainable bamboo and rattan resource base, by consolidating, coordinating and supporting strategic and adaptive research and development.

SRIBS

SRIBS: Established in 1958, Shanghai Research Institute of Building Sciences Group Co., Ltd (SRIBS) is a state-owned science and technology service company, providing services of engineering consulting, testing and design in building, transportation and environment. Developed from a research and development institute, SRIBS has been regarded as a China top 10 green building consultants, China Prefabrication Building Industrial Basement, is among the biggest construction Testing, Inspection and Certification, and environment assessment institutes in Shanghai, and the biggest engineering supervision providers in China. In China, SRIBS ensures the safety, quality and performance of a great number of projects, including Shanghai Center, Shanghai Pudong Airport, Yangshan Port, Shanghai Metro, Hangzhou Bay Bridge, etc.

ICBR
INTERNATIONAL CENTRE FOR BAMBOO AND RATTAN

ICBR: Established in 2000, the International Centre for Bamboo and Rattan (ICBR) is a non-profit research institution affiliated to the National Forestry and Grassland Administration (NFGA) in China. The mission of ICBR is to build direct supports to and cooperate with INBAR, the first intergovernmental international organization headquartered in China, for helping INBAR's better fulfilling its Host Country Agreement, as well as promoting the sustainability of bamboo and rattan industry development both in China and across the world. ICBR mainly carries out six key research fields of bio-resources in preservation, cultivation, improvement, processing and utilization. Since its establishment, ICBR has made a series of achievements on biomass materials, genetic science, bio-resources chemistry, resource cultivation and physiological/ecological studies, green economy, garden flowers and landscape.

Tsinghua University: Tsinghua University is one of the most prestigious and influential universities in China. Since its establishment in 1911, Tsinghua University has developed into a comprehensive, research-intensive university, covering sciences, engineering, humanities, law, medicine, economics, management and art. Through the pursuit of education and research at the highest level of excellence, Tsinghua is developing innovative solutions that will help solve pressing problems in China and the world. Several faculties, such as architecture, civil engineering and environment, focus on frontier issues of future human habitats and support smarter and more sustainable urban–rural development.

University of Pittsburgh: The University of Pittsburgh is a state-related research university, founded as the Pittsburgh Academy in 1787. Pitt is a member of the Association of American Universities (AAU), which comprises 63 preeminent doctorate-granting research institutions in North America. Since 1846, the University of Pittsburgh's Swanson School of Engineering has developed innovative processes and designs that have shaped our state, our country and our world. Swanson School faculty and students are on the forefront of developing solutions to create a better future and continue its founding commitment. The Swanson School focuses on our health, our planet and the ingenuity that keeps us competitive with recognized programs in bioengineering, sustainability and energy. Nanotechnology, manufacturing and product innovation are also critical strategic initiatives.

Contents

About the Authors

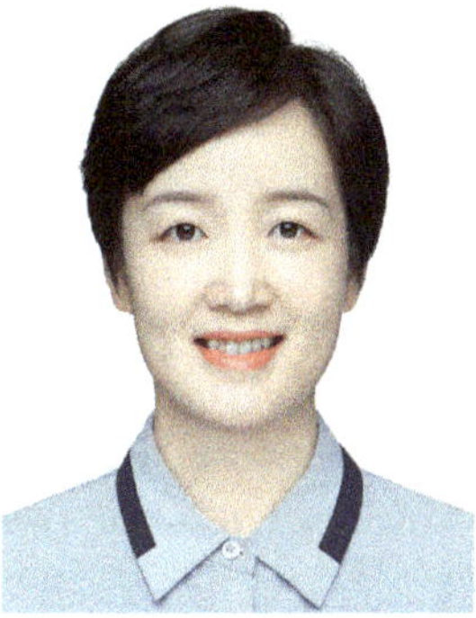

Mrs. K. W. Liu is Coordinator of Global Bamboo Construction Programme at the International Bamboo and Rattan Organization (INBAR). She was born in 1981 in Sichuan, China. She graduated from Beijing Jiaotong University with master degree in Structural Engineering and is now a Ph.D. candidate of the School of Civil Engineering at Tsinghua University. Since 2008, she has been working on promoting bamboo construction around the world. She managed around 20 international bamboo construction projects in Asia, America, Africa and Europe. She is managing the INBAR Bamboo Construction Task Force, comprising more than 30 bamboo construction experts from around 20 countries. As Convener of Working Group 12 on Structural Use of Bamboo within the Timber Structures Technical Committee of International Organization for Standardization (ISO TC165 WG12), she is working on the development and revision of five international standards of bamboo construction currently. She is also actively participating in the development of six technical standards for bamboo construction in China. Her academic achievements include more than 20 peer-reviewed papers, the book *Contemporary Bamboo Architecture in China* in Chinese as the first author, the academic conference proceedings *Modern Engineered Bamboo Structures* in English as one editor-in-chief, and the Chinese-language version of the English book *Sustainable Building in Practice: What the Users Think* as the first translator. She is one of the experts in three committees "Timber and Composite Structures Committee" of China Association for Engineering Construction Standardisation, "Timber and Bamboo Structures Committee" and "Sustainable Civil Engineering Committee" organized by Chinese Society for Urban Studies in China.

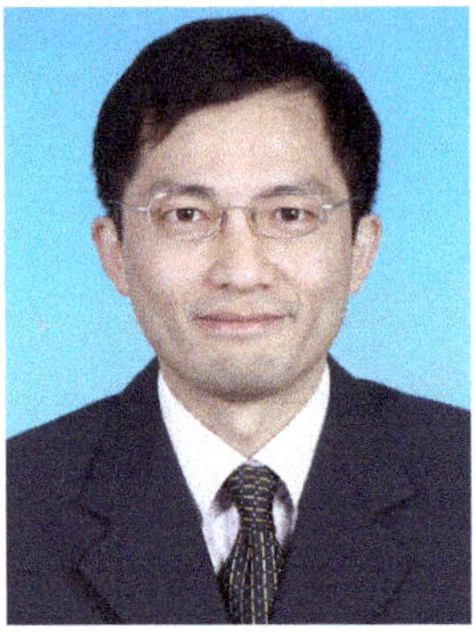

Dr. Q. F. Xu is Chief Engineer of Shanghai Research Institute of Building Sciences Co., Ltd (SRIBS) and Academic Leader of Shanghai Key Laboratory of Engineering Structure Safety. He is Professor of Engineering, Shanghai Subject Chief Scientist, National First-Class Registered Structural Engineer and Registered Consulting Engineer. He was born in 1973 in Jiangsu, China. In 2001, he graduated from Southeast University with a doctoral degree in structural engineering. He has been engaged in the research and technical service of wood and bamboo structure design and construction, performance improvement, maintenance and reinforcement for over 20 years. He is Member of INBAR Construction Task Force and ISO TC 165. His academic achievements include more than 150 academic papers including more than 50 SCI/EI cited papers; six editor-in-chief technical standards and 12 participated technical standards; 14 authorized national invention patents; five Second-Class Prize and two Third-Class Prize of Shanghai Scientific and Technological Progress Award and two First-Class Prize, one Second-Class Prize and one Third-Class Prize of Huaxia Architectural Science and Technology Award.

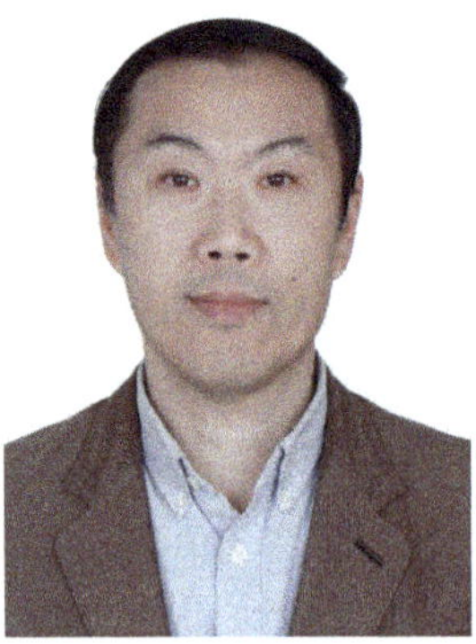

Dr. G. Wang is Professor and Head of Research Group of Bamboo Fiber Composite Materials of New Biomass Materials Institute at International Center for Bamboo and Rattan (ICBR). He is a provincial and ministerial candidate for "Millions of Talents Project in the New Century," and project leader of the 13th Five-Year Plan in China. Born in 1965, Harbin, Heilongjiang Province, he graduated from Northeast Forestry University in 1988 with a bachelor degree and from the Department of Wood Science and Technology at the Chinese Academy of Forestry in 2003 with a doctorate degree in engineering. He has been engaged in processing technology research and product development of bamboo and wood composite materials for a long time. He presided over the completion of more than 30 national projects related to bamboo and wood structural materials and bamboo fiber-based composites. His academic achievements include more than 140 academic papers including more than 50 SCI/EI cited papers; ten editor-in-chief national or industrial standards; one editor-in-chief monograph and three participated monographs, nine provincial and ministerial scientific and technological appraisal and recognition achievements and 20 authorized national patents; one First Prize of National Scientific and Technological Progress Award, two First Prize and two Second Prize of Liang Xi Forestry Science and Technology Award and two Second Prize of Society of Wood Science & Technology.

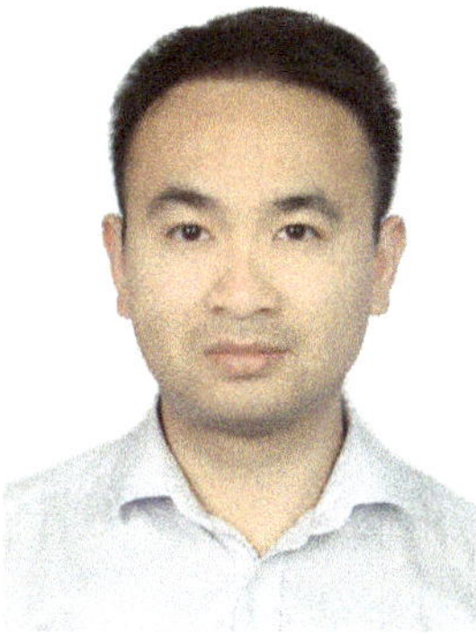

Dr. F. M. Chen is Associate Professor of ICBR, Visiting Scholar at the Swiss Federal Institute of Technology in Zurich and Secretary of National Key Research Projects. He was born in 1985, Anhui, China. He graduated in 2014 from the Department of Science and Engineering of Bamboo-Based Composite Material at the Chinese Academy of Forestry with a doctorate and his tutor is Professor Jiang Zehui. He is engaged in the research and development of new bamboo fiber composite materials and their application in prefabricated buildings. He presided over and participated in six national and provincial-level projects. His academic achievements include more than 30 peer-reviewed papers, including 20 SCI/EI cited papers; co-authored two English and one Chinese monographs; ten authorized national patents for inventions and practical new models; one Scientific and Technological Achievement recognized by the State Forestry and Grassland Administration of China, one Scientific and Technological Achievement of China Society of Forestry, one First Prize of Liang Xi Forestry Science and Technology Award and one Third Prize of Liang Xi Excellent Youth Paper Award.

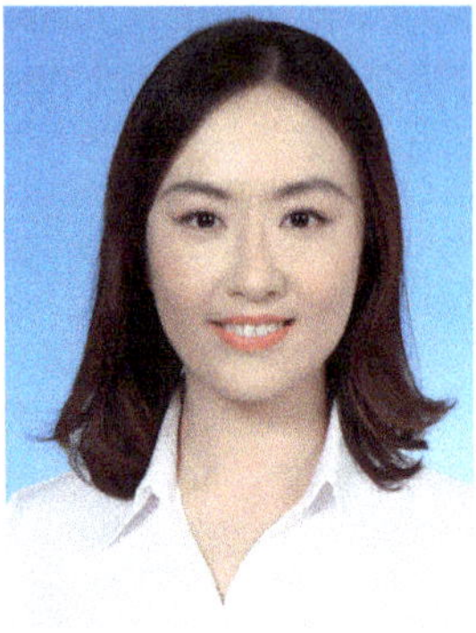

Dr. Y. B. Leng is Senior Engineer of SRIBS. She is National First-Class Registered Structural Engineer and Registered Consulting Engineer. She was born in 1988 in Anhui, China. In 2017, she graduated from Shanghai Jiao Tong University with a doctoral degree in structural engineering. She is currently a researcher of SRIBS, mainly engaged in the research and technical service of wood and bamboo structural design, construction and durability. She has presided over and participated in more than ten national, provincial and ministerial projects, and her academic achievements include more than 40 academic papers including 15 SCI/EI cited papers; six participated technical standards and 15 applied or authorized national invention patents; and sponsored by Shanghai Rising-Star program. She is also a member of ISO TC 165.

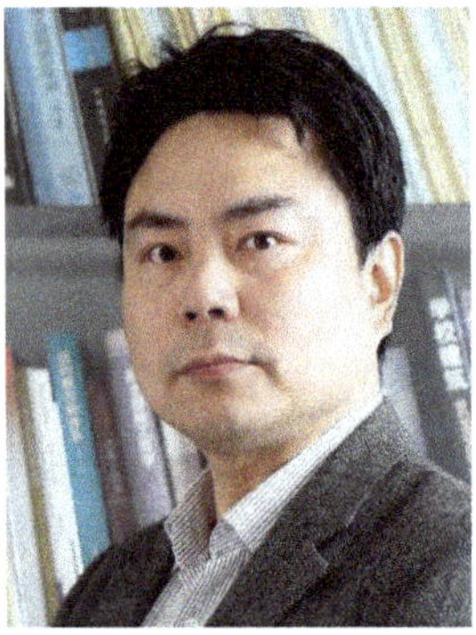

Dr. J. Yang is Professor of Civil Engineering at Tsinghua University. He was born in 1974 in Sichuan, China. He obtained his B.Sc. in Structural Engineering and his Ph.D. in Soil Mechanics at Tsinghua University in 1996 and 2001, respectively. After that, he stayed as a faculty in the School of Civil Engineering to initiate teaching, research and practice on underground space engineering and structural engineering. As a visiting scholar, he participated in the education and research project on wood structures at the University of British Colombia in 2005. He teaches design practices in Architectural Design & Research Institute of Tsinghua University and is also a member of Wood Structure Committee, Architectural Society of China. He has published more than 60 papers, co-authored two books and won one First-Class Prize of China Highway and Transportation Society and one Second-Class Prize of Chinese Ministry of Education.

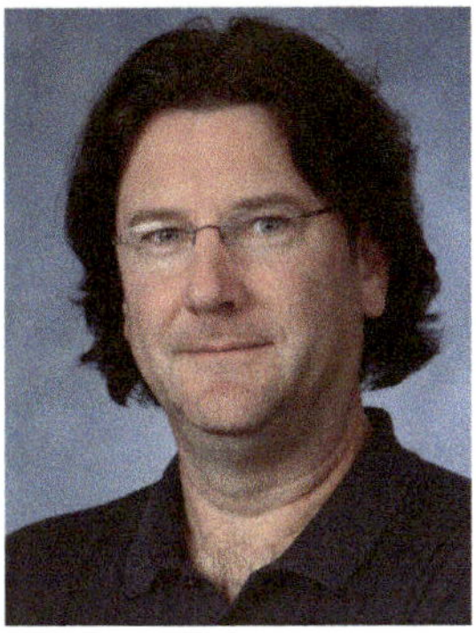

Dr. K. A. Harries is Professor of Structural Engineering and Mechanics at the University of Pittsburgh and is Co-director of the Center for Nonconventional Materials and Alternative Technologies (PITT-NOCMAT). He received his doctorate from McGill University in 1995. Harries' research interests include the use of nonconventional materials—including bamboo—in civil infrastructure. He is the author of over 300 peer-reviewed articles, is the co-editor of *Nonconventional and Vernacular Construction Materials* and is the senior editor of the *Journal of Construction and Building Materials*. He serves on numerous US and international codes and standards development committees and led the effort to revise ISO 22156 *Bamboo Structural Design*. He is Fellow of ASCE, ACI and IIFC and is a professional engineer in Ontario, Canada.

Chapter 1
Distribution of Bamboo Forest Resources and Species for Construction

Bamboo, mainly distributed in tropical, subtropical and temperate regions between N46° and S47°, is one of the most important non-timber products around the world. According to the United Nations Food and Agriculture Organization, bamboo forest area in the world totals 31.5 million ha [1]. The Asia–Pacific region accounts for 55.3%, the Americas 33.2%, and Africa 11.5%. Since the 1990s, global forests have been shrinking steadily, while the area of bamboo forests has increased approximately 3% annually [2], exerting a positive influence on the development of bamboo industries in Asia, Africa and Latin America. China has the greatest bamboo resources in the world. According to data of the Ninth National Forest Resources Inventory of China (2014–2018), the total area of bamboo forests in China covers 6.41 million ha [3], accounting for 20% of the world's total. With the exception of Xinjiang, Inner Mongolia, Heilongjiang and Jilin provinces, all regions and cities in China have bamboo resources. Resources are concentrated mainly in Zhejiang, Jiangxi, Anhui, Hunan, Hubei, Fujian, Guangdong, Guangxi, Guizhou, Sichuan, Chongqing and Yunnan provinces, municipalities or autonomous regions. South of the Yangtze River, Jiangxi, Fujian, Hunan and Zhejiang provinces are the four largest bamboo-producing provinces, producing mainly *Phyllostachys edulis* (moso) bamboo. With 4.68 million ha of moso bamboo resources [3], these regions account for 73% of bamboo resources in China.

In countries and regions where bamboo resources are widely distributed, bamboo has been used as a traditional (or vernacular) construction materials for thousands of years. Although there are 1642 species of bamboo in the world [4], most are unsuitable for construction. There are only about 60 species of bamboo that can be used as construction materials [5] and fewer that can be commercially exploited. Based on data compiled by the International Bamboo and Rattan Organization (INBAR) on the development of bamboo construction in different countries around the world, and on the physical and mechanical properties of different bamboo species, the most common species for construction and their distributions in Asia–Pacific (18 species commonly available for construction), Americas (3 species) and Africa (3 species) are summarised in Table 1.1.

K. W. Liu et al., *Contemporary Bamboo Architecture in China*,
https://doi.org/10.1007/978-981-16-8309-1_1

Table 1.1 Common bamboo species for construction and their geographic distribution [4–10]

Region	Species	Geographic distribution [4]
Asia–Pacific	*Bambusa balcooa**	Bangladesh, India, Laos, Myanmar, Nepal and Vietnam
	*Bambusa bambos**	Bangladesh, India, Laos, Malaysia, Myanmar, Sri Lanka, Thailand and Vietnam
	*Bambusa nutans**	Bangladesh, India, Laos, Nepal, Thailand and Vietnam
	*Bambusa pallida**	Bangladesh, China, India, Laos, Malaysia, Myanmar, Thailand, and Vietnam
	Bambusa pervariabilis	China
	Bambusa polymorpha	Bangladesh, China, Laos, Myanmar and Thailand
	*Bambusa tulda**	Bangladesh, China, India, Laos, Myanmar, Nepal, Thailand and Vietnam
	*Bambusa vulgaris**	Cambodia, China, India, Laos, Myanmar, Thailand and Vietnam
	*Dendrocalamus asper**	Bangladesh, China, Indonesia, Laos, Myanmar, the Philippines, Thailand and Vietnam
	*Dendrocalamus giganteus**	China, India, Laos and Myanmar
	*Dendrocalamus hamiltonii**	Bangladesh, China, India, Laos, Myanmar, Nepal, Thailand and Vietnam
	*Dendrocalamus strictus**	India, Laos, Myanmar, Nepal, Pakistan, Thailand and Vietnam
	*Melocanna baccifera**	Bangladesh, India, Myanmar and Nepal
	Gigantochloa apus	Bangladesh, China, Indonesia, Laos, Malaysia, Myanmar and Thailand
	Gigantochloa atroviolacea	Indonesia
	Gigantochloa atter	Indonesia, Laos, the Philippines and Vietnam
	Gigantochloa macrostachya	Bangladesh, Myanmar
	Phyllostachys edulis	China
Americas	*Guadua angustifolia*	Colombia, Ecuador, Peru and Venezuela
	Guadua aculeata	Costa Rica, El Salvador, Honduras, Guatemala, Mexico, Nicaragua and Panama
	Guadua amplexifolia	Colombia, Costa Rica, El Salvador, Honduras, Mexico, Nicaragua, Panama and Venezuela
Africa	*Oldeania alpine*	Burundi, Cameroon, Congo, Ethiopia, Kenya, Malawi, Rwanda, Sudan, Tanzania, Uganda and Zambia

(continued)

Table 1.1 (continued)

Region	Species	Geographic distribution [4]
	Oxytenanthera abyssinica	Benin, Burkina Faso, Burundi, Cameroon, Central African Republic, Chad, Congo, Ethiopia, Equatorial Guinea, Eritrea, Gambia, Guinea, Guinea Bissau, Kenya, Malawi, Mozambique, Nigeria, Republic of Angola, Republic of Cote d'Ivoire, Republic of Mali, Senegal, Sierra Leone, Sudan, Togo, Tanzania, Uganda, Zambia and Zimbabwe

The Asia–Pacific region is home to the most abundant and diverse bamboo resources suitable for construction. Bangladesh, China, India, Laos, Myanmar, Thailand and Vietnam all have 10 or more species for construction. Among these species, *P. edulis* is most abundant, representing an area of 4.68 million ha, primarily in China [3].

Eleven species, noted with an asterisk (*) in Table1.1, are recommended by India's National Agro-Forestry & Bamboo Mission (NABM) as bamboo species for construction [6]. *B. bambos* and *D. hamiltonii* are defined as the two species suitable for building houses by the Royal Forest Department of Thailand (RFD) [7]. In Hong Kong, *P. edulis* and *B. pervariabilis* are the primary materials of bamboo scaffolding [8]; *D. asper* (Petung in Indonesia) and *G. apus* (Tali) are the main building materials of the famous Green School in Indonesia. *D. strictus* (Tam Vong bao in Vietnam) is often used in the works of Vo Trong Nghia, a famous Vietnamese architect.

In addition to the common species listed in Table 1.1, the physical and mechanical properties of *D. affinis*, *B. oldhamii* and *B. cerosissima* in China can also meet the requirements of building materials [11].

Compared with the Asia–Pacific region, bamboo species suitable for construction in the Americas and Africa are relatively concentrated. In the Americas, *G. angustifolia* generally has the best physical and mechanical properties; this species is also the most commonly used for construction in Latin America. It is called "vegetal steel" [10] by architects. *G. angustifolia* is widely distributed in Venezuela, Colombia, Ecuador and Peru. In addition, *G. angustifolia* is the only bamboo species specified by Colombian's nationalconstruction standard *NSR-10(AIS 2010): Colombian Standards for Seismic Resistant Design and Construction* [12].

In Africa, two species of bamboo, *Ol. alpine* and *Ox. abyssinica*, are suitable for construction, accounting for a total area of about 3.6 million ha [1]. *Ox. abyssinica* is endemic to most sub-Saharan countries, while *Ol. alpine* (called mountain bamboo) is mainly distributed in the uplands of the rift valley in East Africa. Both African species have long been taken as traditional construction materials by local people. For example, in Ethiopia, *Ox. abyssinica* is used for building traditional Amhara bamboo houses, while *Ol. alpine* is used for Sidama bamboo houses [9].

In terms of global distribution and utilization of bamboo resources for construction, China has the advantage of variety, vast resource area and a wide distribution of resources. Such abundant bamboo resources help to ensure a stable supply for the processing and production of bamboo for construction use in China, but also represents a competitive advantage for China's export of bamboo products. In areas rich in bamboo species, the development of a bamboo construction industry can address many issues of social equity: providing many jobs and a secure livelihood while facilitating safe and secure housing.

Different species of bamboo exhibit different material characteristics. Indeed, even the same species sourced from different regions or used in different environmental exposures will vary considerably. The investigation, planning and sustainable management of different bamboo resources and species for construction will therefore also vary and affect the development of local bamboo construction-related industries.

China's leading expertise and technology for bamboo processing can help other regions to fully utilise their resources, thereby developing and driving a truly global bamboo construction industry.

1.1 Chinese and Global Markets for Bamboo Structural Products

In 2018, the reported value of the global export of bamboo products was 2.9B USD [13]. The reported value of Chinese exports of bamboo products is approximately 2.1B USD [13, 14], making China the dominant source of global bamboo exports. Yet this contribution represents only a small proportion of overall trade in bamboo. In China alone, the 2019 output of the bamboo industry was almost 46B USD [14] and the estimated value of global bamboo and rattan production approaches 60B USD [15]. Most bamboo production is for domestic markets. Not surprisingly, at least 40% of global bamboo exports are to the European Union and North America [14]—regions where large bamboo species are not endemic.

Currently, building materials represent a relatively small proportion of the bamboo products market. Chinese exports of bamboo building materials—primarily panels and flooring—were worth approximately 300 M USD (14% of total bamboo exports) in 2018 [13], and 215 M USD (10%) in 2019 [14]. Based in the data from the INBAR Bamboo and Rattan Trade Database [16], the value of flooring products was approximately twice that of panels in 2017, but is roughly the same as panels in 2019. Engineered bamboo materials are presently a very small portion of the market: exports were reported to be only 0.3 M USD in 2018 and 0.5 M USD in 2019 [16]. Nonetheless, as the single largest economic sector in the world, use in construction has the potential to significantly disrupt the bamboo products industry, absorbing any and all industry growth.

References

1. FAO. (2015). *Global forest resources assessment 2015.* Food and Agriculture Organization of the United Nations.
2. Jiang, Z. H. (2007). *Bamboo and rattan in the world.* China Forestry Publishing House.
3. Li, Y. M., & Feng, P. F. (2019). Bamboo resources in China based on the Ninth National Forest Inventory data. *World Bamboo and Rattan, 17*(6), 45–48 (in Chinese).
4. Vorontsova, M., Clark, L., Dransfield, J., et al. (2016). *World checklist of bamboos and rattans.* International Network for Bamboo and Rattan (INBAR).
5. Jayanetti, D., & Follett, P. (1998). *Bamboo in construction: an introduction.* Published jointly by TRADA Technology Limited, International Network for Bamboo and Rattan (INBAR), Department for International Development (DFID).
6. Salam, K., & Pongen, Z. (2008). *Hand book on bamboo.* Cane and Bamboo Technology Centre.
7. RFD & ITTO. (2013). *Physical and mechanical properties of some Thai bamboos for house construction.* Royal Forest Department of Thailand (RFD) and International Tropical Timber Organization (ITTO).
8. Chung, K., & Chan, S. (2002). *Bamboo scaffolds in building construction.* International Network for Bamboo and Rattan (INBAR).
9. Kibwage, J., & Misreave, S. (2011). *The value chain development and sustainability of bamboo housing in Ethiopia.* International Network for Bamboo and Rattan (INBAR).
10. Villegas, M. (2003). *New bamboo architecture and design.* Villages Editores.
11. Yu, Y. L., Huang, X. A., & Yu, W. J. (2014). High performance of bamboo-based fiber composites from long bamboo fiber bundles and phenolic resins. *Journal of Applied Polymer Science, 131*(2), 40371.
12. NSR-10 (AIS 2010). (2010). *Reglamento Colombiano de Construcción Sismo Resistente [Colombian regulations for earthquake resistant construction].* Instituto Colombiano de Normas Técnicas y Certificación (in Spanish).
13. INBAR. (2021). *Trade overview 2018: Bamboo and Rattan commodities in the international market.* https://www.inbar.int/wp-content/uploads/2021/04/Trade-Overview-2018-International——final.pdf.
14. INBAR. (2021). *Trade overview 2019: Bamboo and Rattan commodities in China.* https://www.inbar.int/wp-content/uploads/2021/04/Trade-Overview-2019-China——final-1.pdf.
15. INBAR. (2019). *Trade overview 2017: Bamboo and Rattan commodities in the international market.* https://www.inbar.int/wp-content/uploads/2020/05/1578283314.pdf.
16. INBAR Bamboo and Rattan Trade Database. (2021). http://trade.inbar.int:10444. Last accessed July 17, 2021.

Chapter 2
Types and Characteristics of Bamboo Materials for Construction Uses

Bamboo materials for construction uses can be divided into structural materials, enclosure and decorative materials, and materials for other functions. This chapter will focus on the following:

(1) *Bamboo as load-bearing structural materials*

In terms of a structural material, bamboo can not only be utilised directly in its round, full-culm form, but it may also be processing into different engineered bamboo products including glued laminated bamboo, laminated bamboo sliver lumber and bamboo scrimber. The engineered products are fabricated into beams, columns, shear walls and roof trusses. For structural, load-bearing materials, the physical and mechanical properties are of paramount importance. This chapter presents the physical and mechanical properties of the most popular structural bamboo products for construction in China and the test methods from which they are determined. Additionally, comparisons with common softwood products are presented.

(2) *Bamboo as enclosure and decorative materials*

Engineered bamboo products are widely used for enclosure and decoration. Products such as non-structural glued laminated bamboo, bamboo scrimber, bamboo plywood, bamboo veneer, bamboo particleboard, etc. are used as indoor and outdoor flooring, wallboard, indoor decorative materials (such as bamboo sound-absorbing panels, bamboo perforated-panels, etc.) and outdoor grilles. This chapter will introduce the purposes and use of some materials.

(3) *Bamboo for other functions*

As a construction material, bamboo is also widely used as bamboo scaffolding, in bamboo-reinforced concrete and masonry structures, as a method of soil reinforcement, as well as for strengthening existing structures. The results of researches on these functions will also be discussed in this chapter.

K. W. Liu et al., *Contemporary Bamboo Architecture in China*,
https://doi.org/10.1007/978-981-16-8309-1_2

2.1 Structure and Morphology of Bamboo

The structure of bamboo is composed of ostensibly round culms (poles) with solid transverse diaphragms or 'nodes' separating hollow intermodal regions along the height of the culm.

The microscopic structure of bamboo, shown in Fig. 2.1, consists primarily of vascular bundles of sclerenchyma cells (Fig. 2.1a) surrounded by parenchyma cells (Fig. 2.1b). The vascular bundles can be divided into fibres (Fig. 2.1c), lignified catheters, sieve tubes and cell lumens. Fibres are arranged longitudinally, parallel to the culm axis, giving the bamboo its high strength and stiffness in this direction [2]. A micrograph of a single bamboo fibre is shown in Fig. 2.1d. The size and shape of cells, density of vascular bundles and fibre content vary both along the height of the culm and through the wall thickness. Research shows that the mechanical strength of the upper part of bamboo is greater than that of the lower part [3], and the mechanical strength of the outer part of bamboo wall is greater than that of the inner part [2]. Both phenomena reflect the nature of bamboo growth. Increased strength accompanies vertical taper of the culm resulting in a more uniform section stiffness and capacity [4]. The gradation through the culm wall has the effect of increasing the bending capacity of the culm subject to wind loads [4, 5].

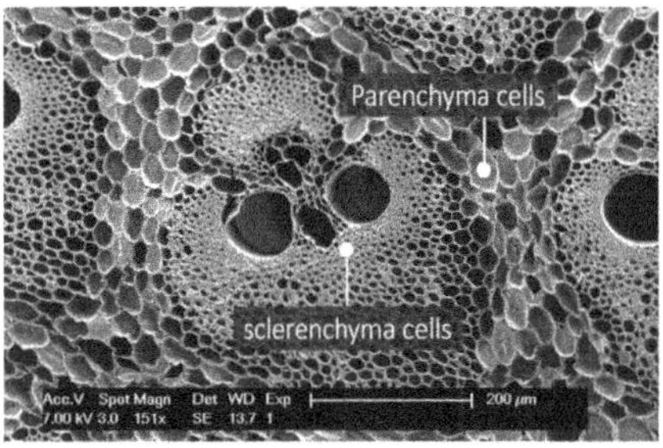

(a) Microstructure of bamboo (width of image is approximately 800 μm)

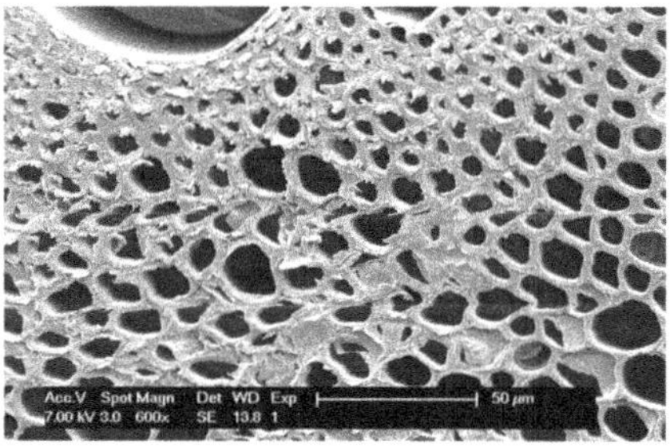

(b) Micrograph of parenchyma cells (width of image is approximately 200 μm)

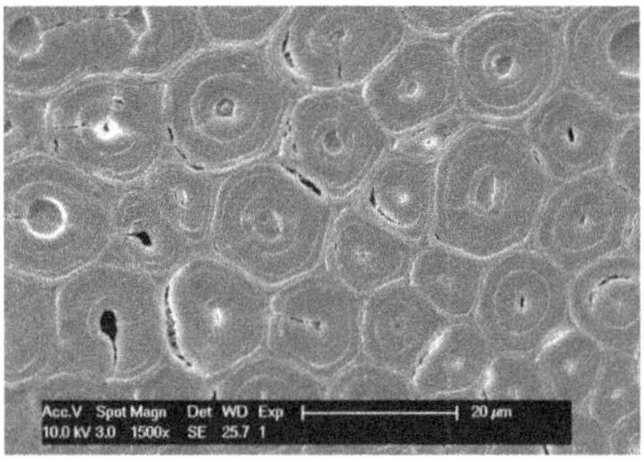

(c) Micrograph of bamboo fibres (width of image is approximately 125 μm)

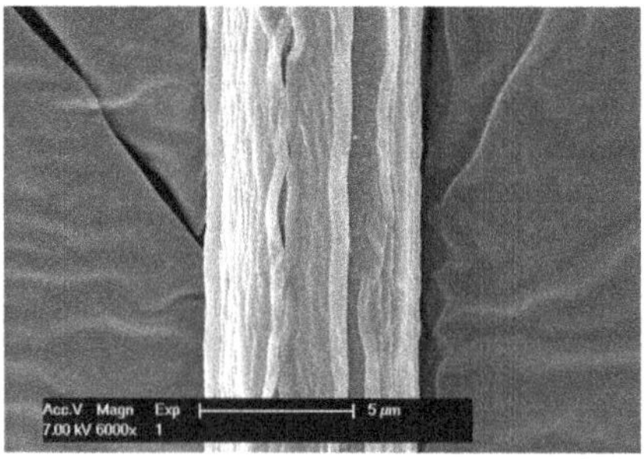

(d) Micrograph of a single bamboo fibre (width of image is approximately 20 μm)

Fig. 2.1 Micro-cytological graphs of bamboo [1]

The bending strength, compressive strength parallel to fibres and tensile strength in the nodal regions are lower than those in intermodal. Due to the less uniform fibre alignment through the node [6], the splitting strength and tensile strength perpendicular to fibres in nodal zone are higher than in the internode [7, 8].

2.2 Bamboo as a Load-Bearing Structural Material

2.2.1 Full-Culm Bamboo

2.2.1.1 Test Methods for Determination of Physical and Mechanical Properties of Bamboo

Bamboo is a naturally occurring, anisotropic materials having graded (varying) structure and material properties; its three-dimensional mechanical properties are complex and the variability can be great. Mechanical properties vary with location along the culm and with bamboo age and moisture content [9]. Necessarily, mechanical properties are determined for a single direction. Most extant studies focus on mechanical properties in the longitudinal direction: parallel to the fibres and along the axis of the culm. Recognising the importance of the weaker properties in the perpendicular direction, more recent research has focused on these [10]. As yet, there are no established models coupling the orthotropic behaviours of bamboo. Bamboo is not a homogeneous orthotropic material [11]; nor is it appropriately considered as a fibre reinforced composite [12].

Most extant research focuses on test methods of mechanical properties and variability of mechanical parameters in small [clear] samples without defects; only a few studies of long-term load and full-scale static, dynamic and vibration tests of structural bamboo materials are available. Some studies report proposed constitutive relationships (i.e., stress versus strain) for bamboo.

The international standard *Bamboo structures—Determination of Physical and Mechanical Properties of bamboo culms—Test Methods* (ISO 22157:2019) [13] provides test methods for measuring the mechanical properties of full-culm bamboo. This standard includes compression strength and modulus of elasticity parallel to fibres (Fig. 2.2a), bending strength and modulus of elasticity (Fig. 2.2b), shear strength parallel to fibres (Fig. 2.2c), and tension strength and modulus of elasticity parallel to fibres (Fig. 2.2d). All but the tension tests use full-culm bamboo specimens. ISO 22157:2019 also provides standard means of determining moisture content and density—both critical parameters to interpreting mechanical test data. In this regard ISO 22157:2019 is aligned with *Bamboo structures—Grading of bamboo culms—Basic principles and procedures* (ISO 19624:2018) [14] which introduces the easily determined "weight per unit length" measure in place of density. ISO

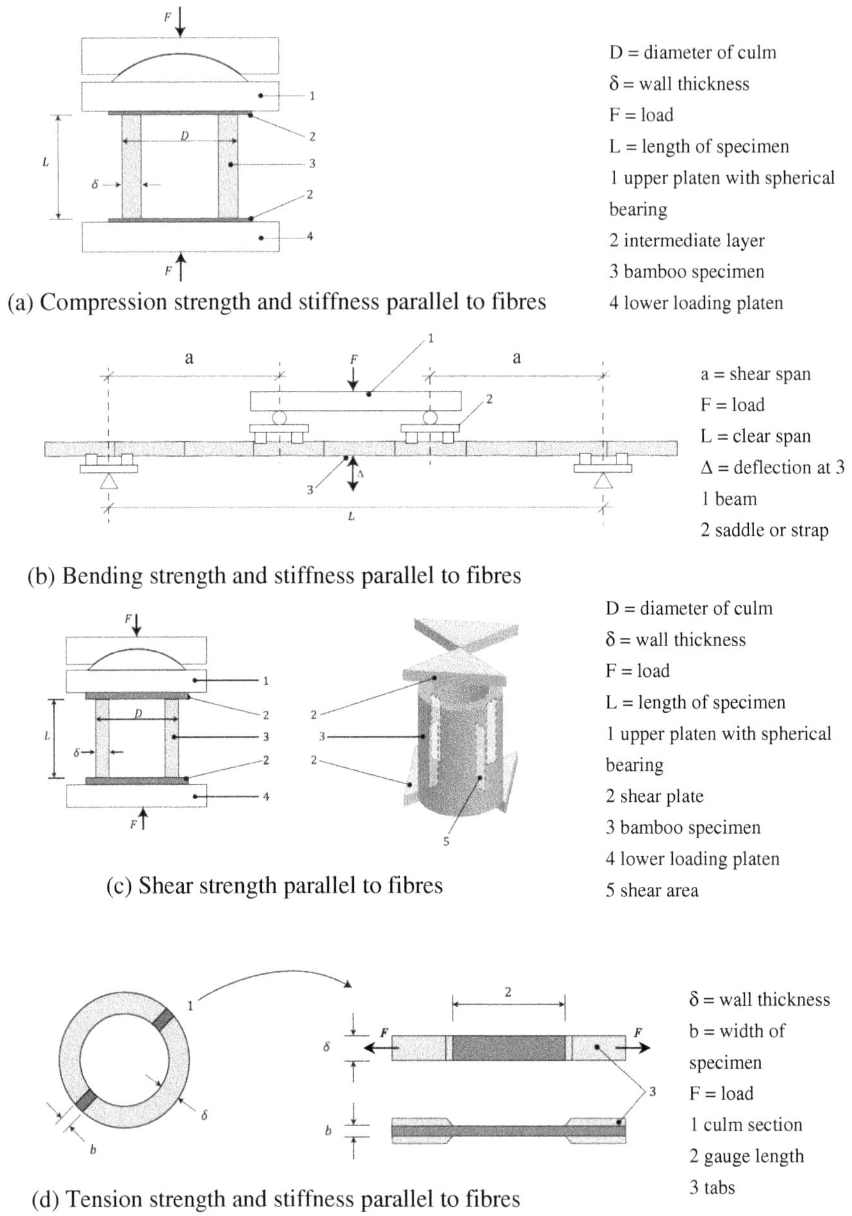

Fig. 2.2 Test methods for mechanical properties of full-culm bamboo [13]

22157:2019 is also permissive in terms of providing multiple methods for determining moisture content and allowing testing to be conducted in ambient (rather than laboratory) conditions when appropriate.

One goal in its development was to ensure that a broad range of laboratories could adopt ISO 22157:2019. The standard needed to be suitable for both state-of-the-art laboratories in the industrialised world as well as austere facilities in the developing world. There are a number of examples in which the standard test has a permissive aspect allowing flexibility in its conduct [15].

The Chinese national standard *Testing methods for physical and mechanical properties of bamboos* (GB/T 15,780-1995) [16] and industry standard *Testing methods for physical and mechanical properties of bamboo used in building* (JG/T 199-2007) [17] specify test methods for mechanical properties of bamboo strips (i.e. "small clear specimens") cut from full-culm bamboo sections. Both standards provide test methods for compression and tension strength parallel to fibres, modulus of elasticity, bending strength, and shear strength parallel to fibres, but the test methods in each standard are slightly different. Liu et al. [18] compared the two Chinese standards and found that the test methods in the industry standard, JG/T 199-2007, were easier to conduct. JG/T 199-2007 also specifies testing methods for compression modulus of elasticity parallel to fibres, compression modulus of elasticity perpendicular to fibres, tension modulus of elasticity parallel to fibres and impact toughness.

Standard test methods allow single dimensional orthotropic material properties to be determined. These are usually adequate for strength-based design of bamboo structures and products. Nonetheless, defining clear constitutive (stress versus strain) relationships remains difficult from standard—relatively easily conducted—tests. Digital image correlation techniques have recently been successfully used in this regard and should be pursued as the norm in bamboo materials testing. From these, it is proposed that not only accurate constitutive relationships can be developed but single and multi-dimensional failure criteria established. For example, Gauss et al. [19] propose and demonstrate the appropriateness of the "limit of proportionality" as a parameter for bamboo structural design rather than the typically used failure strength.

2.2.1.2 Factors Affecting Standard Test Results

Moisture content (MC) greatly affects the mechanical properties of bamboo. Like timber, below the fibre saturation point (about 30–35% MC for bamboo), strength and stiffness typically increase with decreased moisture content (to about 6%). The exceptions are tensile strength parallel to grain and small specimen impact toughness, which both decrease with decreased moisture content [17]. If dried completely (to 0% MC), tested properties fall due to the brittle texture of the culm structure [20]. By convention, properties of bamboo are reported at 12% MC [13, 16, 17].

Above the fibre saturation point, there is little variation in properties of the so-called "green" bamboo [21]. Nonetheless, the compressive, shear and splitting strength of bamboo at its fibre saturation point (MC $\approx$ 30%) is about 75% of that under air-dry conditions (MC $\approx$ 12%) [21].

In addition, age of harvest is known to affect bamboo strength and stiffness [22, 23]. Young bamboo is not strong. With age, lignification improves the bamboo mechanical properties. The optimal age at harvest varies for species but is conventionally reported as being about 3–5 years [23]. Beyond this, strength begins to decrease as the culm becomes more brittle [22].

2.2.1.3 Physical and Mechanical Properties of Full-Culm Bamboo

Table 2.1 lists the physical and mechanical properties of some common bamboo species and reference conifers species used for construction in China. Like timber, bamboo material properties are generally proportional to density. The density of bamboo is greater than that of softwood and, accordingly, material properties are typically superior to those of softwood.

Properties of engineered bamboo are not greatly different than those of their source bamboo. However, engineered bamboo will typically have much less variation in tested properties than full-culm bamboo. Engineered bamboo is described later in this chapter.

2.2.1.4 Long-Term Performance, Durability and Fire Resistance of Full Culm Bamboo

As a construction material, in addition to physical and mechanical properties, long-term performance, durability and fire resistance of bamboo are also very important. Durability of bamboo is a major concern. Starch in bamboo attracts organisms—primarily boring insects—resulting in deterioration of bamboo properties. Bamboo in service conditions in which the moisture content exceeds about 20% are also susceptible to a variety of fungal attacks. The service life of untreated bamboo is roughly 1–3 years in an open air environment or in contact with soil, 4–6 years in a sheltered environment without soil contact, and as many as 10–15 years in a well-conditioned environment. Proper treatment of bamboo can significantly improve its durability [53]. Suggestions for improving bamboo durability include: (a) harvesting in winter, when bamboo has the lowest sugar content; (b) storing vertically for several days after harvesting to keep branches and leaves intact, which can consume some latent sugar; and (c) soaking culms in running water to leach out starch content. In order to reduce microbial damage to bamboo, non-chemical or chemical methods can be used to protect bamboo. Non-chemical methods include fumigation, smoking, brushing lime powder, and painting waterproof and antiseptic coatings; chemical methods include soaking bamboo in chemical solution or a variety of pressurised treatments. In addition, bamboo culms for construction must always be dried (cured) in order to establish a stable equilibrium moisture content prior to being used in a

Table 2.1 Physical and mechanical properties of common full culm bamboo, engineered bamboo and conifers used for construction in China

Species	Tensile strength parallel to fibre (MPa)	Compression strength parallel to fibre (MPa)	Bending strength (MPa)	Shear strength parallel to fibre (MPa)	Elastic flexural modulus (MPa)	Density (kg/m^3)	Moisture content (%)	References
Physical and mechanical properties of common bamboo species used for construction								
Bambusa pervariabilis	/	79	80	/	22,000	/	< 5	[24]
Bambusa tulda	207	79	194	9.9	18,600	910	8.6	[25]
Bambusa vulgaris	118	57	138	/	/	590	/	[26]
Dendrocalamus asper	227	69	84	/	/	767	11.3	[27]
Dendrocalamus giganteus	177	70	193	10.6	16,400	740	8.0	[25]
Dendrocalamus hamiltonii	177	70	89	6.7	9630	590	8.5	[25]
Gigantochloa apus	299	27–49	88	7.5–7.7	/	/	15.1	[28]
Phyllostachys edulis	/	67	181	18.2	10,850	644	/	[29]
Physical and mechanical properties of engineered bamboo used for construction								
Laminated bamboo sliver lumber (along grain)	137	86	161	23	12,200	960	/	[a]

(continued)

Table 2.1 (continued)

Species	Tensile strength parallel to fibre (MPa)	Compression strength parallel to fibre (MPa)	Bending strength (MPa)	Shear strength parallel to fibre (MPa)	Elastic flexural modulus (MPa)	Density (kg/m^3)	Moisture content (%)	References
Laminated bamboo sliver lumber (across grain)	83	35–51	99	16	9407	820–900	/	[30]
Glued laminated bamboo	90–140	40–60	90–120	/	8000–12,000	600–1000	8–12	[31–45]
Bamboo scrimber	120–250	50–80	120–250	/	12,000–13,000	1000–1250	8–10	[45–51]
Physical and mechanical properties of common conifers used for construction								
Larch (northeast China)	130	58	113	8.5	14,500	640	Air-dry	[52]
Cedarwood (Sichuan, China)	83	36	68	6.0	9600	420	Air-dry	[52]
Cypress (Guizhou, China)	102	48	100	10.2	10,000	460	Air-dry	[52]
Fir (Sichuan, China)	97	36	70	4.9	10,000	430	Air-dry	[52]

The data provided was tested under different conditions; Direct comparison is not necessarily appropriate.

[a] Data provided by Professor Wang Zheng, Chinese Academy of Forestry

structure [53]. Regardless of treatment, bamboo should not be used in contact with soil or embedded into cement-based mortars [54, 55].

In addition to durability, long-term performance under sustained load (i.e., creep) affects load-bearing members. Limited study of the creep of bamboo is available. However, creep coefficients determined for small clear *B. stenostachaya* specimens tested in flexure were found to be marginally lower (better) than would be expected from softwood timber [56]. Considering (a) the biological similarity of bamboo and timber and (b) the greater anisotropy of bamboo, creep performance of bamboo loaded parallel to its fibres should be better than timber. In the absence of additional data, it is recommended that creep of bamboo loaded such that sustained stress is parallel to the fibres may be addressed using the same coefficients appropriate for softwood construction [56, 57].

Like timber, bamboo is combustible. There is little existing data on the fire-safety performance of bamboo. Limited testing indicates that the combustibility and charring behaviours of bamboo are marginally superior to softwood timber [58, 59]. At present, existing research shows that low-rise residential full-culm bamboo construction can meet the requirements of fire safety performance when proper fire design measures (e.g., egress, suppression, protection) are adopted. There is presently no data available to support performance-based fire design methods for full-culm bamboo construction.

2.2.2 Engineered Bamboo

Beginning in the late 1970's, researchers began to slice or reconstitute full-culm bamboo using modern gluing methods to produce bamboo-based engineered building materials that meet the modern construction industry requirements and standards. Engineered bamboo is a composite material which combines various bamboo in some form—strips, splits, veneers, fragments, fibres, multiple-culm bundles, etc. —with adhesive to achieve elements having uniform geometry and more stable mechanical properties. Based on composition and processing method, there are three categories of structural engineered bamboo product: laminated bamboo sliver lumber, glued laminated bamboo and bamboo scrimber, as shown in Fig. 2.3 and described in the following paragraphs.

Laminated bamboo sliver lumber is fabricated from large diameter bamboo culms that are split into strips of fixed width. After the processing steps of cutting, weaving, drying and gluing, the resulting bamboo curtains are hot pressed into boards in which the curtains may be uniformly oriented or cross laminated. The resulting boards can be used directly as wall, floor and roof panels. The boards can be further processed to produce beams, columns and other structural components through cutting, gluing, cold pressing and finger jointing. The manufacturing process is summarised in Fig. 2.4. Some researchers and producers describe cross-laminated bamboo sliver lumber as "Glubam".

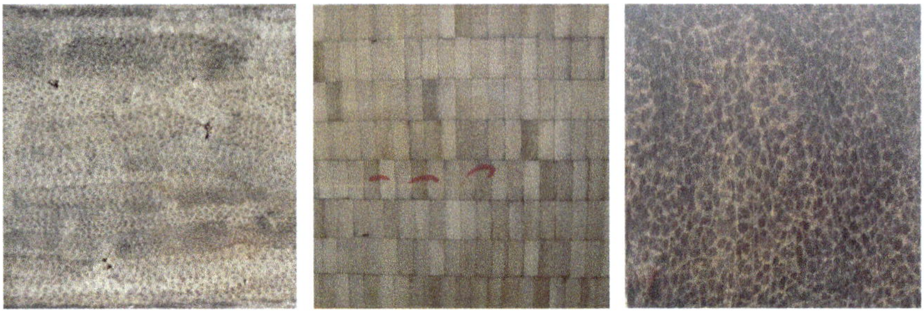

(a) Laminated bamboo sliver lumber. (*photo by Chinese Academy of Forestry*)
(b) Glued laminated bamboo. (*photo by Nanjing Forestry University*)
(c) Bamboo scrimber. (*photo by Nanjing Forestry University*)

Fig. 2.3 Classifications of engineered bamboo product for structural uses

Glued laminated bamboo consists of bamboo slices having fixed width and thickness are hot pressed into boards in which all fibres are oriented in the same direction. Relatively large diameter culms are required; primarily *P. edulis* (moso) is used. The manufacturing process includes cutting, stripping, raw slicing, boiling, caramelisation, drying, finish planning, color separation, assembling, pressing and sanding. Glued laminated bamboo can be fabricated as single or multiple layer flat-pressed, side pressed or cross-laminated panels. Glued laminated bamboo can also form more complex shapes such as I-beams.

Bamboo scrimber is a kind of board or dimensional (square-edged) lumber composed of bamboo bundles or fibrous veneers impregnated with water-soluble phenolic resin. Members are assembled by dispersing the fibres in a mould in a unidirectional fashion, impregnating the fibres and compressing the product in a hot (or cold) pressing operation. For scrimber, raw materials can be obtained from smaller culms (generally greater than 5 cm diameter). Compared with 25–30% material utilization for laminated bamboo products, bamboo scrimber can achieve as high as 90% raw material utilisation.

In addition to engineered bamboo produced in forms familiar to the construction industry, bamboo, with its high axial tension strength and good flexibility, has been proposed for a variety of composite applications. Lightweight bamboo winding composites—composed of bamboo sliver tape windings and resin—have been proposed for pipeline and other applications conducive to filament winding [61]. Similar applications may be suitable for other construction applications such as curved facia and so forth.

2.2.2.1 Physical and Mechanical Properties of Engineered Bamboo

Many studies of addressed the properties of glue-laminated engineered bamboo [31–45] and bamboo scrimber [45–51] products. Summarised in Table 2.1, the density of these engineered products is typically greater than the constituent bamboo;

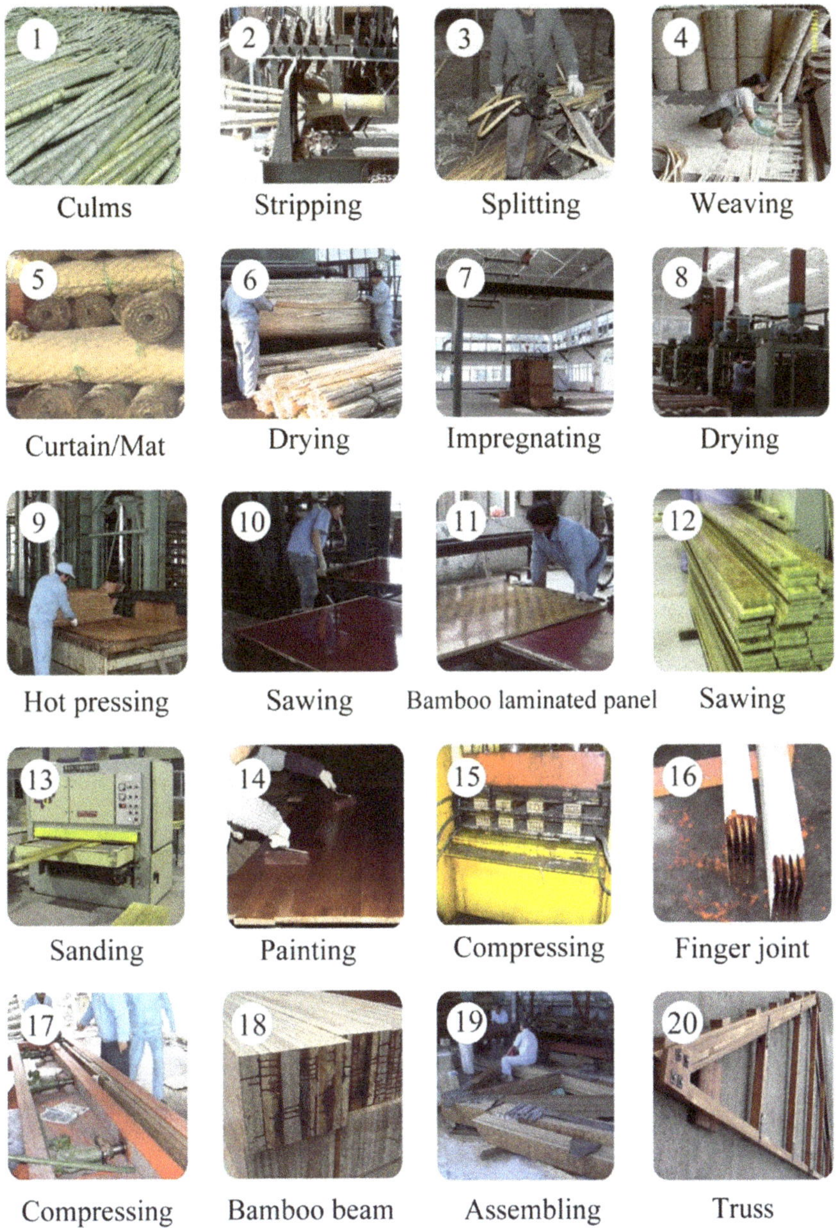

Fig. 2.4 The manufacturing process of laminated bamboo sliver lumber for structural uses [60]

this is especially true for scrimber which is densified or formed in a compression moulding process. Mechanical properties are comparable or marginally improved with processing and variability is reduced. Geometric variation is virtually eliminated and more controlled moisture properties (affecting dimensional stability and durability) can be achieved with engineered products. The strength-to-weight ratio of engineered bamboo can exceed that of steel and the material is generally suitable for any structural application in which timber may be used. Like other engineered products, mechanical properties measured using small samples do not fully reflect strength characteristics of in situ use. It is necessary to study the performance of engineered bamboo materials and products in full-scale tests. The associated issues related to cost and standard compliance still need to be addressed before wide spread adoption of engineered bamboo in construction is viable.

At present, there is no standard test methods for determining the physical and mechanical properties of engineered bamboo. In this case, researchers mostly refer to related test methods for timber and composite lumber products, such as ASTM D5456-19 *Standard Specification for Evaluation of Structural Composite Lumber Products* [62]. The size and shape of test specimens corresponding to each standard used vary. As a result, properties obtained from different tests should not be compared. The macroscopic constituents (slivers, laminates, etc.), adhesive area/volume and presence and distribution of defects also affect test results and must be considered when selecting a specimen size and appropriate test method. Manufacturing process and environment can also affect mechanical properties. Therefore, it is very important to combine the physical and mechanical properties measured in the laboratory with the anticipated in situ conditions to obtain mechanical parameters that can be used in structural design. Considerable effort is presently ongoing to develop appropriate measures of engineered bamboo properties and performance and to establish appropriate constitutive models and failure criteria. At present, the properties of engineered bamboo are usually compared with those of engineered timber materials.

2.2.2.2 Durability of Engineered Bamboo

The durability of engineered bamboo is superior to that of unprocessed full-culm bamboo although it remains susceptible to insect and fungal attack as well as to deterioration from environmental stressors such as UV exposure and humidity. At present, the evaluation of bamboo durability mostly refers to appropriate standards for laboratory [63, 64] and/or field [64, 65] tests for wood. These address susceptibility to fungal and microorganism as well as termite attack.

Studies incorporating simulated accelerated aging and natural outdoor aging have been conducted. Performance is measured in terms of appearance, degradation of physical and mechanical properties and material loss following after aging or exposure to specific environments. Engineered bamboo, while still susceptible, is resistant to brown and white fungal rot [66–69]. On the other hand, engineered bamboo is not resistant to mildew and additional anti-mildew processing is necessary [68]. A variety of water-borne preservative treatments, including Alkaline Copper Quaternary (ACQ) and Copper Azole (CuAz), have been demonstrated to significantly improve the durability of glue-laminated bamboo [69]. Likely due to the resin-rich exterior, bamboo mat composites have been observed to be more resistant to attack than scrimber products [69]. Accelerated aging tests of glue-laminated bamboo identify degradation of the adhesive glueline as affecting mechanical properties [70]. Sealing the edges of such materials was recommended. In general, the durability of scrimber, being denser, having a greater resin content and not being reliant on the performance of individual thin gluelines is superior to laminated bamboo products.

2.2.2.3 Fire Resistance of Engineered Bamboo Structures

Testing of engineered bamboo at elevated temperatures and for fire resistance is relatively limited. Like durability testing, test methods and evaluation standards for timber structures are mostly used. Prescribed fire protection measures for engineered bamboo structures are similarly referenced to timber structures. Multiple studies have shown the ignition parameters, flame spread and charring rate of engineered bamboo to be marginally better than those of conifer timber commonly used for structural purposes [58, 59, 71, 72]. Shah et al. [73] established the correlation of engineered bamboo density and thermal conductivity. Models for the degradation of mechanical properties [72, 74] and changes in thermophysical properties [71] have been proposed.

Following the 2008 Wenchuan Earthquake, glubam 'resettlement' structures were constructed for the Guangyuan Beijie and Nanying Primary Schools. Zhou et al. [75] studied the fire performance of such prefabricated glubam structures and compared this with comparable light gauge steel construction. Results indicated that the glubam structures provided improved safety and fire resistance. Xiao et al. [76] demonstrated the efficacy of treating glubam with a fire retardant coating and conducted a test on a full-scale light-frame glubam house which achieved a fire resistance of one hour.

2.3 Bamboo as Enclosure and Decoration Materials

Bamboo has many non-structural applications. The dominant markets are flooring and decorative materials.

2.3.1 Bamboo Flooring

Bamboo flooring is attractive (having the texture and consistent colour of bamboo), dimensionally stable, durable, resistant to abrasion and harder than most wood flooring. Over 40 years, bamboo flooring has evolved from simple integrated single-colour bamboo slices to products integrating different materials and finally to modern reconstituted materials. Like structural bamboo, flooring can be laminated or scrimber bamboo products and can be all-bamboo or composed of a bamboo-wood composite. Colour ranges from bleached bamboo to natural to caramelised[1] [77, 78]. Bamboo flooring suitable for indoor or outdoor use is available. Its high strength and superior toughness and wear resistance make bamboo flooring well suited for use in public spaces. Due to its toughness and relative light weight, bamboo flooring is also commonly use in shipping containers.

For use as flooring, resistance to wear and water, resistance to environmental degradation (typically only a concern for outdoor use), dimensional stability and combustion performance are considerations for bamboo floor selection. Physical and mechanical properties of flat-pressed bamboo floors are found to be superior to side-pressed bamboo, bamboo scrimber or bamboo-wood composite floors [79]. Natural bamboo was additionally found to better maintain material characteristics than bleached or caramelised materials [79]. Bamboo scrimber flooring, while denser and having a much higher raw material utilisation, was shown to be the least dimensionally stable of bamboo flooring materials [79]. Side-pressed bamboo flooring was shown to be susceptible to changes in temperature and humidity [79]. Painted coatings have been shown to improve durability (especially against water) and wear resistance of bamboo flooring [80]; although penetrating stains offer little protection or improvement [80].

In an outdoor environment, bamboo scrimber coated with UV-resistant paint or weather-resistant wood oil showed good performance. 24-h water uptake was about 4.5% and through thickness swelling limited to about 1.7% [81]. Following accelerated environmental conditioning, material property retention remained above 85% [81]. Bamboo scrimber treated with other coatings, such as a UV-curing varnish did not perform as well [81].

In terms of fire resistance, bamboo scrimber was shown to perform better than flat- and side-pressed bamboo in terms of time to ignition, rate of heat release and effective heat of combustion, and smoke production [82]. Scrimber was recommended as a favourable flooring material for fire egress routes [82].

2.3.2 *Indoor Bamboo Decorative Materials*

In recent years, indoor bamboo decorative materials, such as ceilings, sound absorption board and perforated boards, have become popular because they are natural, environmentally friendly and differ from wood. A number of large public building

[1] Caramelisation is a chemical process known as the Maillard reaction. In the bamboo industry, the term 'carbonisation' is often mistakenly used to refer to the process of caramelisation. Carbonisation occurs at much higher temperature and is part of the process of making charcoal.

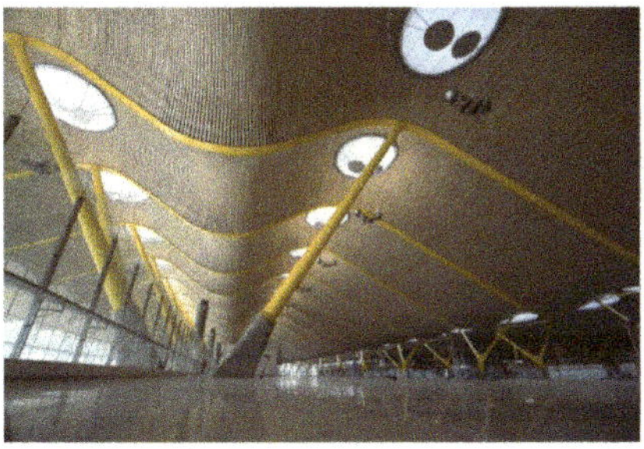

(a) Terminal 4 of Barajas Airport, Madrid, Spain

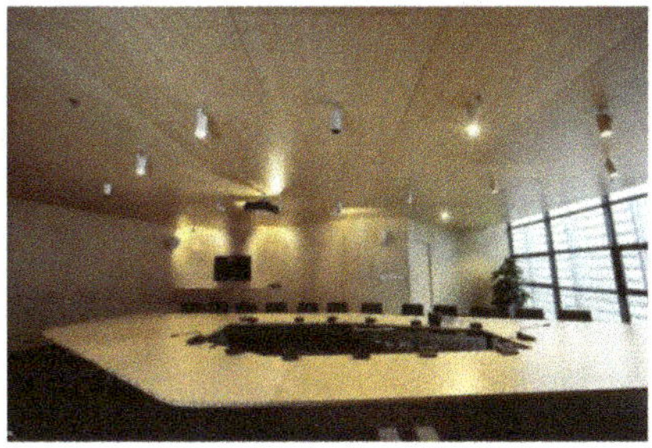

(b) Vanke Headquarters, Shenzhen, China

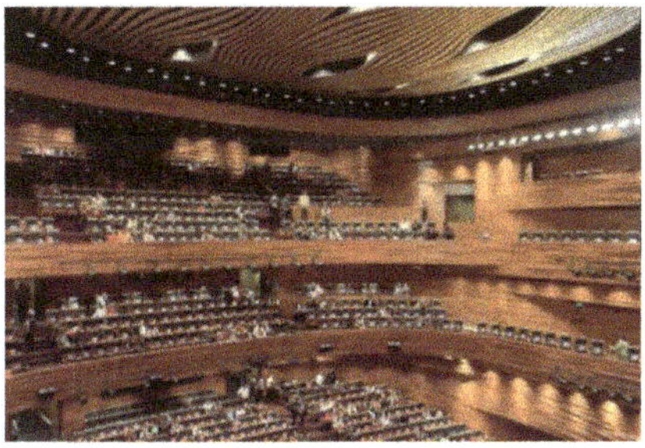

(c) Wuxi Grand Theatre, Jiangsu, China

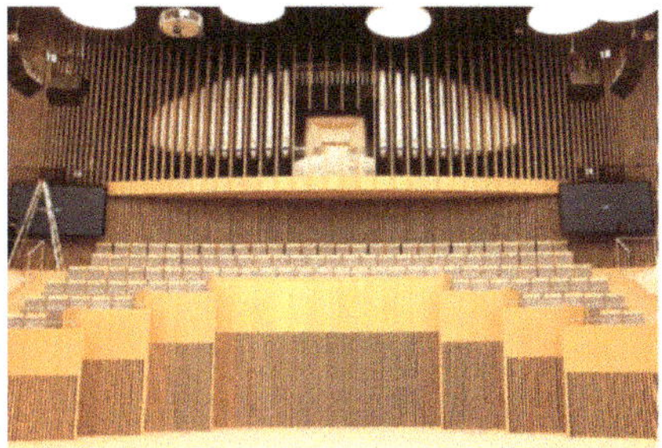

(d) Fuzhou Strait Culture and Art Center, Fujian, China

Fig. 2.5 Application of indoor bamboo decorative materials (*photos by Hangzhou Dasso Technonogy Co., Ltd.*)

projects have used such materials. Perhaps most recognisable, the 230,000 m^2 fire-resistant bamboo ceiling in Terminal 4 of Madrid's Barajas Airport (Fig. 2.5a) is the largest single application of bamboo in the world. The Barajas Airport has had a great influence on the perception of bamboo for construction and represents a significant breakthrough for Chinese bamboo products in international construction. The project is also a milestone in terms of demonstrating fire-proof bamboo technology. Vanke Headquarters was completed in Shenzhen, China in 2011. This LEED Platinum certified project contains 120,000 m^2 of decorative bamboo (Fig. 2.5b), including doors, panels, wallboards and even ceilings made of bamboo. Using 200 cubic meters of renewable environmentally friendly bamboo materials, the Wuxi Grand Theatre was completed in 2012 in Jiangsu, China (Fig. 2.5c). A variety of bamboo materials were used including ceiling and sound-absorbing wallboard. The 150,000 square meter Fuzhou Strait Culture and Art Center (Fig. 2.5d), built in 2018 in Fujian, China, adopts bamboo throughout the building. More than 40,000 m^2 of engineered bamboo, including floors, sound-absorbing wallboard, fire-proof panels and grilles, and outdoor weather-resistant bamboo was used.

The synergistic effect of bamboo colour and texture produces its unique decorative characteristics. In general, indoor decorative bamboo may be green, yellow or caramelised. Green is primarily for full-culm bamboo decorative elements. While invoking nature, the colour is too bright and cold, and not suitable for large-scale indoor application. Most decorative bamboo-based panels are yellow. The light colour having low brightness and saturation provides a sense of elegance. Yellow bamboo is well suited to large areas such as ceilings and walls (Fig. 2.5). Caramelised bamboo is darker and less saturated, invoking a restrained atmosphere. Caramelised bamboo is often used for dividing space as flooring. The low saturation and good colour inclusiveness of bamboo creates a soft warm atmosphere that is suitable for living space [83].

2.4 Bamboo in Other Structural Load-Bearing Applications

2.4.1 Bamboo Scaffolding

In South and East China, because of its widespread availability, bamboo is often used to build construction scaffolds (Fig. 2.6). Although the use of bamboo scaffolding is extensive, related research is relatively scarce. Chung et al. [84] carried out experimental research on the physical and mechanical properties of *P. edulis* and *B. pervariabilis* commonly used in scaffolds and recommended design values, material selection conditions and preservation methods. The effects of moisture content—particularly that higher than the fibre saturation point—in scaffold elements has been reported [21]. Experimental studies of single culm buckling have led to proposed design methods [85]. Multi-culm arrangements over multiple floors have been tested and requirements for lateral bracing recommended [86]. The redundancy inherent in scaffold construction is critical to their safe use: the loss of one or more vertical members should not lead to collapse of the scaffold [21, 86].

Due to the lack of attention bamboo scaffolding design, accidents still occur frequently. In recent years, various documents have been issued to normalise and/or restrict the use of bamboo scaffolding. However, in Hong Kong, due to the importance attached to the safety of bamboo scaffolding and strict compliance with standards, bamboo scaffolding use is more common than in other areas of China. In Hong Kong,

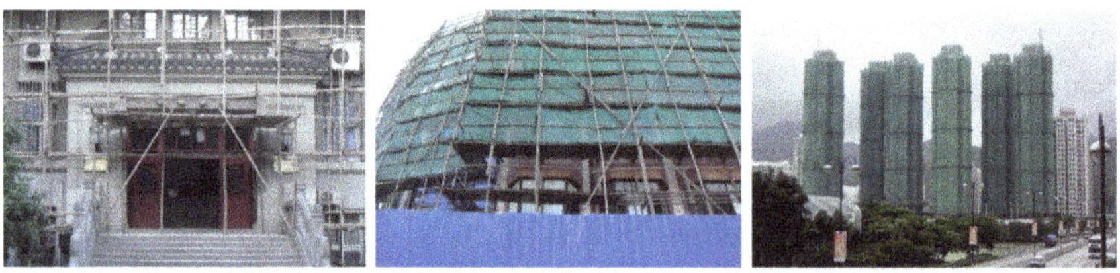

(a) Over portal having little redundancy (*photo by co-author Xu*)

(b) Mid-rise construction (*photo by co-author Xu*)

(c) High-rise construction (six towers clad in bamboo scaffolding) (*www.globalphotos.org*)

Fig. 2.6 Examples of bamboo scaffolding [21]

bamboo scaffolding has been used for 60-storey buildings. It is reinforced every 6 to 8 storeys, and a comprehensive safety inspection is required for every interlace for subsequent use. Therefore, bamboo scaffolding can still be widely used on the basis of a full understanding of the relationship between bamboo and scaffolding system to ensure standardised practices.

2.4.2 Bamboo-Reinforced Concrete Structures

In the early 1900's, bamboo bars were used to replace steel bars as reinforcement in concrete construction [87, 88]. Typically, however, the strength and durability of bamboo-reinforced concrete members were found to be unsatisfactory. A primary reason for poor performance was that without waterproof treatment, bamboo bars would absorb water during concrete placement. Once the concrete cures, drying shrinkage affects the bamboo resulting in the bamboo separating from the surrounding concrete resulting in very poor bond performance [89].

In the 1950s, interest in bamboo-reinforced concrete resurged as a potential means of rebuilding marine facilities throughout the Pacific theatre following the Second World War [90]. Researchers from all over the world began to study means of treating and preserving bamboo bars for use in concrete [91]. Experimental evaluation of the performance of bamboo-reinforced beams, slabs and walls was found to be improved using effective waterproofing measures [89, 92, 93]. Despite some bamboo-reinforced concrete construction in China in the mid-twentieth century, bond between the [typically asphalt-coated] waterproofed bamboo and concrete was poor resulting in lower-than expected capacities [94] although the bamboo itself remained in good shape, showing little degradation [86]. Bamboo-reinforced concrete construction was eventually banned in China [95, 96].

More recently, stiffer water-proof coatings (typically epoxy) that incorporate a sand bond-enhancement layer have been proposed and tested [93]. Anchorage length [97] and bond remain an issue however and no know studies have achieved bond strength approaching that of deformed steel reinforcing bars [55]. Heat-treated, densified engineered-bamboo composite bars having a sand coating have been demonstrated to provide both good dimensional stability when embedded in concrete and bond that reaches about 80% of the minimum achieved by steel bars [98]. Nonetheless, the modulus of bamboo remains at least an order of magnitude lower than steel making design using conventional approaches for reinforced concrete wholly inappropriate [55].

In a recent review of bamboo-reinforced concrete, Archilla et al. [55], echoing Li's paper title "*Bamboo-reinforced concrete members are not suitable for load carrying structures*" [95] conclude: "*bamboo-reinforced concrete is an ill-considered concept, having significant durability, strength and stiffness issues, and does not meet the environmentally friendly credentials often attributed to it*". Furthermore, bamboo-reinforced concrete is often justified as being an 'environmentally friendly' alternative to steel-reinforced concrete; it is not. To be safe, bamboo-reinforced concrete members must be considerably larger than steel-reinforced members, requiring more concrete and formwork [55]. The treatment required to both protect the bamboo and ensure adequate bond is both costly and often toxic [55]. Vo and Navard [99] draw a very prescient conclusion: "*A large proportion of* [the methods used to overcome issues of biomass durability when embedded in concrete] *are effectively helpful in easing the concrete preparation and leading to better final materials. However, most of them, if not all, have little practical value since they are either impossible to be implemented because of the use of chemicals which are not environmentally-friendly or much too expensive.*"

This is in no means to say that bamboo acting in a composite manner with concrete may not make an efficient structural system. Light cement bamboo frame (LCBF) panels, known colloquially as bahareque construction [100], are well established. LCBF construction is a modern technique, recognised and promoted by ISO 22156 [57], utilising composite shear panels composed of a wall matrix of bamboo or metal lath nailed onto a bamboo framing system, plastered with cement or lime mortar render. This system works well because the stresses in the wall matrix are very low.

Korde et al. [101] have demonstrated a bamboo-concrete composite tied arch system spanning 4.5 m and propose spans of 10 m.

2.4.3 Bamboo-Reinforced Masonry Structures

Because of the different role masonry reinforcement plays (as different from concrete reinforcement), bamboo reinforcement may be suitable for hollow-core masonry in non-seismic environments [55]. Studies have demonstrated the use of small bamboo culms used as in-plane reinforcement for concrete masonry walls [102] and simple design calculations have been proposed [103]. Nonetheless, durability concerns remain and bamboo used in such an application will require waterproof and bond-enhancing treatment [102] which draws into question its efficacy [99].

2.4.4 Application of Bamboo Bars in Soil Reinforcement Engineering

The application of bamboo as a method of reinforcing soil engineering has a long history. Examples include Fujian earth building in China, in which a wall made of raw soil mixed with fine sand, lime, bamboo sheets, wood strips, etc., is constructed by repeated rubbing, kneading and pressing in an approach similar to cob construction. Bamboo bars have been proposed to reinforced earth engineering, including slope treatment [104], soft soil foundation reinforcement [105, 106] and deep foundation pit consolidation [107]. The experimental studies, numerical analyses and engineering practices show that bamboo bars can effectively reinforce the soil although the method of bamboo treatment to ensure durability is critical. A unique approach results in adequate soil consolidation before the bamboo deteriorates in situ [105, 106]. Although bamboo piles have been shown to adequately stabilise a deep foundation pit, large deformations are present making the approach of limited utility [107].

The application of bamboo bars in reinforced soil engineering is still in its infancy. Common wisdom is that regardless of treatment, bamboo should not be used in contact with soil or embedded into cement-based mortars [54, 55]. New methods of treatment may make such applications feasible but need to be considered in light of the conclusion of Vo and Navard [99], reported in Sect. 2.4.2.

2.4.5 Application of Bamboo in Strengthening of Existing Structures

Engineered bamboo laminates have been proposed for use as post-applied repair materials, especially well-suited to timber structures. Bolted [108] and adhesively bonded [109] applications have been proposed. Xu et al. [109] reinforced timber beams using adhesive-applied bamboo sheets (obtained by simple processing of full-culm bamboo) and demonstrated increases in capacity of up to 83% in addition to increases in ultimate deflection capacity. The adhesive bonding technique was demonstrated to work well. Zhu et al. [110] conducted tests of reinforced concrete beams strengthened with bamboo sheets reporting an approximately 25% increase in ultimate capacity. The adhesive-applied bamboo was also shown to controlled crack widths well.

Bhattacharya et al. [111] tied bamboo culms to the outside of masonry walls to reinforce the wall and buttress columns and ring beams. The experimental results demonstrated that the bearing capacity of the wall can be doubled using this reinforcement method.

In order to improve the buckling resistance of steel members, Feng et al. [112] wrapped bamboo strips around steel members, using steel wires to bind the bamboo strips. Then fibre-reinforced polymer (FRP) sheets were intertwined around the bamboo strips to restrain the member horizontally and vertically. The experimental results showed that this strengthening method can significantly improve the bearing capacity and ductility of cruciform and I-shaped steel members.

References

1. Chen, F. M., Wang, G., Cheng, H. T., et al. (2016). Development of advanced bamboo fiber based composite material. *Journal of Northeast Forestry University, 44*(02), 80–85. (in Chinese).
2. Abdul Khalil, H. P. S., Bhat, I. U. H., Jawaid, M., et al. (2012). Bamboo fibre reinforced biocomposites: A review. *Materials and Design, 42*, 353–368.
3. Qi, J. Q., Xie, J. L., Huang, X. Y., et al. (2014). Influence of characteristic inhomogeneity of bamboo culm on mechanical properties of bamboo plywood: Effect of culm height. *Journal of Wood Science, 60*, 396–402.
4. Harries, K., Bumstead, J., Richard, M., et al. (2017). Geometric and material effects on bamboo buckling behaviour. *ICE Structures and Buildings, 170*(4), 236–249.
5. Janssen, J. (1981). *Bamboo in building structures. PhD dissertation.* Eindhoven University of Technology.
6. Liese, W. (1998). *The anatomy of bamboo culms* (204 pp). Brill.
7. Taylor, D., Kinane, B., Sweeney, C., et al. (2015). The biomechanics of bamboo: Investigating the role of the nodes. *Wood Science and Technology, 49*(2), 345–357.
8. Qi, J. Q., Xie, J. L., Yu, W. J., et al. (2015). Effects of characteristic inhomogeneity of bamboo culm nodes on mechanical properties of bamboo fiber reinforced composite. *Journal of Forestry Research, 26*, 1057–1060.
9. Janssen, J. J. A. (2012). *Mechanical properties of bamboo.* Springer Science & Business Media.
10. Harries, K. A., Sharma, B., & Richard, M. (2012). Structural use of full culm bamboo: The path to standardization. *International Journal of Architecture, Engineering and Construction, 1*(2), 66–75.
11. Akinbade, Y., Nettleship, I., Papadopoulos, C., et al. (2021). Modelling round bamboo as a naturally varying functionally graded material. *Wood Science and Technology, 55*, 155–179.
12. Akinbade, Y., & Harries, K. (2021). Is the rule of mixture appropriate for assessing bamboo material properties? *Construction and Building Materials, 267*, 120955.
13. ISO 22157:2019. (2019). *Bamboo structures—Determination of physical and mechanical properties of bamboo culms—Test methods.* Switzerland: International Organization for Standardization.
14. ISO 19624:2018. (2018). *Bamboo structures—Grading of bamboo culms—Basic principles and procedures.* Switzerland: International Organization for Standardization.
15. Trujillo, D., & Harries, K. (2020). *Developing ISO Standards for Bamboo*, Timber 2020, September 2020, London [virtual].
16. GB/T 15780-1995. (1995). *Testing methods for physical and mechanical properties of bamboos.* Standards Press of China (in Chinese).
17. JG/T 199-2007. (2007). *Testing methods for physical and mechanical properties of bamboo used in building.* Standards Press of China (in Chinese).
18. Liu, B., Chen, Z. Y., Yin, Y. F., et al. (2008). Comparison of two standards for evaluating physical and mechanical properties of bamboo materials. *China Wood Industry, 22*(4), 26–29. (in Chinese).
19. Gauss, C., Savastano, H., & Harries K. (2019). Use of ISO 22157 mechanical test methods and the characterisation of Brazilian P. edulis bamboo. *Construction and Building Materials, 228*, 116728.
20. Wang, H., Wang, H. K., Li, W. J., et al. (2013). Effects of moisture content on the mechanical properties of moso bamboo at the macroscopic and cellular levels. *BioResources, 8*(4), 5475–5484.
21. Xu, Q. F., Harries, K. A., Li, X. M., et al. (2014). Mechanical properties of structural bamboo following immersion in water. *Engineering Structures, 81*, 230–239.
22. Correal, D. J. F., & Arbeláez, C. J. (2010). Influence of age and height position on Colombian Guadua angustifolia bamboo mechanical properties. *Maderas. Ciencia y tecnología, 12*(2), 105–113.

23. Saito, Y., Ishii, M., & Sato, M. (2013). The suitable harvesting season and the part of moso bamboo (Phyllostachys pubescens) for producing binderless boards. *Wood Science and Technology, 47*(5), 1071–1081.
24. Chung, K., & Chan, S. (2002). *Bamboo scaffolds in building construction.* International Network for Bamboo and Rattan (INBAR).
25. Naik, N. (2005). *Mechanical and physico-chemical properties of bamboos carried out by aerospace engineering department.* Indian Institute of Technology-Bombay.
26. Mbuge, D. O. (2000). *Mechanical properties of bamboo (Bambusa vulgaris) grown in Muguga, Kenya.* MSc of University of Nairobi.
27. RFD & ITTO. (2013). *Physical and mechanical properties of some Thai bamboos for house construction.* Royal Forest Department of Thailand (RFD) and International Tropical Timber Organization (ITTO).
28. Seethalakshmi, K. K., & Muktesh Kumar, M. S. (1998). *Bamboos of India: a compendium.* Peechi Bamboo Information Centre, India, Kerala Forest Research Institute.
29. Gao, L., Wang, Z., Lin, T., et al. (2010). A comparative study of main physical and mechanical properties of Arundinaria alpine and Phyllostachys pubescens. *World Bamboo and Rattan, 8*(4), 20–22. (in Chinese).
30. Xiao, Y., & Shan, B. (2013). *Glubam structures.* China Architecture & Building Press. (in Chinese).
31. Jiang, Z. H., Chang, L., Wang, Z., et al. (2005). Physical and mechanical properties of glued structural laminated bamboo. *China Wood Industry, 19*(4), 22–24. (in Chinese).
32. Zhang, Y. T., & He, L. P. (2007). Comparison of mechanical properties for glued laminated bamboo wood and common structural timbers. *Journal of Zhejiang Forestry College, 24*(1), 100–104. (in Chinese).
33. Sulastiningsih, I., & Nurwati. (2009). Physical and mechanical properties of laminated bamboo board. *Journal of Tropical Forest Science, 21*(3), 246–251.
34. Mahdavi, M., Clouston, P., & Arwade, S. (2010). Development of laminated bamboo lumber: Review of processing, performance, and economical considerations. *Journal of Materials in Civil Engineering, 23*(7), 1036–1042.
35. Mahdavi, M., Clouston, P., & Arwade, S. (2012). A low-technology approach toward fabrication of laminated bamboo lumber. *Construction and Building Materials, 29*, 257–262.
36. Correal, J., Echeverry, J., Ramírez, F., et al. (2014). Experimental evaluation of physical and mechanical properties of glued laminated guadua angustifolia kunth. *Construction and Building Materials, 73*, 105–112.
37. Correal, J., Ramirez, F., Gonzalez, S., et al. (2010). Structural behavior of glued laminated guadua bamboo as a construction material. In *Proceedings of the 11th World Conference on Timber Engineering*, Trentino, Italy.
38. Xiao, Y., Yang, R. Z., & Shan, B. (2013). Production, environmental impact and mechanical properties of glubam. *Construction and Building Materials, 44*, 765–773.
39. Verma, C., & Chariar, V. (2012). Development of layered laminate bamboo composite and their mechanical properties. *Composites Part B: Engineering, 43*, 1063–1069.
40. Verma, C., Sharma, N., Chariar, V., et al. (2014). Comparative study of mechanical properties of bamboo laminae and their laminates with woods and wood based composites. *Composites Part B: Engineering, 60*, 523–530.
41. Li, H. T., Zhang, Q. S., Huang, D. S., et al. (2013). Compressive performance of laminated bamboo. *Composites Part B: Engineering, 54*, 319–328.
42. Li, H. T., Zhang, Q. S., & Wu, G. (2015). Stress-strain model of side pressure laminated bamboo under compression. *Journal of Southeast University (Natural Science Edition), 45*(6), 1130–1134. (in Chinese).
43. Li, H. T., Wu, G., Zhang, Q. S., et al. (2016). Experimental study on side pressure LBL under Tangential eccentric compression. *Journal of Hunan University (Natural Sciences), 43*(5), 90–96. (in Chinese).
44. Sharma, B., Gatóo, A., & Ramage, M. H. (2015). Effect of processing methods on the mechanical properties of engineered bamboo. *Construction and Building Materials, 83*, 95–101.

45. Sharma, B., Gatóo, A., Bock, M., et al. (2015). Engineered bamboo for structural applications. *Construction and Building Materials, 81*, 66–73.
46. Guan, M. J., & Zhu, Y. X., & Zhang, X. A. (2006). Comparison of bending properties of scrimber and bamboo scrimber. *Journal of Northeast Forestry University, 34*(4), 7 (in Chinese).
47. Zhang, J. Z., Ren, H. Q., Zhong, Y., et al. (2012). Analysis of compressive and tensile mechanical properties of recombinant bamboo. *Journal of Nanjing Forestry University (Natural Science Edition), 36*(4), 107–111. (in Chinese).
48. Huang, D. S., Zhou, A. P., & Bian, Y. L. (2013). Experimental and analytical study on the nonlinear bending of parallel strand bamboo beams. *Construction and Building Materials, 44*, 585–592.
49. Huang, D. S., Bian, Y. L., Zhou, A. P., et al. (2015). Experimental study on stress-strain relationships and failure mechanisms of parallel strand bamboo made from phyllostachys. *Construction and Building Materials, 77*, 130–138.
50. Li, H. T., Su, J. W., Wei, D. D., et al. (2016). Comparison study on parallel bamboo strand lumber under axial compression for different directions based on the large scale. *Journal of Zhengzhou University (Engineering Science), 37*(2), 67–72. (in Chinese).
51. Chen, L. B. (2015). Study on control of Phenolic Resin in production process of bamboo recombinant material. *Anhui Agricultural Science Bulletin, 21*(20), 98–101. (in Chinese).
52. Long, W. G., Yang, X. B., Wang, Y. W., et al. (2005). *Design manual for timber structures (The third edition).* China Architecture & Building Press. (in Chinese).
53. Jayanetti, D., & Follet, P. (1998). *Bamboo in construction: an introduction.* International Network for Bamboo and Rattan (INBAR).
54. Kaminski, S., Lawrence, A., Trujillo, D., et al. (2016). Structural use of bamboo. Part 2: durability and preservation. *Structural Engineer, 94*(10), 38–43.
55. Archila, H., Kaminski, S., Trujillo, D., et al. (2018). Bamboo reinforced concrete—a critical review. *Materials and Structures, 51*, 102.
56. Gottron, J., Harries, K. A., & Xu, Q. F. (2014). Creep behaviour of bamboo. *Construction and Building Materials, 66*(1), 79–88.
57. ISO 22156:2021. (2021). *Bamboo structures—Bamboo culms—Structural design.* International Organization for Standardization.
58. Mena, J., Vera, S., Correal, J. F., et al. (2012). Assessment of fire reaction and fire resistance of guadua angustifolia kunth bamboo. *Construction and Building Materials, 27*(1), 60–65.
59. Xu, Q. F., Chen, L. Z., Harries, L., et al. (2017). Combustion performance of engineered bamboo from cone calorimeter tests. *European Journal of Wood and Wood Products, 75*, 161–173.
60. Liu, K. W., & Oliver, F. (2013). An overview of global bamboo architecture: Trends and challenges. *World Architecture, 12*, 27–34. (in both English and Chinese).
61. Chen, F., Li, M., Xin, W., et al. (2019). Development of bamboo winding composite pipe (BWCP) and its compression properties. *BioResources, 14*(3), 5875–5882.
62. ASTM D5456-19. (2019). *Standard specification for evaluation of structural composite lumber products.* ASTM International.
63. GB/T 13942.1-2009. (2009). *Durability of wood—Part 1: Method for laboratory test of natural decay resistance.* Standards Press of China (in Chinese).
64. GB/T 18261-2013. (2014). *Test method for anti-mildew agents in controlling wood mould and stain fungi.* Standards Press of China (in Chinese).
65. GB/T 13942.2-2009. (2009). *Durability of wood—Part 2: Method for field test of natural durability.* Standards Press of China (in Chinese).
66. Huang, Y. X., Ji, Y. H., & Yu, W. J. (2019). Development of bamboo scrimber: A literature review. *Journal of Wood Science, 65*(1), 1–10.
67. Yuan, Z. R., Wu, X. W., Wang, X. Z., et al. (2020). Effects of one-step hot oil treatment on the physical, mechanical, and surface properties of bamboo scrimber. *Molecules, 25*(19), 4488.
68. Yang, K., Li, X. G., Wu, Y. Q., et al. (2021). A simple, effective and inhibitor-free thermal treatment for enhancing mold-proof property of bamboo scrimber. *European Journal of Wood and Wood Products* , 1–7.

69. Jin, J. W., Wu, C., Qin, D. C., et al. (2016). Decay resistance of bamboo oriented strand board pretreated with copper-based preservatives. *BioResources, 11*(1), 1541–1553.
70. Yan, X., Li, Z., Wang, R., et al. (2012). Mechanical properties of Glubam sheets after artificial accelerated aging. *Key Engineering Materials, 517*, 43–50.
71. Xu, M., Cui, Z., Tu, L., et al. (2019). The effect of elevated temperatures on the mechanical properties of laminated bamboo. *Construction and Building Materials, 226*, 32–43.
72. Xu, M., Cui, Z., Chen, Z., et al. (2017). Experimental study on compressive and tensile properties of a bamboo scrimber at elevated temperatures. *Construction and Building Materials, 151*, 732–741.
73. Shah, D., Bock, M., Mulligan, H., et al. (2016). Thermal conductivity of engineered bamboo composites. *Journal of Materials Science, 51*(6), 2991–3002.
74. Zhong, Y., Wen, L. L., & Zhou, H. B. (2014). Study on bending performance of laminated bamboo board in and after exposure to high temperature. *Journal of Building Materials, 17*(6), 1115–1120. (in Chinese).
75. Zhou, Q., She, L. Y., Xiao, Y., et al. (2011). Fire-resistance simulation and test of prefabricated bamboo house. *Journal of Building Structures, 32*(7), 60–65. (in Chinese).
76. Xiao, Y., & Ma, J. (2012). Fire simulation test and analysis of laminated bamboo frame building. *Construction and building materials, 34*, 257–266.
77. Jiang, S. X., Zhang, Q. S., & Jiang, S. H. (2002). On structure, production, and market of bamboo-based panels in China. *Journal of Forestry Research, 13*(2), 151–156.
78. Yu, X., Zeng, L., Zhang, G., et al. (2019). Environmental impact of bamboo laminated flooring and bamboo scrimber flooring investigated via life cycle assessment. *BioResources, 14*(4), 9132–9145.
79. Zhang, W. F., Wang, J., Wang, G., et al. (2015). Comparative analysis of physical-mechanical properties of different kinds of bamboo flooring. *Forestry Machinery & Woodworking Equipment, 5*, 14–16. (in Chinese).
80. Deng, J. C., Chen, F. M., Wang, G., et al. (2014). The implication of painting treatments on water and wear resistance of bamboo flooring. *China Forest Products Industry, 41*(2), 50–52. (in Chinese).
81. Chen, J. P., Guagliano, M., Shi, M. H., et al. (2020). A comprehensive overview of bamboo scrimber and its new development in China. *European Journal of Wood and Wood Products*, 1–17.
82. Liu, Z. J., Wang, J. X., Fei, B. H., et al. (2017). Investigating fire performance of heat-treated bamboo scrimber flooring. *General Technical Report-Forest Products Laboratory, USDA Forest Service, 2017*(FPL-GTR-249), 349–358.
83. Vân, N. T. B. (2018). Bamboo-the eco-friendly material–one of the material solutions of the sustainable interior design in Viet Nam. *MATEC Web of Conferences. EDP Sciences, 193*, 04014.
84. Chung, K., & Yu, W. (2002). Mechanical properties of structural bamboo for bamboo scaffoldings. *Engineering Structures, 24*(4), 429–442.
85. Yu, W., Chung, K., & Chan, S. (2003). Column buckling of structural bamboo. *Engineering Structures, 25*(6), 755–768.
86. Yu, W., Chung, K., & Chan, S. (2005). Axial buckling of bamboo columns in bamboo scaffolds. *Engineering Structures, 27*(1), 61–73.
87. Chow, H. K. (1914). *Bamboo as a material for reinforcing concrete.* Massachusetts Institute of Technology. http://hdl.handle.net/1721.1/81516.
88. Datta, K. (1936). *Versuche Uber die Verwendung von Bambus im Beton bau [Experiments on the use of bamboo in concrete construction]* (Vol. 17, no. 1). Der Bauingenieur (BerlinWilmersdorf).
89. Hu, X. F. (1957). Waterproof treatment of bamboo (waterproof treatment of bamboo in bamboo-reinforced concrete). *Journal of Tongji University, 4*(1), 185–192. (in Chinese).
90. Glenn, H. E. (1950). *Bamboo reinforcement in Portland cement concrete.* Engineering Experiment Station, Clemson Agricultural College, Clemson, South Carolina, Bulletin No. 4.

91. JTJ 025-86. (1986). *Specifications for design of steel structure and timber structure highway bridges and culverts*. China Communications Press (in Chinese).
92. Khatib, A., & Nounu, G. (2017). Corrugated bamboo as reinforcement in concrete. *Proceedings of the Institution of Civil Engineers-Structures and Buildings, 170*(4), 311–318.
93. Ghavami, K. (1995). Ultimate load behaviour of bamboo-reinforced lightweight concrete beams. *Cement and Concrete Composites, 17*(4), 281–288.
94. Janssen, J. (2000). *Designing and building with Bamboo*. International Network for Bamboo and Rattan (INBAR).
95. Li, D. H. (1986). Bamboo-reinforced concrete members are not suitable for load carrying structures. *Architecture Technology, 4*(1), 52–53. (in Chinese).
96. Notice of the State Construction Committee on the suspension of the use of bamboo-reinforced concrete. *The Bulletin of the State Council of the People's Republic of China, 1957*(19), 354–355 (in Chinese).
97. Schneider, N., Pang, W., & Gu, M. (2014) *Application of bamboo for flexural and shear reinforcement in concrete beams* (pp. 1025–1035). Structures Congress 2014, American Society of Civil Engineers, Boston, Massachusetts, United States.
98. Javadian, A., Wielopolski, M., Smith, F. C., et al. (2016). Bond-behavior study of newly developed bamboo-composite reinforcement in concrete. *Construction and Building Materials, 122*, 110–117.
99. Vo, L. T. T., & Navard, P. (2016). Treatments of plant biomass for cementitious building materials—a review. *Construction and Building Materials, 121*, 161–176.
100. Gonzalez, G., & Gutierrez, J. (2005). Structural performance of bamboo 'bahareque' walls under cyclic load. *J Bamboo Rattan, 4*(4), 353–368.
101. Korde, C., West, R., Gupta, A., et al. (2015). Laterally restrained bamboo concrete composite arch under uniformly distributed loading. *Journal of Structural Engineering, ASCE, 141*(3), B4014005.
102. Moroz, J., Lissel, S., & Hagel, M. (2014). Performance of bamboo-reinforced concrete masonry shear walls. *Construction and Building Materials, 61*(1), 125–137.
103. Lyer, S. (2002). *Guidelines for building bamboo-reinforced masonry in earthquake-prone areas in India*. University of Southern California.
104. As'ad Munawir, M. D., Zaika, Y., & Agoes Soehardjono, M. D. (2013). Bearing capacity on slope modeling with composite bamboo pile reinforcement. *Jurnal Terpublikasi: International Journal of Engineering and Advanced Technology (IJEAT)*.
105. Suyuti, M. A. S., & Zulkarnain, K. M. (2019). Bearing capacity of soil bags on soft ground reinforced by bamboo pile. *International Journal of GEOMATE, 16*(53), 32–39.
106. Hakim, R. (2020). Bamboo pile installed into soft ground to increase required reinforcement force for coastal dike. *Journal of Physics: Conference Series, 2020, 1569*(4), 042042.
107. Dai, Z., Chen, Y., Zheng, G., et al. (2014). *Numerical analysis on the mechanism of bamboo soil nails and bamboo piles in rows for retaining deep foundation pit* (pp. 720–730). GeoShanghai International Congress: Tunneling and Underground Construction, American Society of Civil Engineers (ASCE), Shanghai, China.
108. Platt, S., & Harries, K. (2015). Strength of bolted bamboo laminate connections. In *Proceedings 15th International Conference Non-conventional Materials and Technologies (NOCMAT 15)*, Winnipeg, Canada.
109. Xu, Q. F., Chen, J. F., & Li, X. M. (2012). Study on timber beams strengthened with bamboo strips. *Journal of Sichuan University (Engineering Science Edition), 44*(1), 36–42. (in Chinese).
110. Zhu, L., Xu, Q. F., & Chen, J. F. (2012). Strengthening RC beams with externally bonded bamboo strips. *Structural Engineers, 28*(3), 141–146. (in Chinese).
111. Bhattacharya, S., Nayak, S., & Dutta, S. (2014). A critical review of retrofitting methods for unreinforced masonry structures. *International Journal of Disaster Risk Reduction, 7*(1), 51–67.
112. Feng, P., Zhang, Y., Bai, Y., et al. (2013). Combination of bamboo filling and FRP wrapping to strengthen steel members in compression. *Journal of Composites for Construction, 17*(3), 347–356.

Chapter 3
Research and Development Status of Different Types of Bamboo Structures

This chapter summarises the history and current research and development status of both full culm bamboo structures and engineered bamboo structures. Although it is difficult to map research and development, insight into both the depth and breadth of a field can be obtained by looking at the record of academic publishing in the field. As will be discussed in this chapter and is shown in Fig. 3.1, bamboo research is on a dramatic upward trend, especially in the twenty-first century. A recent search of the *Scopus* indexing data base[1] revealed 7928 scholarly articles attributed to bamboo "engineering" or "materials science". The majority of these articles have only appeared in the last six years and an overwhelming majority have their country of origin listed as China. The modern bamboo research community is thriving. A more refined review of publication metrics is described in Chap. 5.

3.1 Full Culm Bamboo Structures

The use of round, full culm, bamboo has a long history. About 7000 years ago, at Hemudu, ancestors of Chinese people began to build houses using bamboo with other natural materials [1]. Today, people in many countries still keep the traditions of living in bamboo-constructed buildings. Bahareque houses in Colombia (Fig. 3.2a), Quincha houses in Ecuador (Fig. 3.2b), Sidama houses in Ethiopia (Fig. 3.2c) and Ganlan houses in China (Fig. 3.2d) are all examples of contemporary wide-spread bamboo houses [2]. Most traditional bamboo houses were designed and constructed without the participation of architects and engineers. They were built by local people according to experience passed down from their ancestors. Nonetheless, as seen in Fig. 3.2, such construction is far from being 'informal'.

[1] search ran 2 July 2021 for the word "bamboo" in the article title, abstract or keywords. The search was limited to journal articles and conference articles in the fields of engineering and materials science.

K. W. Liu et al., *Contemporary Bamboo Architecture in China*,
https://doi.org/10.1007/978-981-16-8309-1_3

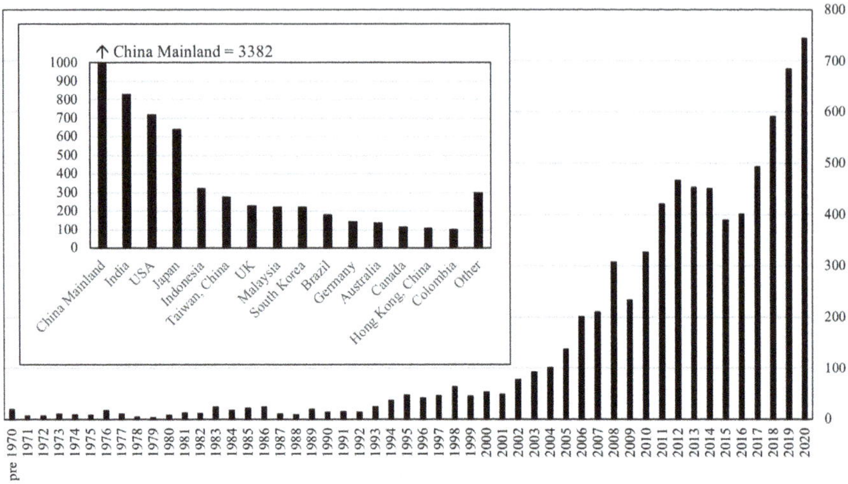

Fig. 3.1 Scopus search results for "bamboo" through 2020

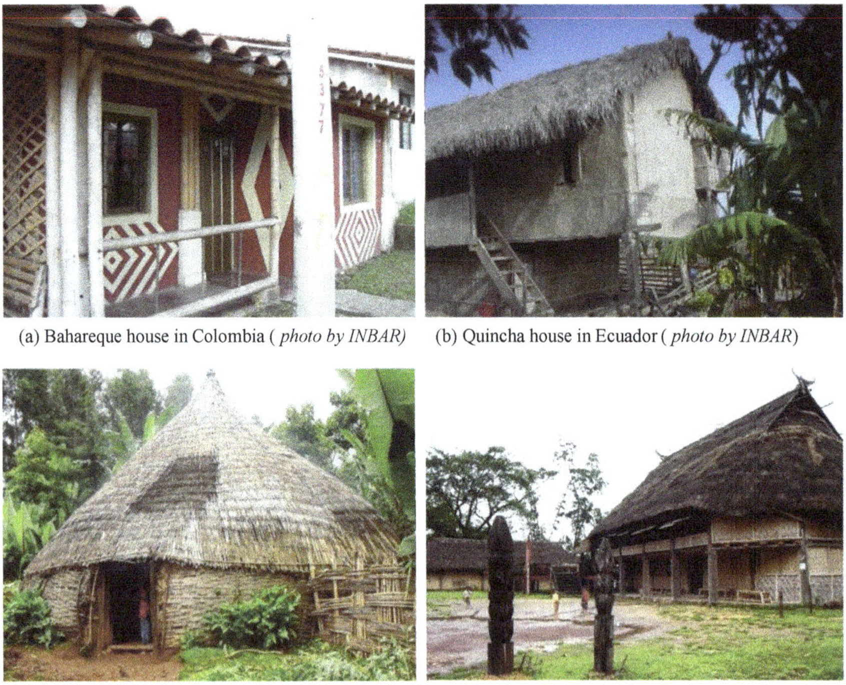

(a) Bahareque house in Colombia (*photo by INBAR*) (b) Quincha house in Ecuador (*photo by INBAR*)

(c) Sidama house in Ethiopia (*photo by INBAR*) (d) Ganlan house in China (*photo by Huang Wenkun*)

Fig. 3.2 Traditional full-culm bamboo houses around the world

Despite the development of modern construction technologies and the gradual encroachment of modern architectural theory, traditional bamboo construction has important roles to play: technically, culturally and socially. It is crucial that the development of modern construction technologies integrate and consolidate the inheritance of traditional materials skills.

Since the 1970s, modern bamboo structures having architects' participation in design and construction have developed in an unprecedented manner. A group of innovative architects, represented by Mr. Simón Veléz from Colombia, not only solved the bearing capacity problem of hollow full-culm bamboo joints, but also overcame the low strength of traditional bamboo joints by pouring cement mortar into the culms in order to anchor mechanical connections such as through-culm bolts. By improving the connection capacity, the potential application of full culm bamboo construction is expanded in addition to permitting construction of larger public buildings [2]. Nonetheless, from ancient times to the present, no formal system of design and construction of full culm bamboo structures has been established.

In 1999, a 6.4 magnitude earthquake struck the coffee-producing area of Colombia. As many as 1000 people died as a result of the collapse of older concrete and masonry structures. However, the traditional Bahareque bamboo houses in the area survived. Subsequently, the Colombian Earthquake Engineering Association (AIS) carried out research on the seismic performance of Bahareque construction. Over the next few years, with the efforts of AIS, Colombia added a new chapter, *One and Two Storey Houses Made of Mortared Bahareque*, to its 2002 issue of *NSR-98: Colombian Standards for Seismic Resistant Design and Construction* (AIS 2002) [3]. In 2004, the International Organization for Standardization (ISO) issued the world's first design standard for bamboo structures (ISO 22156:2004: *Bamboo—Structural Design*). Colombia adapting the intent of the ISO standard, augmented the available research resulting in the inclusion of Chapter G *Guadua Structures* in the revised NSR-10 *Colombian Standards for Seismic Resistant Design and Construction NSR-10* (AIS 2010) [4]. With this, Colombia has become the first country in the world to incorporate the design of full culm bamboo structures into national standards, which has greatly promoted the development of bamboo structures across Latin America. Later, in addition to Colombia, a large number of bamboo structures were also built in Ecuador, Mexico, Jamaica and Panama as shown in Fig. 3.3. It is noted, however, that the Colombian standards consider only a single bamboo species: *Guadua angustifolia.*

In Asia, where bamboo is plentiful, a number of excellent full culm bamboo structures have emerged in the past two decades, such as those shown in Figs. 3.4, 3.5, 3.6, 3.7 and 3.8.

Bamboo has been a traditional building material in southern China for millennia. By the 1980s, however, rapid economic growth drove 'modern' preferences toward concrete, masonry and steel construction. Consideration and research on bamboo stagnated. With the twenty-first century bringing an emphasis on sustainability,

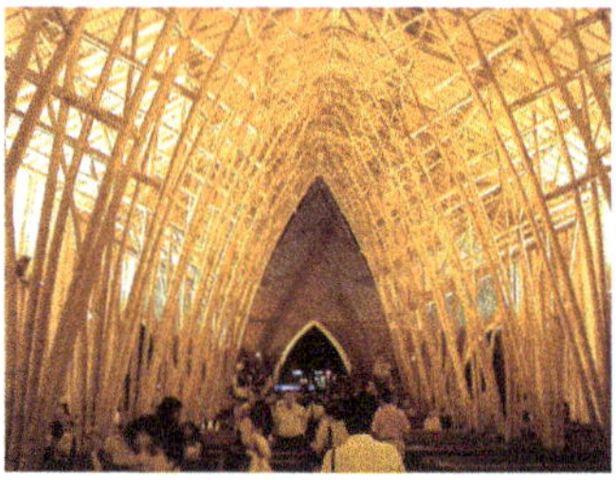

(a) Pereira Church in Colombia *(photo by Simón Veléz Architects)*

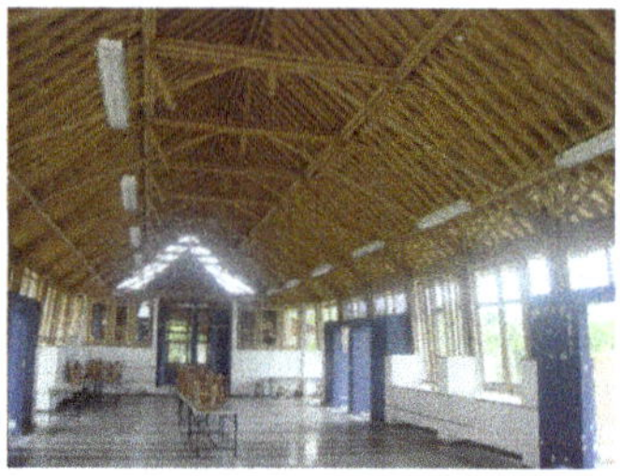

(b) A teaching building in Colombia *(photo by co-author Wang)*

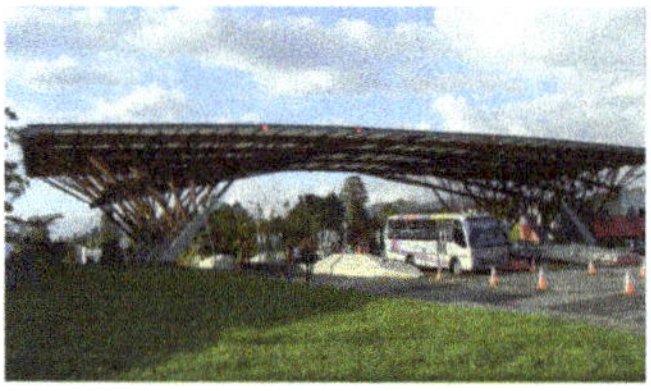

(c) Traffic toll station in Colombia *(photo by co-author Liu)*

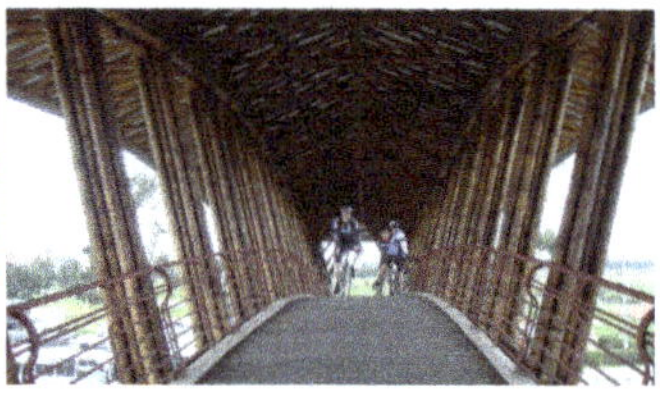

(d) Jenny Garzon Bridge in Colombia *(photo by co-author Liu)*

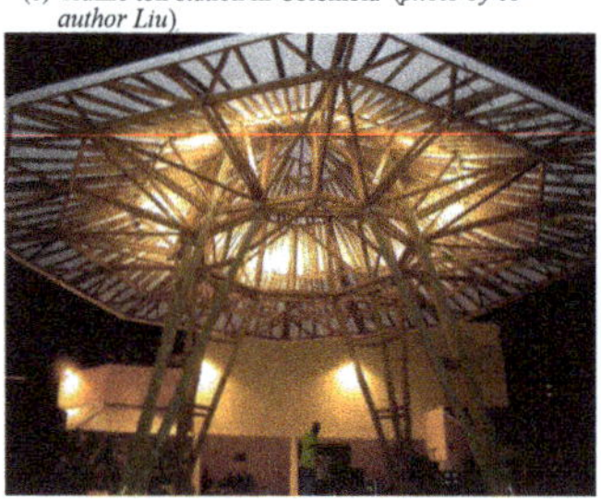

(e) Bob-Marley Memorial Stage in Jamaica *(photo by Simón Veléz Architects)*

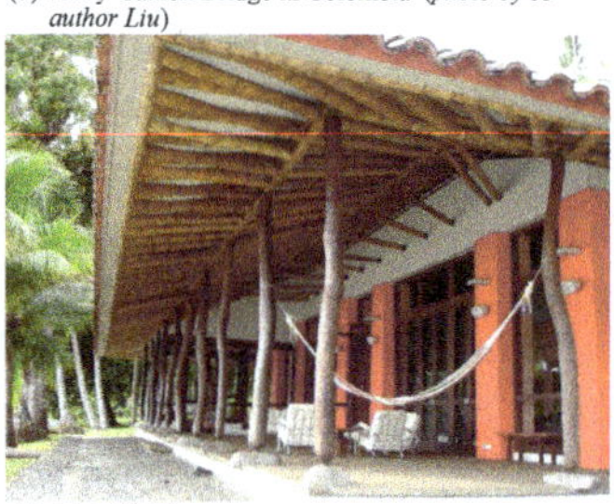

(f) A hotel in Panama *(photo by Simón Veléz Architects)*

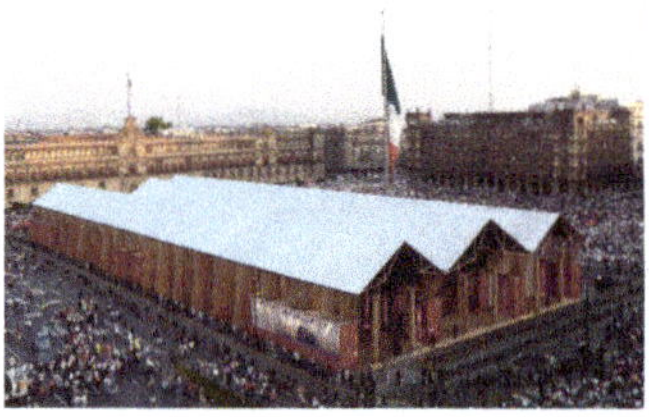

(g) Nomadic Art Museum in Mexico *(photo by Simón Veléz Architects)*

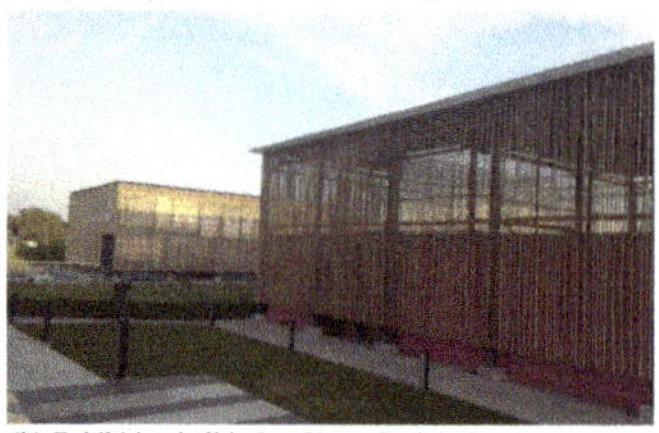

(h) Exhibition hall in Las Voces Park in Peru *(photo by Pablo Jacome)*

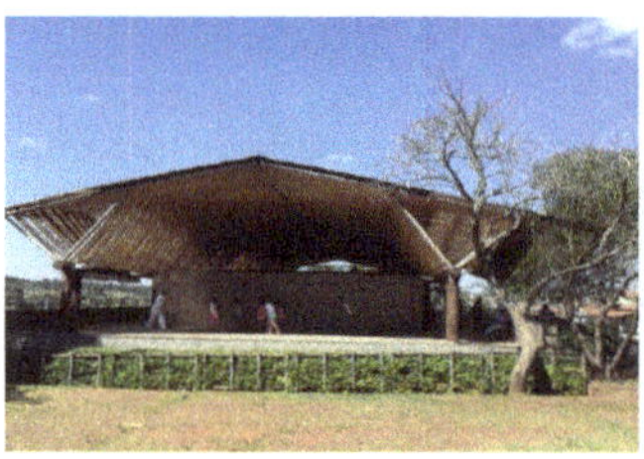

(i) Max Feffer Cultural Center in Brazil *(photo by Pablo Jacome)*

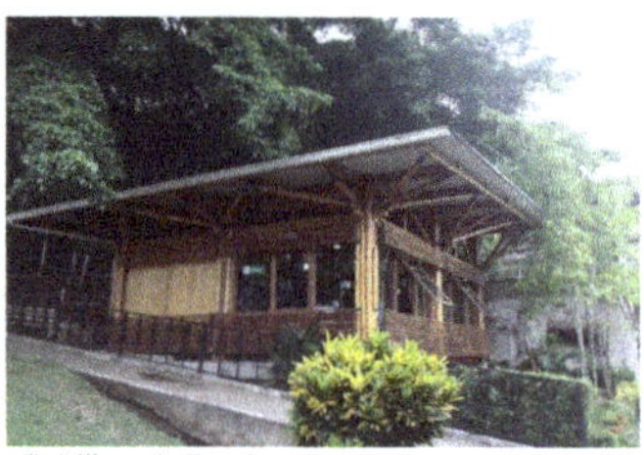

(j) A library in Ecuador *(photo by Pablo Jacome)*

Fig. 3.3 Examples of bamboo structures in Latin America

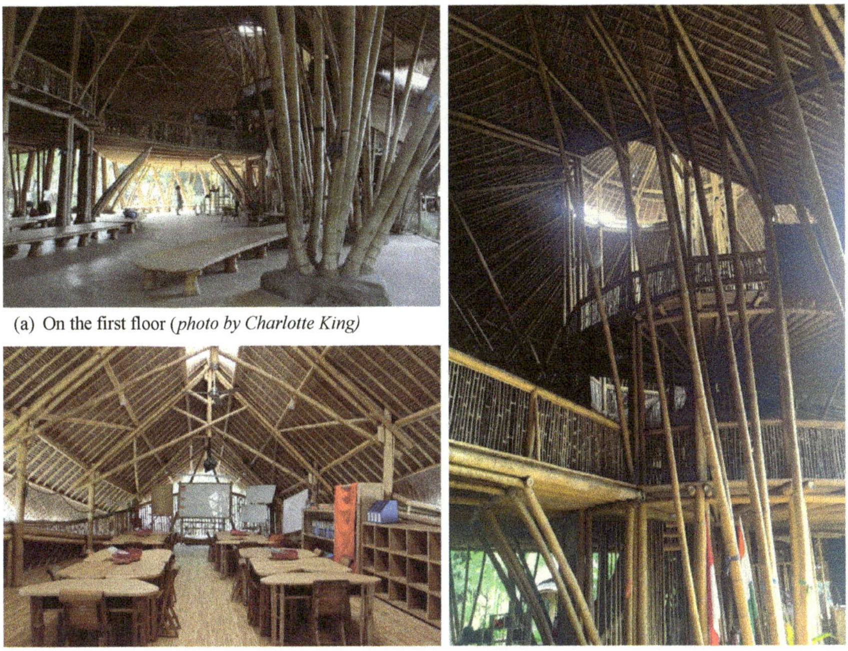

(a) On the first floor (*photo by Charlotte King*)

(b) In the classroom (*photo by Charlotte King*)

(c) View from the first floor to the roof (*photo by Hans Friederich*)

Fig. 3.4 Green School in Indonesia

interest in bamboo has rebounded. Through cultural exchange programs, many international architects—recognising the deep cultural significance of bamboo in China—executed a range of full-culm bamboo structures in China. Only then did bamboo as a construction material return to the vision of the Chinese people.

In 2006, construction began on the Cross Water Ecotourism Resort in the Nankun mountain ecotourism district of Huizhou, Guangdong, China. Many international designers including Simón Veléz were invited to participate in the design. Sr Veléz built a bamboo corridor bridge using construction technologies common in Colombia (Fig. 3.9a, b). The bridge was constructed using native moso (*P. edulis*) bamboo from Nankun mountain and employed traditional Chinese tiles as the roof. In addition, many buildings built with bamboo (Fig. 3.9c, d) were included in the project.

In 2010, a large number of internationally renowned architects gathered at the Shanghai World Expo. A total of nine pavilions were integrated with bamboo elements, including many large pavilions. Among them, the Indian Pavilion [5] (Fig. 3.10) has the world's largest bamboo dome, with a diameter of 35 m and a height of 18 m. The dome consists of 36 bamboo quarter arc-shaped ribs, each consisting of nine 100 mm diameter bamboo culms arranged in a triangular section. More than 500 culms, having a weight of over 41 tons were used.

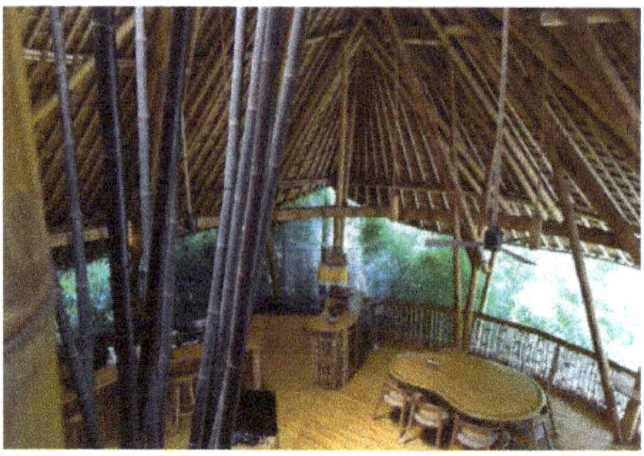

(a) In the restaurant

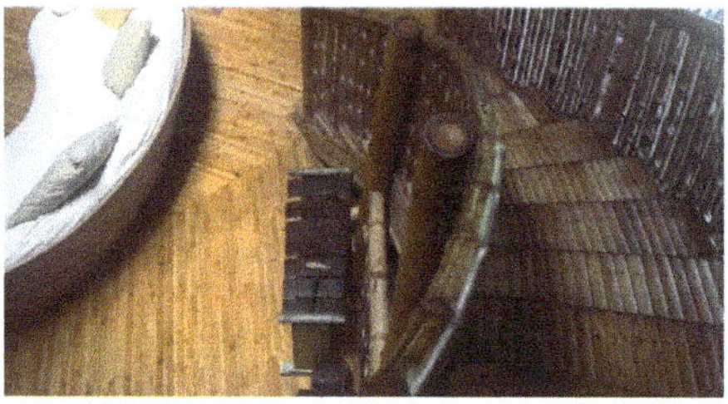

(b) Bamboo stairs

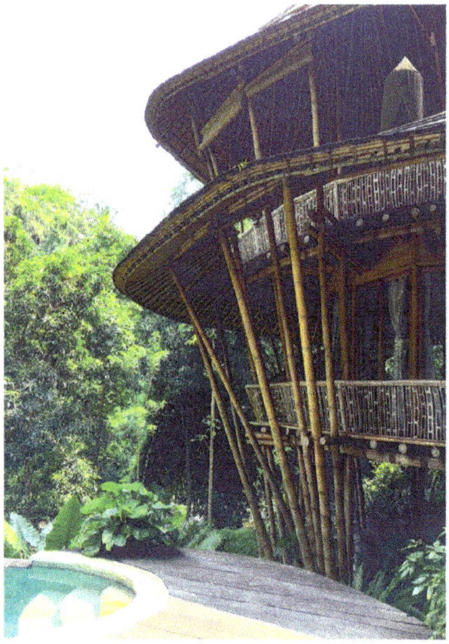

(c) Exterior view

Fig. 3.5 Green Village in Indonesia (*photos by Hans Friederich*)

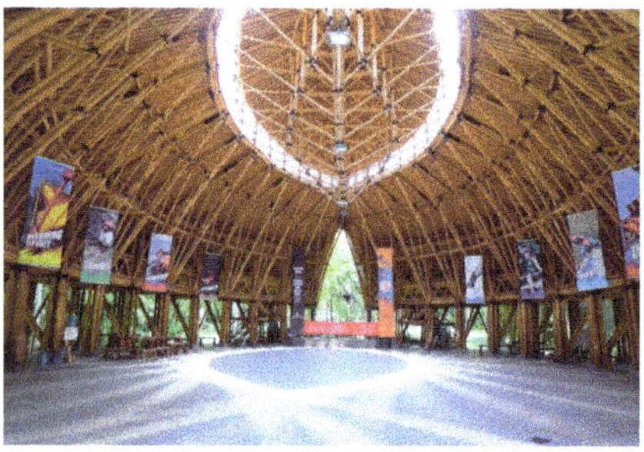

(a) Interior view of OBI Great Hall

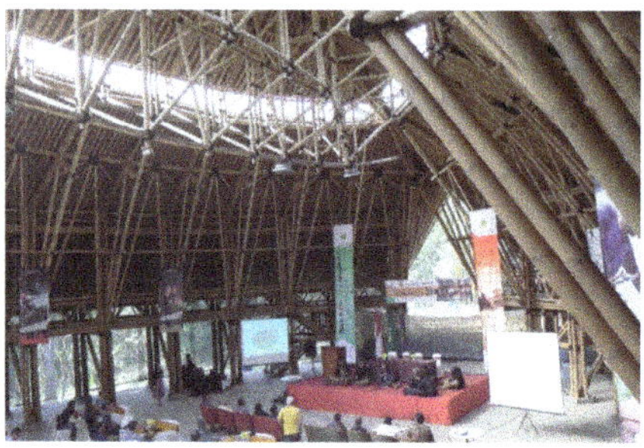

(b) Activity held at OBI Great Hall

Fig. 3.6 OBI Great Hall in Java, Indonesia (*photos by Andry Widyowijatnoko*)

In the Germany-Chinese House at Shanghai World Expo (Fig. 3.11), the bamboo structure was covered with translucent panels which highlighted the beauty and elegance of the full culm and modern glued laminated bamboo used. The two-storey exterior supporting structure of the pavilion was fabricated of 200 mm diameter, 8 m long culms (4-year old *D. giganteus* from Yunnan, China). All bamboo underwent special fire protection treatment before construction and passed a required fire test [6]. The metal connections of Germany-Chinese House are exquisitely designed. One end is embedded in a bamboo culm filled with cement mortar, and the other end is connected by bolts with other components (Fig. 3.11). Inside the building,

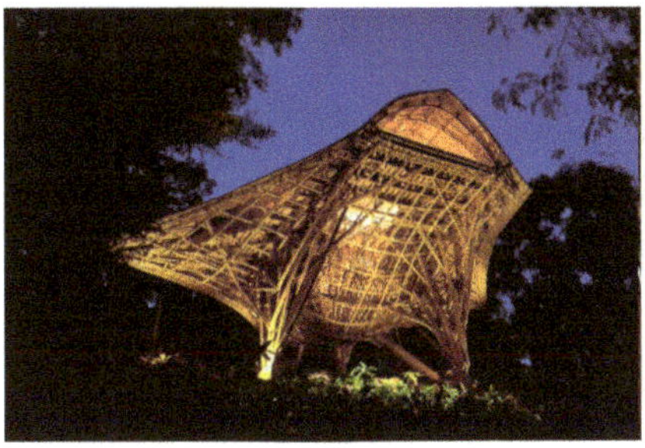

(a) Exterior view of Eco-Children's Activities and Education Center (*photo by Boris Zeisser*)

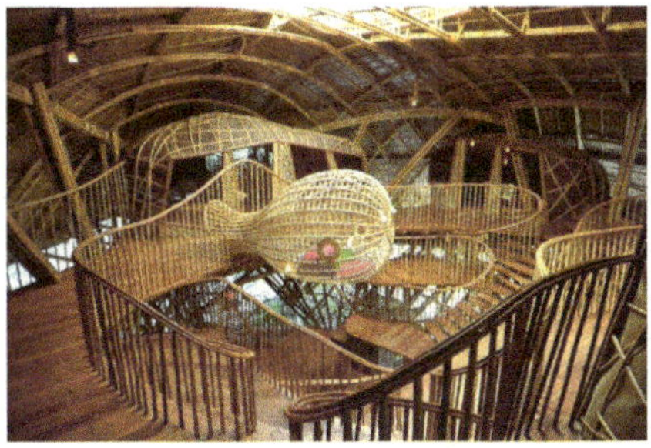

(b) Interior view of Eco-Children's Activities and Education Center (*photo by Boris Zeisser*)

The Eco-Children's Activities and Education Center was awarded Second Prize of the International Bamboo and Rattan Product Design Competition organised by INBAR at the 2010 Shanghai World Expo.

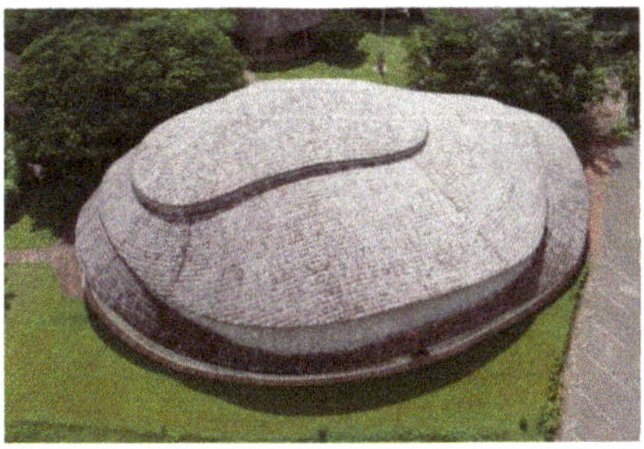

(c) Exterior view of Bamboo Sports Hall at Panyaden International School (*photo by Markus Roselieb*)

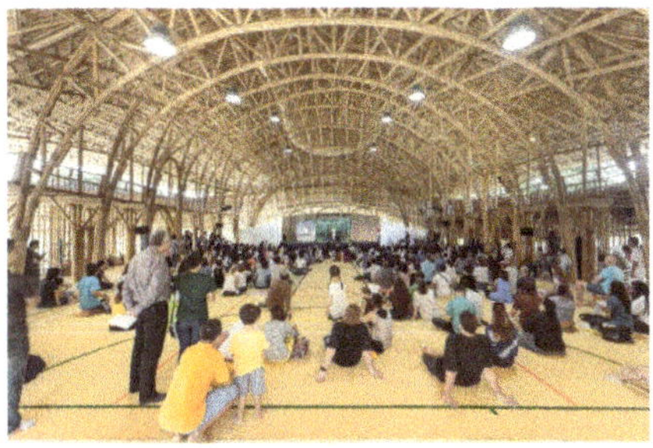

(d) Interior space of Bamboo Sports Hall at Panyaden International School (*photo by Alberto Cosi*)

Fig. 3.7 Full culm bamboo buildings in Thailand

eight portal-shaped glued laminated bamboo frames are arranged parallel to each other to support the second floor. For the first time, this building successfully applied advanced prefabrication technology to the whole process of construction and later non-destructive disassembly [2]. In 2017, the Pavilion was successfully relocated to Hangzhou, China.

In September 2016, the first International Bamboo Architecture Biennale was held in the peaceful village of Xitou, Baoxi Township in Zhejiang, China. Curated by local artist Ge Qiantao and international architect George Kunihiro to illustrate the curatorial theme of "Spirit of Place and Rural Construction", the event showcased the different-styled construction of 18 bamboo-based buildings designed by 12 renowned architects. The biennial exhibition took "on-site" as its way, villages as its carrier, and "bamboo" as its theme. Using natural materials such as local rammed earth, barrier stone and porcelain chips to inspire people's wisdom with "low technology" familiar to the natives, the exhibition explored a path for sustainable development for rural areas in China [7]. The reception centre, designed by Vietnamese architect Vo Trong Nghia (Fig. 3.12a), was built with a two-storey full culm bamboo structure. The bamboo trusses were arranged along the axis to form the main load-bearing structure. Each bamboo truss consists of a bunch of bamboo

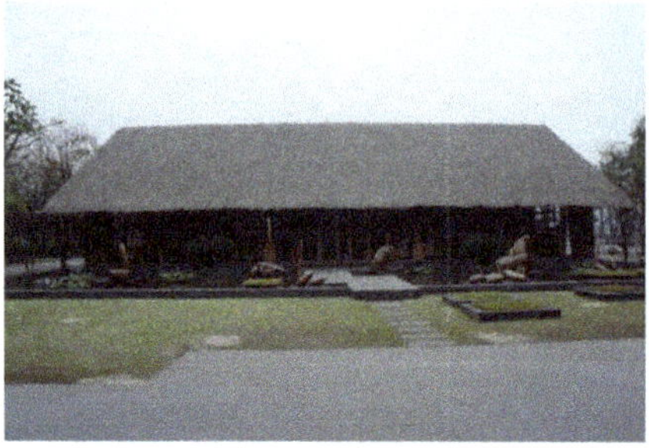
(a) Front façade

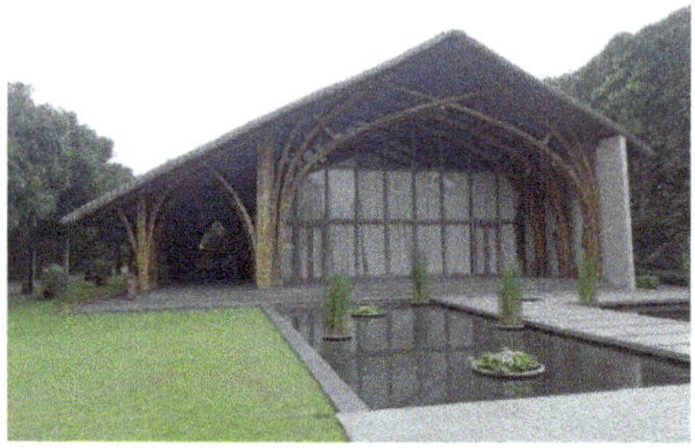
(b) Side façade

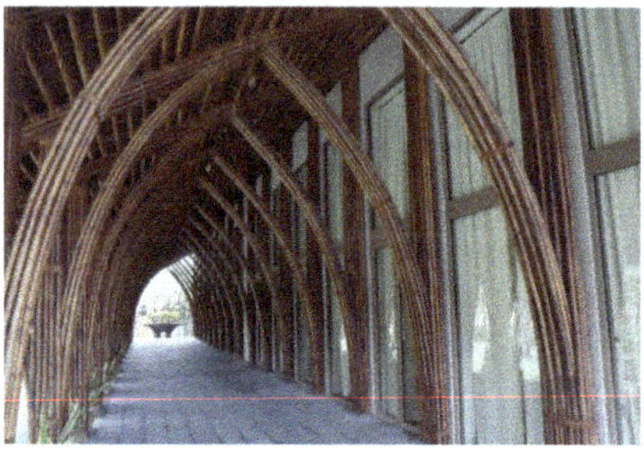
(c) Under the corridor

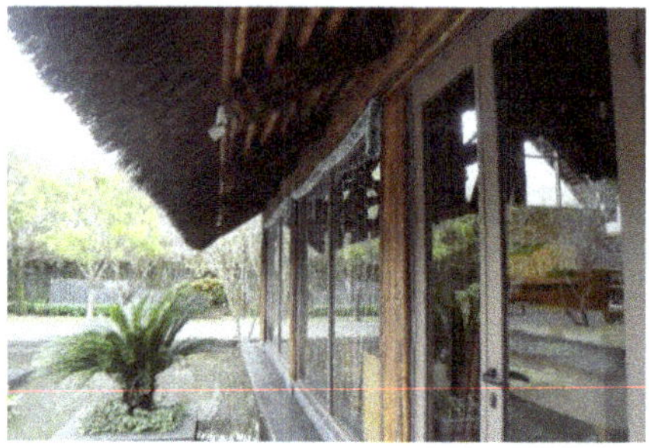
(d) View under eaves

Fig. 3.8 Naman retreat in Vietnam (*photos by Cai Wei*)

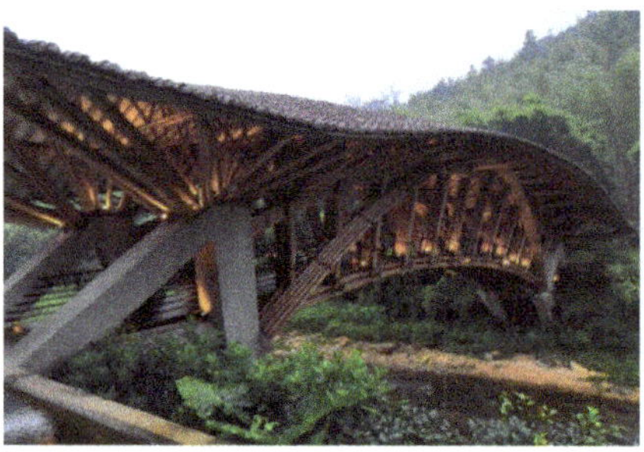
(a) Side view of Bamboo Corridor Bridge

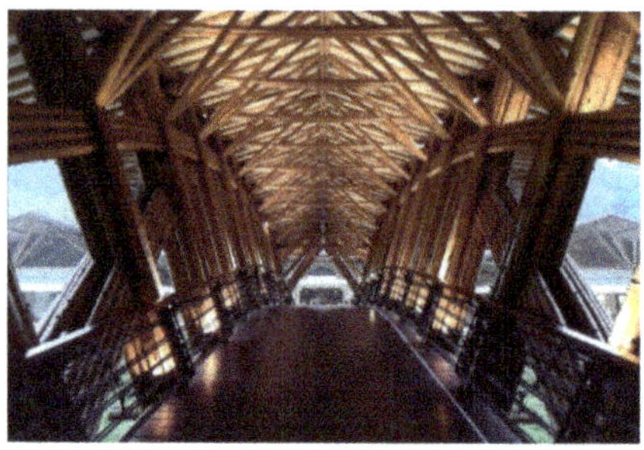
(b) On the bridge

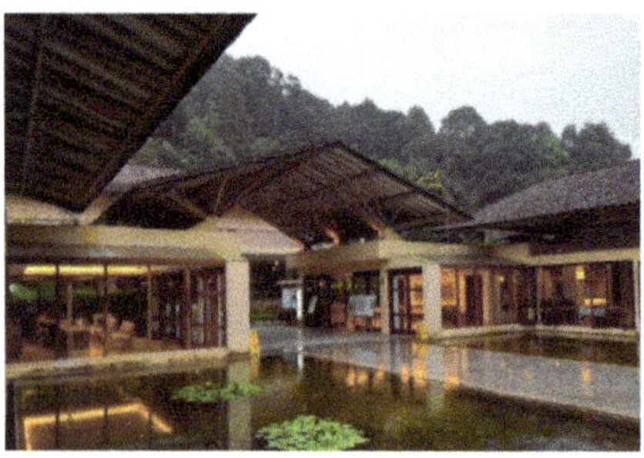
(c) Bamboo buildings surrounding pools

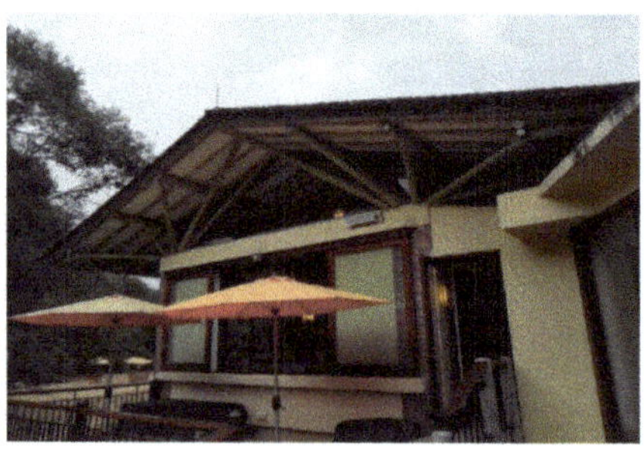
(d) Bamboo buildings aside the mountains

Fig. 3.9 Full culm bamboo structures in Cross Water Ecotourism Resort, Guangdong, China (*photos by Jin Yongzhi*)

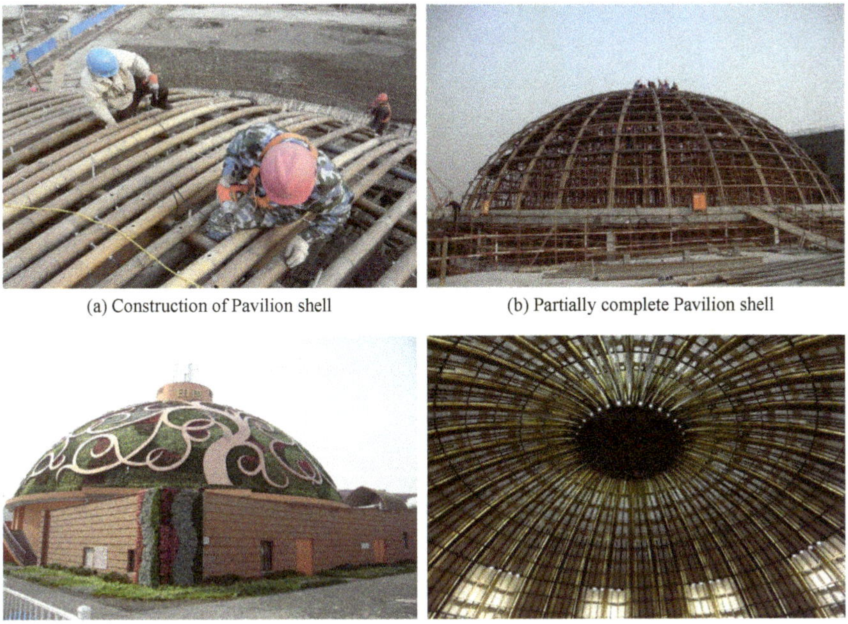

(a) Construction of Pavilion shell (b) Partially complete Pavilion shell

(c) Complete Pavilion (d) Interior of Pavilion shell

Fig. 3.10 Indian Pavilion at Shanghai World Expo (*photos by China Jingye Engineering Technology Company*)

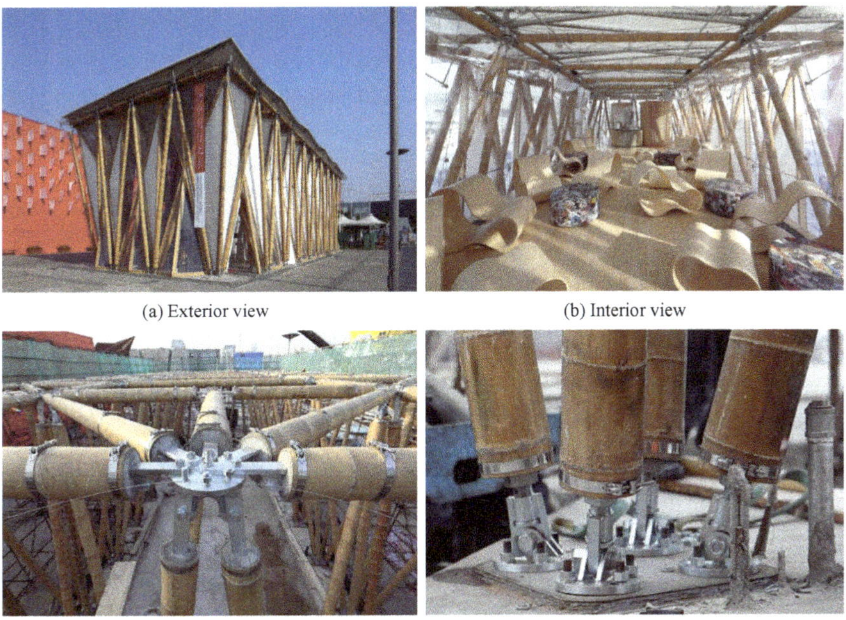

(a) Exterior view (b) Interior view

(c) Framing connection at roof level (d) Base of column elements

Fig. 3.11 Germany-Chinese House at Shanghai World Expo (*photos by Markus Heinsdorff*)

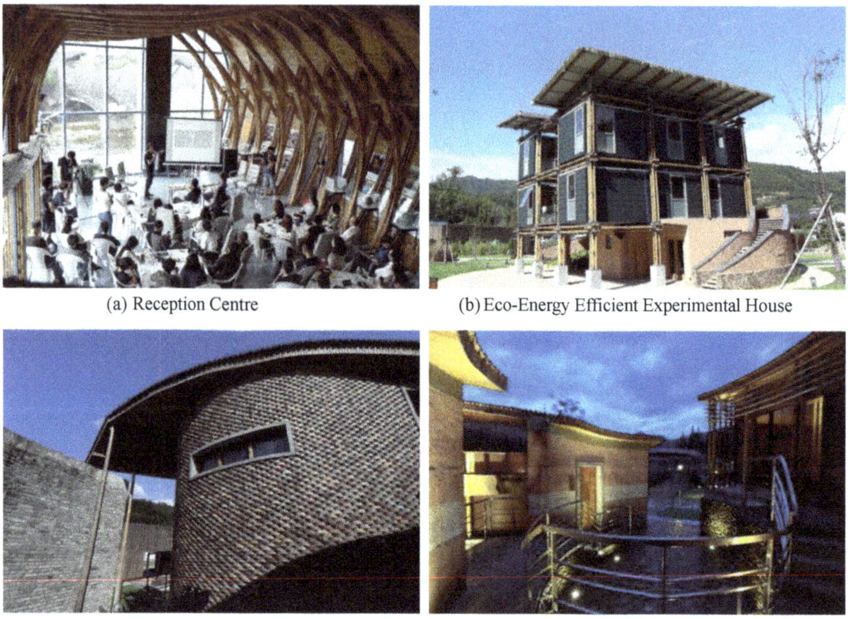

(a) Reception Centre (b) Eco-Energy Efficient Experimental House

(c) Art Hotel – Living in the Water (d) Art Hotel – Living in the Flowers

Fig. 3.12 International bamboo architecture biennale (*photos by Qiantao Studio*)

culms. The Eco-Energy Efficient Experimental House, designed by Italian architect Mauricio Cardenas Laverde (Fig. 3.12b), applied the concept of prefabrication to the full culm bamboo structure. The cleverly-designed metal joints connected individual culms to form the columns and beams of the structure in three directions. Two hotels, Art Hotel—Living in the Flowers and Art Hotel—Living in the Water, designed by Chinese architect Yang Xu (Fig. 3.12c, d), combined bamboo with local rammed earth and celadon to create unique local architectural art.

This biennale was a revolutionary "on-site" practice. Its phenomenological, sociological, ecological and cultural values show the infinite possibility of artistic and humanistic intervention in the countryside to release productivity. It was also the best interpretation of "spirit of place". With the integration of architectural art, means of production, mode of production and productivity "on-site" was transformed and upgraded. Returning home, starting a business, working in the dragon kiln and opening a homestay have now become the foci of Xitou Village, Baoxi, where the biennial exhibition is located. Once an unknown small village has become a well-known rural tourism destination, and Baoxi National 4A-level tourist attraction with Xitou Village at its core area was successfully created. In 2017, Baoxi received more than 200,000 tourists from China and abroad, and its tourism income amounted to 35 million yuan ($\approx$ 5 million USD; 4.6 million euro). In 2018, Xitou Village of Baoxi was selected as "Ten Most Beautiful Forest Households in China". In 2019, Baoxi was named "Provincial Tourism Town". Baoxi has now become the most attractive artistic landscape in China's rural areas, showing the world the interaction between man and nature, human and environment, and human and architecture [7].

Under the influence of internationalization, more and more Chinese architects have begun to rethink and reuse traditional local materials in recent years, trying to reestablish lost traditional skills and culture. In 2016, the architect Cui Kai, a Fellow of Chinese Academy of Engineering, designed the Kunqu Opera Theater in Kunshan, Jiangsu, China (Fig. 3.13a). He made full use of local bamboo materials and ingeniously combined the tension and rhythm of full culm bamboo with Kunqu opera performances by respecting the original pattern of residential houses and courtyards. In 2017, the Bamboo Village Lounge project in Shangcun Village, Anhui, China explored a sustainable development path for the protection of traditional villages. Architect Song Yehao's team chose bamboo as the main material of the building. They designed "six bamboo umbrellas" and "three black tents" (Fig. 3.13b), and constructed a bamboo shelter shared by villagers and tourists. In 2017, the Bamboo Exhibition Corridor of Ancient Kiln in Jinshan Mountain in Shangyu, Zhejiang, China also used full culm bamboo as the main material for a shed to protect and display celadon kilns. Architect Zhong Dekun combined the celadon dragon kiln built 2000 years ago with nature to create a harmonious symbiotic architectural

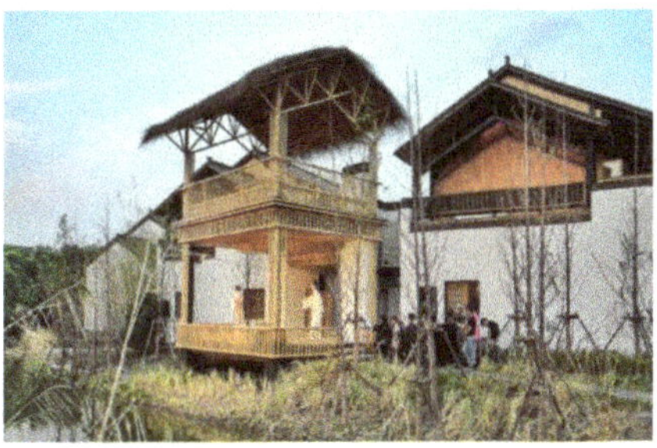

(a) Kunqu Opera Theater in Kunshan, Jiangsu, China (*photo by China Architecture Design & Research Group*)

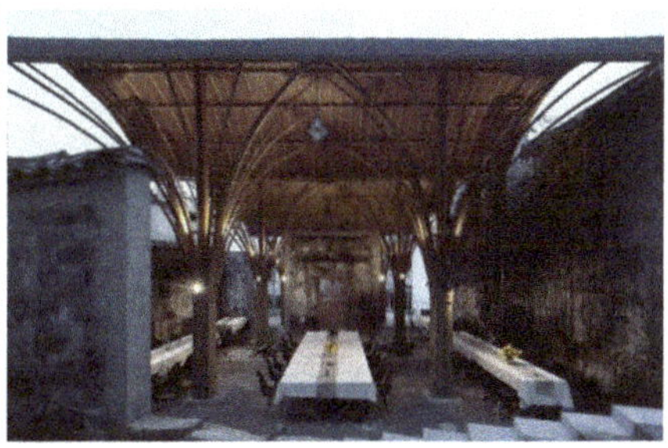

(b) Bamboo Village Lounge in Shangcun Village, Anhui, China (*photo by SUP Atelier*)

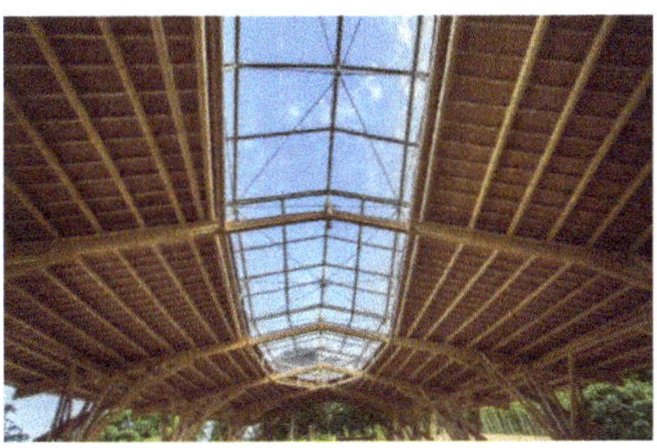

(c) Bamboo Exhibition Corridor of the Ancient Kiln in Jinshan Mountain in Shangyu, Zhejiang, China (*photo by Anji Zhujing Bamboo Industry Technology Co.,Ltd.*)

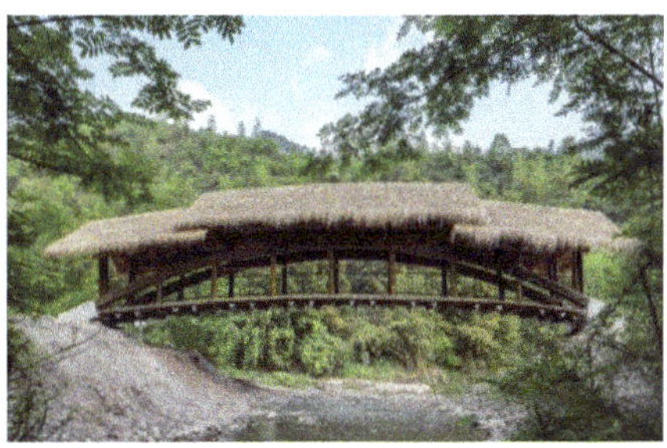

(d) Yixin Bridge in Dujia Village of Xinglong Town, Chongqing, China (*photo by Tsinghua University*)

Fig. 3.13 Full culm bamboo structures in China

environment hidden in the mountains and forests (Fig. 3.13c). In 2017 and 2018, a team from Tsinghua University built two bamboo bridges in Xiaowu and Dujia Villages of Xinglong Town, Chongqing, China (Fig. 3.13d). These bridges spanned 13.5 and 21 m, respectively, and were built to address a lack of safely river-crossing facilities in the region.

There are two gardens at the 2019 Beijing International Horticultural Exhibition (2019 Beijing Expo) that choose full culm bamboo as a structural material for their pavilions, including a large-scale garden pavilion and an outdoor landscape. Both are very unique and demonstrate high level architectural art and construction technology of contemporary bamboo architecture in China. Figure 3.14 shows the INBAR Pavilion at INBAR Garden. The total area of INBAR Garden is about 3600 m^2, including the 1600 m^2 INBAR Pavilion exhibition hall and 2400 m^2 outdoor landscape and green roof area. The main structure of the INBAR Pavilion is made up of nine huge bamboo arches. The exhibition space is delicately hidden under the green garden supported by the arches. The green space of the garden spreads to the roof along the arch feet and gradually disappears. Architecture and landscape were interwoven, forming a vivid organic whole on the ground. Figure 3.15 shows the UNESCO Garden. The exhibition hall is an open parabolic landscape gallery

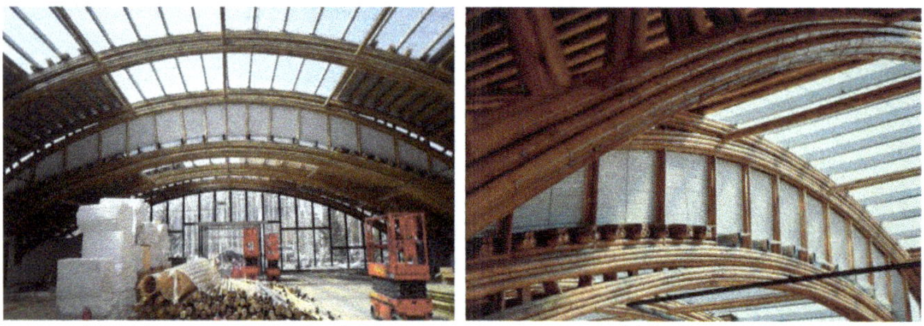

(a) Interior view under construction (b) Bamboo arches

Fig. 3.14 INBAR Pavilion at 2019 Beijing Expo (*photos by Jin Yongzhi*)

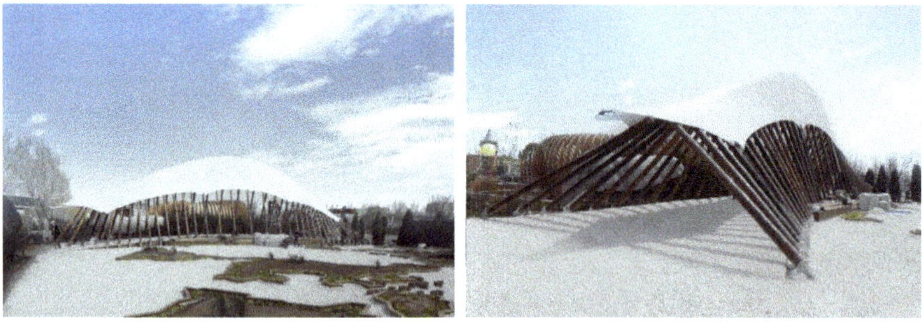

(a) Landscape at UNESCO Garden (b) Bamboo Landscape Gallery

Fig. 3.15 UNESCO Garden at 2019 Beijing Expo (*photos by Nicolas Godelet*)

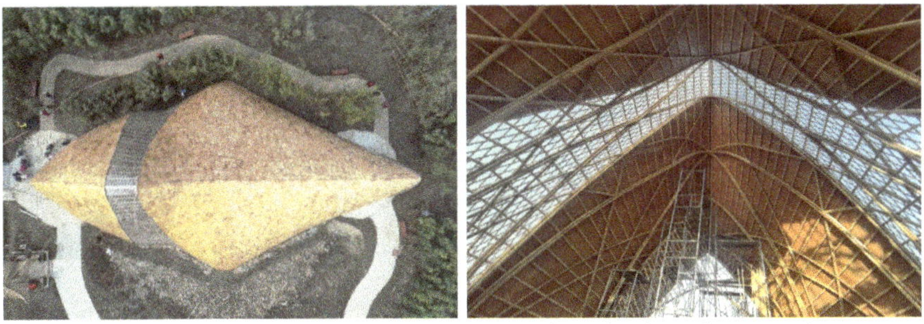

(a) Aerial view (b) Interior view under construction

Fig. 3.16 INBAR Pavilion at 2021 Yangzhou International Horticultural Exhibition (*photos by Liu Gang and Tu Yanfei*)

with large bamboo components supporting each other, forming a complex spatial geometry. The garden surrounded by bamboo represents an earth space that will be planted by young tourists. In time, it will become a lush landscape.

In 2021, a modern Fish-Shaped Bamboo Pavilion (Fig. 3.16), covering an area of about 280 m^2, was constructed using full-culm bamboo in the INBAR Garden at the 2021 Yangzhou International Horticultural Exhibition. The design of the building is inspired by the beautiful phrase: "Celebrating with gold and jade". The internal structure of full-culm bamboo structure is magnificent, and fully demonstrates the characteristics and charm of traditional bamboo in modern architectural design.

With the increasing popularity of full-culm bamboo, Yibin International Bamboo Products Trading Centre (Fig. 3.17) applied full bamboo culms to its large hybrid structure in an innovative way. The 2021 project is the first and largest single building in China to use a steel and full culm bamboo hybrid structure for load-bearing. In order to address the fire-resistance of full culm bamboo materials and meet requirements of China's National Code for Fire Protection Design of Buildings, and following rigorous experiments and expert evaluation, the project employs a fire spray cooling system for the fire protection of steel bamboo hybrid trusses. As shown in Figs. 3.18 and 3.19, two fire experiments with 1:5 model and 1:1 model were carried out respectively.

More details of the described full culm bamboo structures are presented in Chap. 6.

3.1.1 Research Status of Full Culm Bamboo Structures

In terms of realising the potential of bamboo constructions, connection design is critical and represents the greatest impediment to broader adoption of bamboo construction techniques. Force transfer into a culm (or between culms) is hampered by the highly anisotropic nature of the material; connection details must navigate this complex space. Common connection forms include: lashing with ropes, bolted

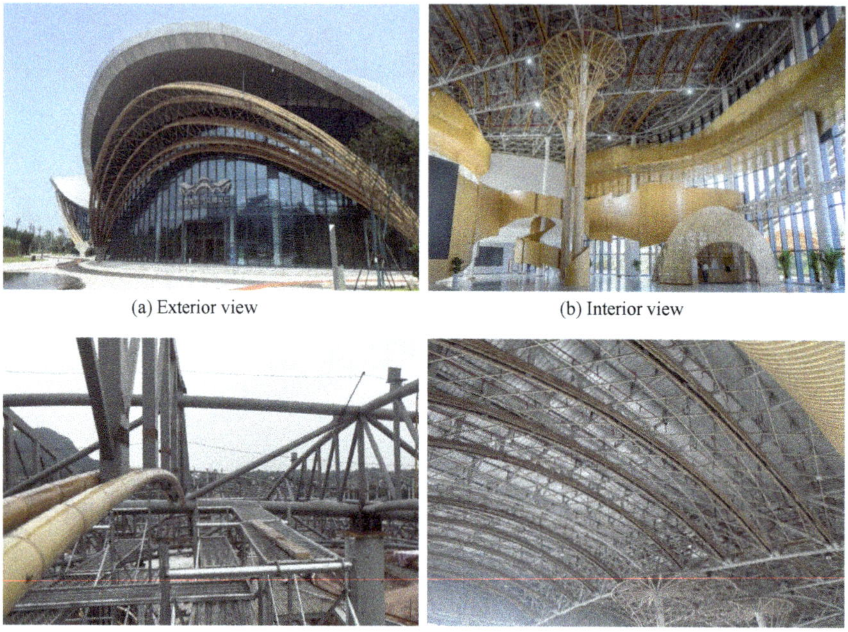

(a) Exterior view (b) Interior view

(c) Detail of hybrid steel bamboo truss (d) Hybrid steel bamboo vault trusses

Fig. 3.17 Yibin International Bamboo Products Trading Centre (*photos by Xi'an Construction Engineering Green Construction Group Co., Ltd. and Xi'an Architectural Design and Research Institute Co., Ltd.*)

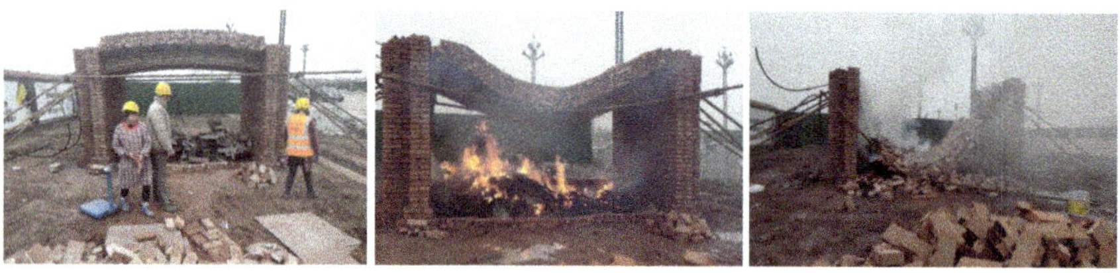

(a) Preparation for the test (b) Large deformation of the arch (c) Bamboo arches collapse

Fig. 3.18 Fire experiment for Yibin International Bamboo Products Trading Centre (1:5 model). (*photos by Xi'an Construction Engineering Green Construction Group Co., Ltd. and Xi'an Architectural Design and Research Institute Co., Ltd.*)

(doweled) connections, steel plate (gusset) connections and sleeve connections [8]. Widyowijatnoko and Harries [9] provide an extensive English-language review of traditional and modern means of connecting full culm bamboo.

Lashed joints are not appropriate for large-scale bamboo structures: they have relatively low capacity and are susceptible to deterioration. Bolts and other mechanical fasteners are commonly used although their capacity can also be limited. To overcome this, the culm cavity is often filled with cement mortar or grout. Although

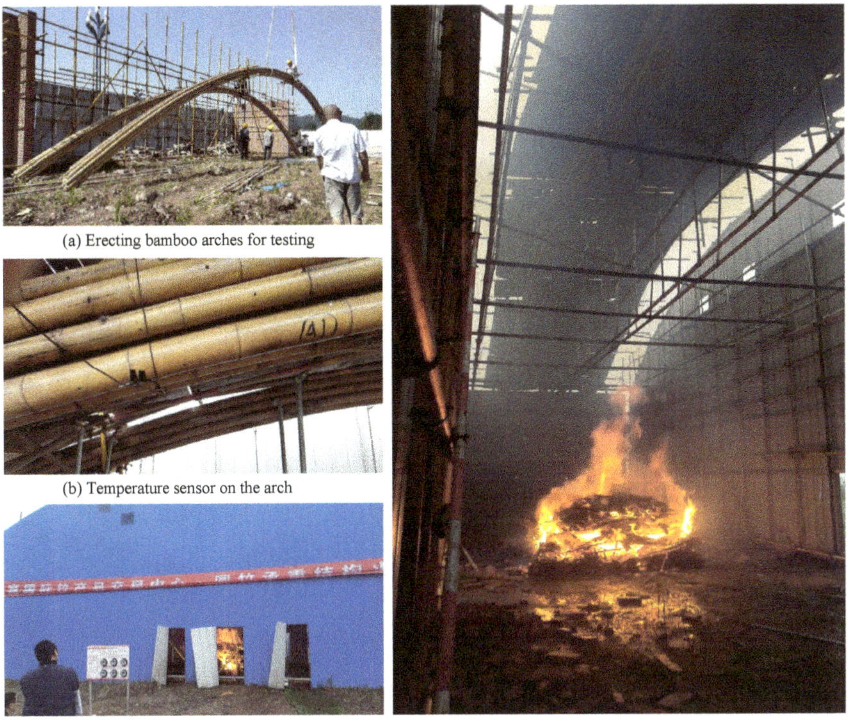

(a) Erecting bamboo arches for testing

(b) Temperature sensor on the arch

(c) Observe temperature data on the panel

(d) Fire experiment under the bamboo arch

Fig. 3.19 Fire experiment for Yibin International Bamboo Products Trading Centre (1:1 model). *(photos by Xi'an Construction Engineering Green Construction Group Co., Ltd. and Xi'an Architectural Design and Research Institute Co., Ltd.)*

this improves the connection capacity, it introduces other inherent drawbacks and effects the sustainability associated with bamboo construction.

When several culms converge at one point, a variety of connections can be made: simple bolted gusset plate connections and nodal connections using internal embedment into the culm cavity (see Fig. 3.11) or external sleeves can be used [10], the latter approaches mitigate the need to penetrate the culm wall—potentially causing splitting—at the culm end. Means of fabricating high-capacity bamboo connections without the need for culm wall penetration or costly hardware is an important area of new research.

Bamboo structural forms that leverage its superior properties while mitigating undesirable behaviour are also an important area of research development. Combinations of structural systems with simple joint assemblies that reduce the forces on individual connections and provide redundancy continue to be investigated [11–13]. The use of bamboo shell structures, particularly for rapidly deployable emergency shelters [13] is also a verdant area for research at this time.

3.1.2 Summary of Full Culm Bamboo Structures

Generally speaking, the use of modern full culm bamboo structures in China is still in its infancy. Although more and more Chinese architects have begun to use this natural material, most architects and structural engineers lack the necessary knowledge of bamboo and other raw materials and their construction techniques. On the other hand, for consumers, the impressions of bamboo architecture remain those of 'traditionally built' bamboo houses. They do not know that modern full culm bamboo architecture can be fashionable, comfortable and safe, and can fully meet the needs of modern work and life. Finally, China needs to establish design methodologies and construction technology for Chinese bamboo species. On the one hand, the experiences using bamboo for construction from other countries can be leveraged. On the other hand, systematic research on Chinese bamboo species should be carried out to provide reliable fundamental data on which to build a system of design and construction.

3.2 Engineered Bamboo Structures

In the late 1970s, China began to study and produce engineered bamboo, although initially intended almost entirely for non-structural uses. Since the 1990s, Chinese material scientists have gradually extended engineered bamboo materials into the field of structural load-bearing applications. Nonetheless, it was not until 2004 that engineered bamboo was first used as a structural material in practical engineering. A primary school building (Fig. 3.20a) was built in Pingbian County, Yunnan Province, China in cooperation with INBAR, the Research Institute of Wood Industry of the Chinese Academy of Forestry (CAF), and the Chinese Academy of Building Research. Eight roof trusses with spans of more than 6 m were fabricated of glued laminated bamboo; the roof panels and inner and outer wall sheathing were bamboo

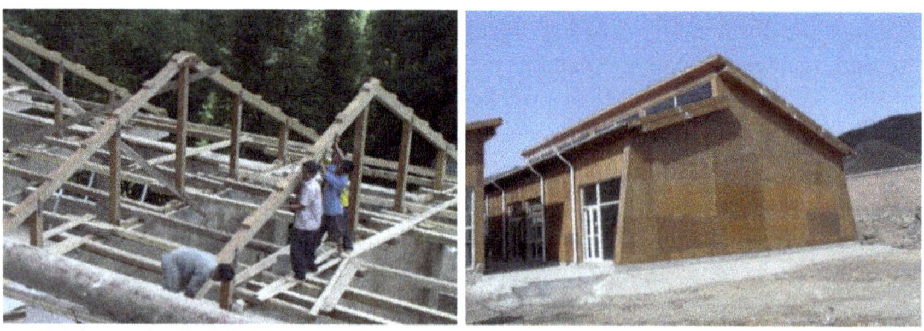

(a) Primary school buildings in Pingbian County, Yunnan, China *(photo by CAF)*

(b) Demonstration bamboo house in Nankou, Beijing, China *(photo by INBAR)*

Fig. 3.20 Demonstration bamboo houses in Yunnan and Beijing, China plywood [14]. On the basis of this project, from 2006 to 2007, INBAR, CAF and the Beijing Institute of Architectural Design cooperated to construct a demonstration structure of engineered bamboo (Fig. 3.20b) in Nankou, Changping, Beijing. The structural members of the building frame were all made of laminated bamboo sliver lumber.

Later, more universities and institutions joined in the research, development and application of engineered bamboo structures. In 2007, the Institute of Modern Bamboo, Wood and Composite Structure of Hunan University built model modern bamboo villa on the campus of Hunan University. A light frame glued laminated bamboo structural system (glubam) was adopted. The wall cavity was filled with thermal insulation, and the surface of external wall was rendered with cement mortar to prolong the service life of the house [15]. Following the 2008 Wenchuan earthquake, Nanjing Forestry University and Southeast University jointly developed a modern seismic resistant bamboo house (Fig. 3.21a). The house adopted a post-and-beam structure with joist and studs embedded in the wall. Bamboo plywood was used for floors and roofs, and glued laminated bamboo and bamboo scrimber were used for all structural members. Because of requirements for heat preservation and insulation, double-layer bamboo plywood wallboards were used [16]. In 2009, with the support of the USA-based Blue Moon Fund, INBAR cooperated with Hunan University to modularise the design, manufacture and construction of a light frame bamboo structure system in the Black Bamboo Park, Beijing (Fig. 3.21b). All the members were processed in Changsha, China and assembled after being transported to the construction site. Seven workers completed the construction of the main structure in eight days and completed final delivery in one month. Later, the Hunan University team applied the concept of a light frame glued laminated bamboo structure to many projects [15].

In 2017, a team from the Nanjing Forestry University built a three-storey bamboo frame structure (Fig. 3.22) at the Baima Branch Campus of Nanjing Forestry University using glued laminated bamboo. In 2020, another team from the International

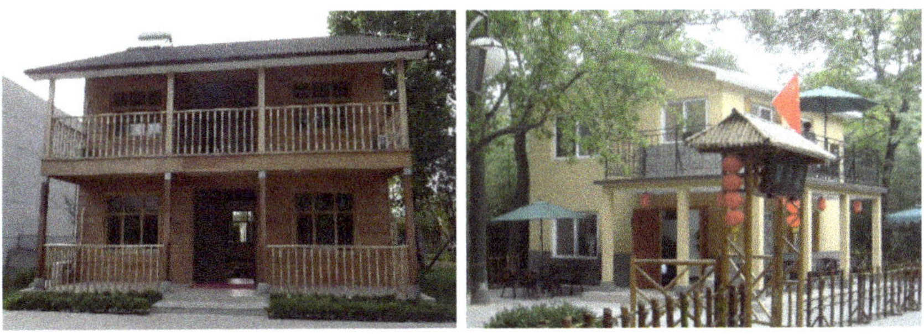

(a) Demonstration bamboo house at Nanjing Forestry University, Nanjing, China (*photo by Lv Qingfang*)

(b) Bamboo teahouse in Black Bamboo Park, Beijing, China (*photo by co-author Liu*)

Fig. 3.21 Demonstration bamboo houses in Nanjing and Beijing, China

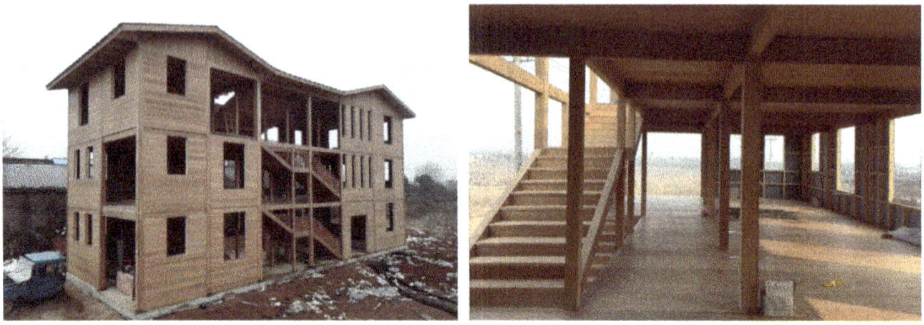

(a) Exterior view (b) The first floor

Fig. 3.22 Demonstration bamboo house at Nanjing Forestry University, Nanjing, China (*photos by Huang Dongsheng*)

Research Center for Bio-composite Building Materials and Structures of Nanjing Forestry University collaborated with Ganzhou Sentai Bamboo & Wood Co., Ltd. to build a two-storey office building (Fig. 3.23) made of glued laminated bamboo for the company in Ganzhou, Jiangxi, China.

In addition, since 2006, Ningbo University has conducted research on steel-bamboo composite structures, including properties of composite floor, wall, beam, column and other steel-bamboo composite members, in addition to the construction of model houses [17, 18].

In addition to the application of engineered bamboo in buildings, China's research teams have extended it to bridge structures. In 2006, the team at Hunan University designed and built the world's first glued bamboo pedestrian bridge in the campus (Fig. 3.24a) [19]. The bridge is 5 m long and 1.5 m wide. In 2007, Hunan University and Leiyang Agricultural Comprehensive Development Office cooperated to construct the world's first modern bamboo bridge capable of carrying heavy trucks (Fig. 3.24b, c) [20]. The 3.5 m wide bridge is supported on nine 10 m long beams. The diaphragms between the beams was also made from bamboo, all of which were

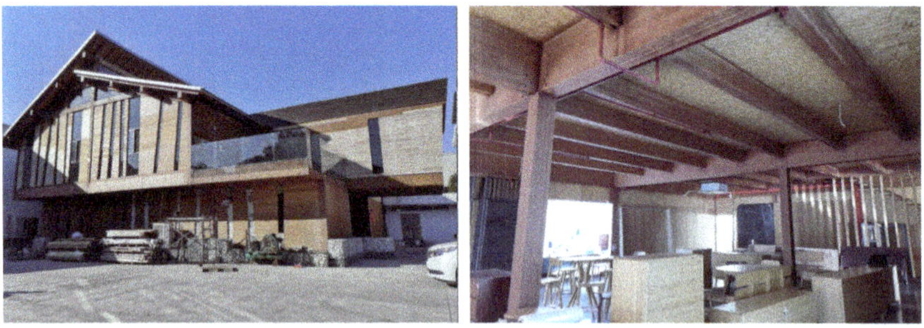

(a) Exterior view (b) Interior view

Fig. 3.23 Demonstration bamboo office building in Ganzhou, Jiangxi, China (*photos by Li Haitao*)

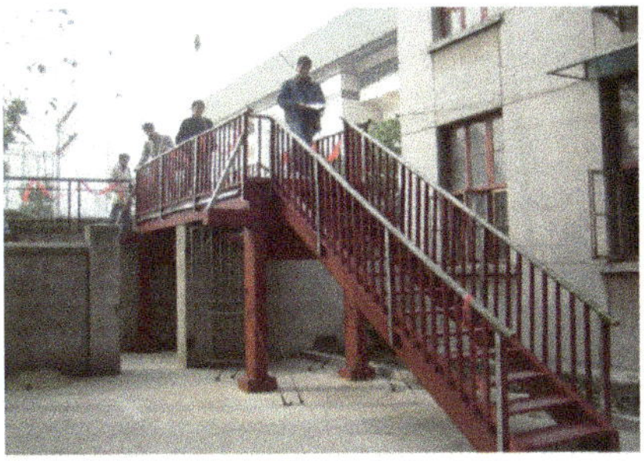

(a) Glued bamboo pedestrian bridge in Hunan University, Hunan, China [3-19] *(designed and photo by Y. Xiao and B. Shan)*

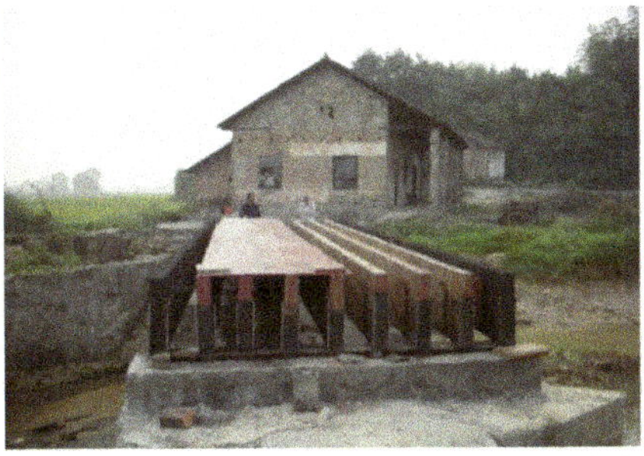

(b) Bamboo bridge capable of carrying heavy trucks in Leiyang, Hunan, China [3-20] *(designed and photo by Y. Xiao and B. Shan)*

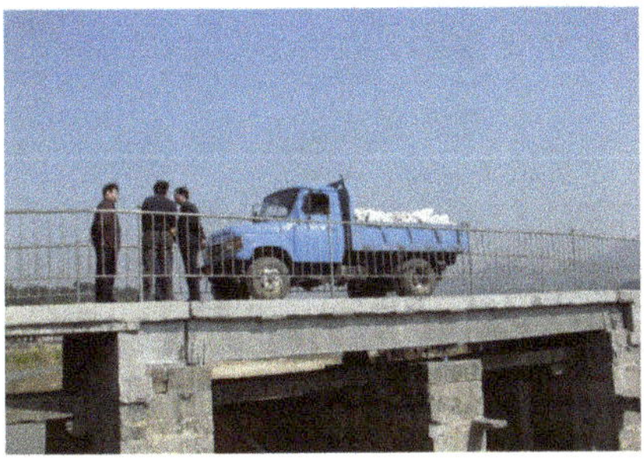

(c) Bamboo bridge capable of carrying heavy trucks in Leiyang, Hunan, China [3-20] *(designed and photo by Y. Xiao and B. Shan)*

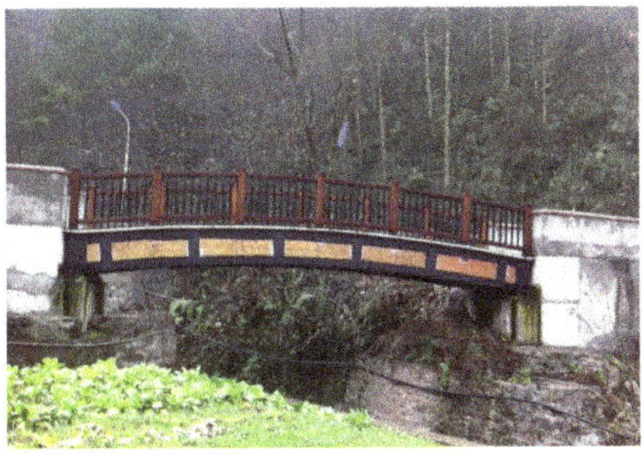

(d) Mao Yisheng glued bamboo demonstration public-benefit bridge in Liutang Township, Chongqing, China *(photo by Zhang Xiubiao)*

Fig. 3.24 Examples of engineered bamboo bridges

connected by bolts. This bridge was awarded the 2008 Best Engineering Innovation Award by *Popular Science* [15].

In 2015, the International Bamboo and Rattan Center, Chongqing Jiaotong University and Mao Yisheng Foundation cooperated to build a 12 m long by 3 m wide glued laminated bamboo demonstration bridge in Liutang Township, Shizhu County, Chongqing, China (Fig. 3.24d). The bridge was designed based on the *Technical Specification for Urban Pedestrian Bridges and Pedestrian Tunnels* (CJJ 69–95) for a 20-year design life.

The Engineering Research Center of Bamboo Winding Composite Materials of the National Forestry and Grassland Administration of China, established in 2016, carried out research work on filament-wound bamboo composite water pipes (Fig. 3.25a) and bamboo composite pipe for urban municipal tunnels (Fig. 3.25c). A residential house was even made using the filament-wound process (Fig. 3.25b). The wound material incorporates bamboo and straw with other biological waste as filler materials.

In a particularly innovative application, a team of Yale University researchers have demonstrated the use of additive manufacturing (so called 3D printing) to build high capacity bamboo structures and new connections by 'printing' a mixture of bamboo

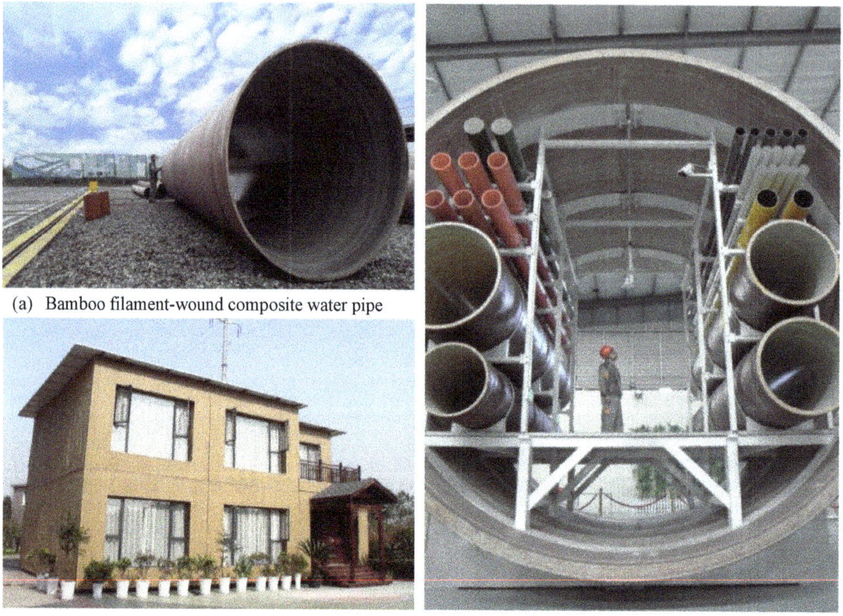

(a) Bamboo filament-wound composite water pipe

(b) Bamboo filament-wound integral composite residential house

(c) Bamboo filament-wound composite pipes for urban municipal tunnels

Fig. 3.25 Bamboo filament-wound composite products (*photos by Engineering Research Center for Bamboo Winding Composites of the National Forestry and Grassland Administration*)

powder and corn starch. The structure, shown in Fig. 3.26, was exhibited during the United Nations Environment Conference in March 2019.

In addition to research projects, with the deepening of architects' understanding of engineered bamboo, some commercial construction projects have begun to use engineered bamboo as structural materials in public buildings or outdoor landscapes. In a renovation project at Niubeishan Mountain in Sichuan Province, architect Li Daode created a youth hostel through digital design and generation logic (Fig. 3.27a). The rolling roof of the hostel, echoing it environment, uses bamboo scrimber made from *D. affinis*, which is abundant in Sichuan Province. For the design of the new Zhaojun Museum (Fig. 3.27b), architect Cao Xiaoxin tried to trace back to the ancient Chinese civil construction method. He used stabilised rammed earth as a new "soil" and bamboo scrimber as a new "wood" to translate the traditional construction methods, and to express his respect to the wisdom of the ancients. Architects Liang Jingyu and Ye Siyu, in the renovation and expansion project of the Zen Center of Baoguo Temple in Lezhi, Sichuan Province (Fig. 3.27c), tried to find light, easy-to-assembly small members to form a spatial grid structure and apply them to the large-span roof of the Zen Hall on the sites of the Sui Dynasty (581 A.D–618 A.D) buildings. For this project, bamboo scrimber was processed into 20 mm by 40 mm rods, which addressed the challenge that the site could only rely on human transportation and small equipment. In the Sanhe Canteen Project (Fig. 3.27d), the team led by architect

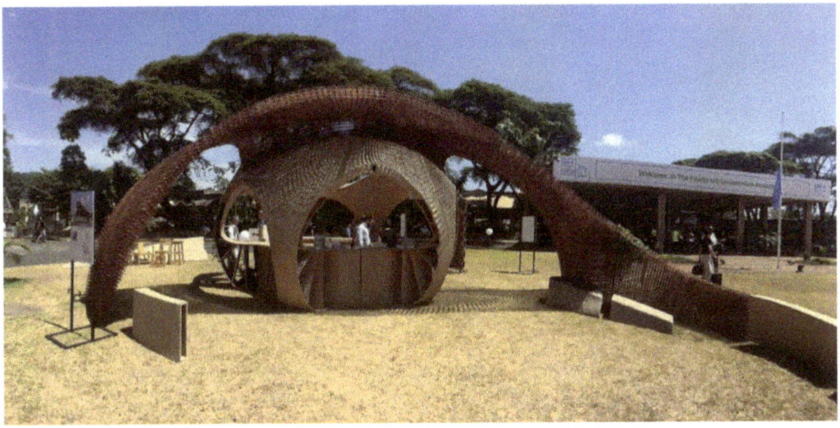

Fig. 3.26 3D printed bamboo structure displayed at the United Nations Environment Assembly (UNEA) in 2019 (*photo by Borja De La Peña Escardó*)

Wang Gang, boldly adopted the excellent material properties of bamboo scrimber to create a 17.8 m span tree crown structural system. An outdoor eave up to 6 m deep was formed by taking advantage of the bamboo structure resulting in a durable, light and flexible rural building.

At the China House Vision exhibition in 2018, the water-drop landscape architecture "Living Garden" (Fig. 3.27e), created by MAD Architects and Hanneng Group, integrated thin-film solar energy with a bamboo structure and outdoor environment, which reflected the integration of science and technology with nature. The outdoor heteromorphic bamboo landscape of the Shanghai Qinsen Group at the 2019 Beijing International Horticultural Exposition brought the application of glued laminated bamboo in an outdoor landscape into full play (Fig. 3.27f). In this structure, the design inspiration came from the curl and undulating three-dimensional space of wood shavings. These wonderful cases will be illustrated in greater detail in Chap. 6.

In addition to permanent structures, engineered bamboo has been employed as temporary shelter following disasters and some Chinese scientific research institutions and universities have made corresponding explorations. In 2005, the Research Institute of Wood Industry of CAF and INBAR cooperated to build the first temporary bamboo house in Tong County, Beijing. In 2007, with the technical support of Beijing Chengdong International Modular Housing Corporation, CAF and INBAR combined the improved modular bamboo wallboard with light steel frame for the construction of temporary bamboo houses. Following the 2008 Wenchuan earthquake, ICBR used bamboo plywood wall and roof panels to build 25 temporary houses in Wolong (Fig. 3.28a) and 20 in Dujiangyan for the resettlement of affected residents. In June 2008, Hunan University donated more than 2000 m^2 earthquake-resistant temporary resettlement houses to Guangyuan Beijie Primary School in Sichuan Province with the support of the Blue Moon Fund (Fig. 3.28b). The structures comprised a glubam frame enclosed with bamboo plyboard. All components were prefabricated

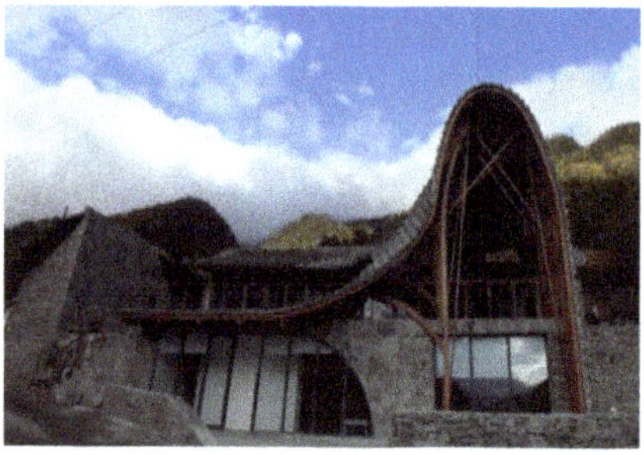

(a) Youth Hostel in Niubeishan Mountain, Sichuan, China (*photo by Hongyazhuyuan Science and Technology Company*).

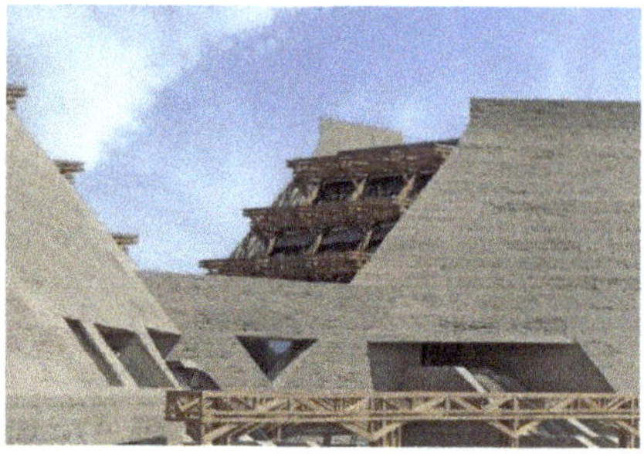

(b) Zhaojun Museum, Huhhot, Inner Mongolia, China (*photo by Zhang Guangyuan*)

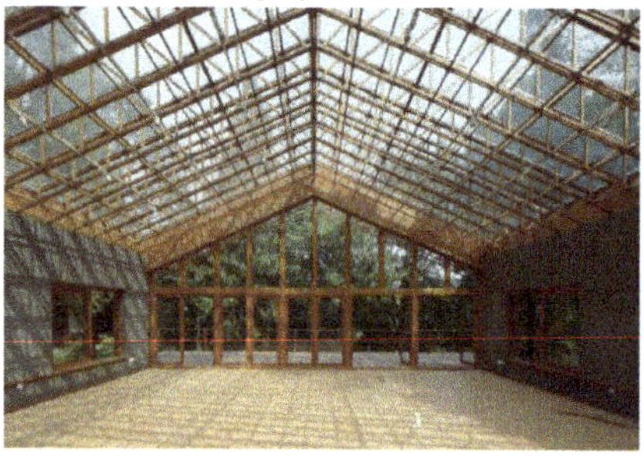

(c) Zen Center of Baoguo Temple in Lezhi, Sichuan, China (*photo by Approach Architecture Studio*)

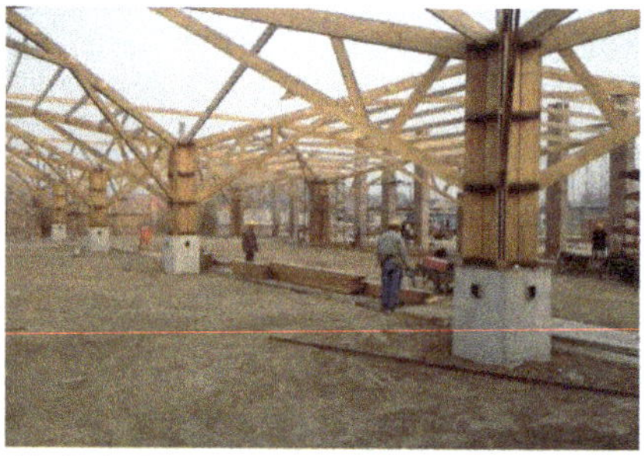

(d) Sanhe Canteen, Hebei, China (*photo by Master Studio of Beijing University of Civil Engineering and Architecture*)

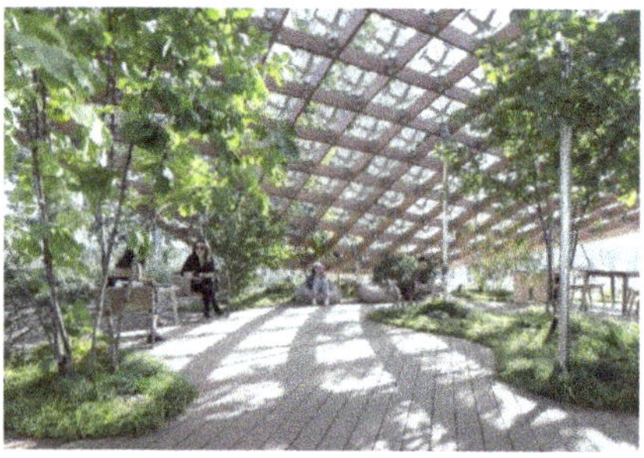

(e) Living Garden, Beijing, China. (*photo by Tian Fangfang and provided by MAD Architects*)

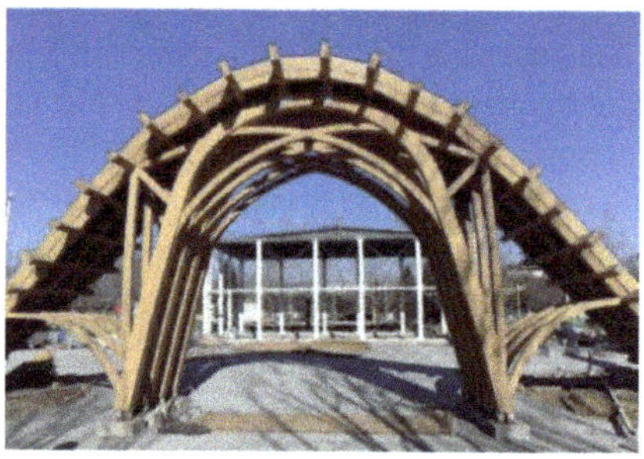

(f) Outdoor Heteromorphic Bamboo Landscape of Shanghai Qinsen Group at 2019 International Beijing Horticultural Exposition (*photo by Ganzhou Sentai Bamboo & Wood Co., Ltd*)

Fig. 3.27 Examples of engineered bamboo structures in China

and assembled on site [15]. In 2011, INBAR conducted a survey of users of ordinary light steel temporary houses and bamboo temporary houses, and assessed their living requirements, physical environment and transportation. Bamboo temporary houses showed obvious advantages over ordinary light steel houses in heat preservation, sound insulation, fire prevention and cost [21].

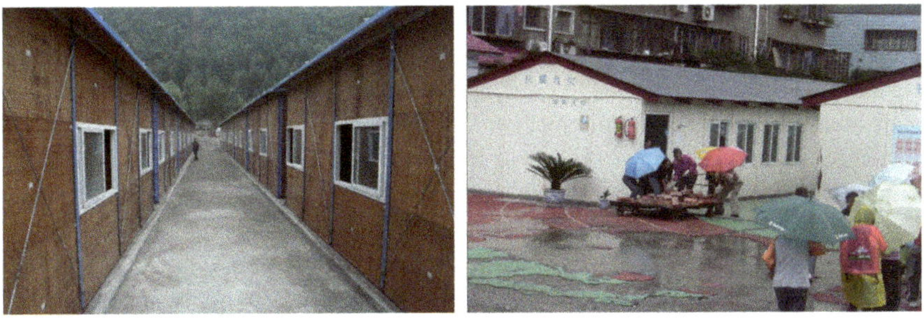

(a) Wolong earthquake-stricken area, Sichuan, China (*photo by co-author Wang*)

(b) Guangyuan Beijie Primary School in Sichuan, China (*photo by Guangyuan Beijie Primary School*)

Fig. 3.28 Temporary bamboo houses

3.2.1 Research Achievements in Engineered Bamboo Structures

Post-and-beam and light frame glued laminated bamboo construction are the two main types of engineered bamboo structures. In the study of mechanical properties of engineered bamboo members, the bending properties of beams and the compressive properties of columns are the two most basic considerations.

Lv et al. [22, 23] studied the flexural properties of laminated bamboo lumber beams and the compressive properties of bamboo scrimber columns. In their tests failure of the beams was mainly caused by tensile fracture following delamination and diagonal tearing of the outermost tension laminate. This is an ideal behaviour since it shows that the beams respect the plane sections assumption (Bernoulli beam theory) and there is clear warning prior to failure. The bamboo scrimber columns showed excellent deformation recovery under axial compression, which was considered important in maintaining ductility energy dissipation capacity and reducing residual deformation following earthquakes. Xiao et al. [24–27] studied the influence of different connection methods, manufacturing methods, and the presence of finger joints on flexural properties of glubam beams and axial compressive properties of glubam columns. Calculation methods for bearing capacities and failure criterion of glubam beams and columns were established. The fatigue performance of glubam beams was also studied.

Varela et al. [28] studied the lateral load resisting capacity of wood-frame shear walls sheathed with glued laminated *Guadua* bamboo panels; the walls showed good energy dissipation capacity. Sinha et al. [29] studied the mechanical properties of glued laminated bamboo beams, and confirmed that these possessed higher tension and bending capacities than beams fabricated of commonly used softwood, such as Douglas fir. Huang et al. [30] and Zhou et al. [31] studied the bending performance of bamboo scrimber beams made from five-year-old moso (*P. edulis*) bamboo from which the fibres were taken from different regions of the culms. An analytical model to evaluate the bending capacity of scrimber was proposed based on experimental studies. Sharma et al. [32, 33] conducted bending tests on glued laminated bamboo and bamboo scrimber beams and compared the effects of processing methods.

Li et al. [34–36] studied the mechanical properties of glued laminated bamboo beams and columns made from moso bamboo. Through bending tests of simply supported glued laminated bamboo beams with different shear span-to-depth ratios, the plane sections assumption was confirmed. Through axial compression tests of glued laminated bamboo columns having different slenderness ratios, the influence of slenderness on compression capacity and failure modes was analysed, and a coefficient capturing the observed behaviour was proposed. Through eccentric compression tests of laminated bamboo columns, the influence of eccentricity was also analysed.

The connections of engineered bamboo structures are similar to those of glued laminated timber structures. Formal research on connections for timber structures goes back more than a century [37]; and has informed modern international design standards. Many such timber connections are believed to be equally applicable for engineered bamboo structures.

Fei et al. [38] studied the bearing capacity of bolted connections for glued laminated bamboo structures, and confirmed their high strength, stiffness and ductility. Yang et al. [39, 40] carried out compressive and tensile tests of bolted connections for Glubam structures, analysed failure modes of the connections, and established relevant bearing capacity calculation models. Wang et al. [41] studied the shear capacity of bolted connections of bamboo plywood through symmetrical double-shear tests. They analysed the influence of ply orientation and density on bearing capacity, and compared this to bolted connections for timber structures. It was found that the bearing capacity of bolted connections of bamboo structures was greater than that of timber structures made of larch. Bolted connections of glued laminated bamboo can achieve comparable bearing capacities to those given in timber design standards. Feng [42] applied Johanson yield theory to bolted connections of engineered bamboo structures to establish predictions of shear capacity of bolted connections in Glubam; these predictions were validated through testing a Glubam frame structure.

Due to the high hardness and poor workability of bamboo scrimber, it is difficult to use conventional timber connections such as nail plates; bolted gusset plate and slotted-in plate connections, are preferred. Li [43] studied the influence of several parameters—including end spacing, thickness of the main connected members and diameter, spacing and number of bolts—on the mechanical performance of bolted connections in bamboo scrimber. The applicability of the design of bolted connections in timber was evaluated for bolted connections of bamboo scrimber, and formulations for dowel-bearing strength and bearing capacity were derived. Zhou et al. [44] studied the parallel-to-fibres tensile capacity of bamboo scrimber members connected with bolted connections in a slotted-in steel plate, and the compressive and deformation capacity of a single bolt.

Zhang et al. [45] carried out reversed cyclic tests on connections between laminated bamboo beams and bamboo scrimber columns as well as a shaking table test of an engineered bamboo frame structure. The dynamic characteristics, energy dissipation and seismic performance of the engineered bamboo frame structure were studied, and the connections were optimised.

3.2.2 Summary of Engineered Bamboo Structures

The history of engineered bamboo structures is not long, and development remains mostly in the exploratory stage. However, compared with full culm bamboo structures, it is relatively easy to promote engineered bamboo structures in China. Firstly, because the properties of engineered bamboo are similar to those of laminated timber and plywood, research methods and design bases can be rapidly adapted from timber research. Secondly, at present, there are many institutions in China engaged in the research on engineered bamboo structures, and the number is increasing year by year. In addition to a large amount of basic research on processing technology and material properties of engineered bamboo, areas such as component performance, structural systems, seismic performance, durability, fire protection and performance and thermal insulation performance are being explored. Thirdly, for the mainstream construction market and consumers, because construction methods and forms are similar to those of modern industrial buildings, engineered bamboo is relatively easy to accept. Fourthly, major timber exporters, such as Canada, the United States and Finland, have promoted policies of timber construction in China. Once a stable market is established, engineered bamboo structures will also develop stably. Finally, engineered bamboo is more competitive than timber in terms of global policies of protecting forest resources and reducing wood cutting.

References

1. Liu, L., & Chen, X. (2012). *The archaeology of China: From the late paleolithic to the early bronze age*. Cambridge University Press.
2. Liu, K. W., & Frith, O. (2013). An overview of global bamboo architecture: Trends and challenges. *World Architecture, 12*, 27–34.
3. Kaminski, S., Lawrence, A., & Trujillo, D. (2016). *Design guide for engineered bahareque housing*. International Network for Bamboo and Rattan (INBAR).
4. NSR-10 (AIS 2010). (2010). *Reglamento Colombiano de Construcción Sismo Resistente*. Instituto Colombiano de Normas Técnicas y Certificación.
5. Naidu, D. R., & Liu, K. W. (2013). India Pavilion, Expo 2010 Shanghai, China, 2010. *World Architecture, 12*, 80–85.
6. Li, Z. J. (2010). German-Chinese house at the Expo 2010. *Architecture Technique, 2010*(zl), 154–157 (in Chinese).
7. Liu, K. W., Ge, Q. T., & Song, Y. H. (2019). Eco-cultural value of Chinese Modern Bamboo Architecture. *Ecological Civilization World, 2*, 18–33 (in Chinese).
8. Zhang, N., & Bai, W. F. (2008). Analyzing and improving the construction of bamboo houses node. *Science Technology and Engineering, 8*(18), 5318–5322, 5326 (in Chinese).

9. Widyowijatnoko, A., & Harries, K. A. (2019). Chapter 20: Joints in bamboo construction in nonconventional and vernacular construction materials. In K. A. Harries & B. Sharma (Eds.), *Characterisation, properties and applications* (2nd ed.). Woodhead (Elsevier) Publishing Series in Civil and Structural Engineering.
10. Albermani, F., Goh, G., & Chan, S. (2007). Lightweight bamboo double layer grid system. *Engineering Structures, 29*(7), 1499–1506.
11. Sharma, B., Mitch, D., Harries, K. A., et al. (2011). Pushover behaviour of bamboo portal frame structure. *International Wood Products Journal, 2*(1), 20–29.
12. Richard, M., & Harries, K. A. (2012). Experimental buckling capacity of multiple-culm bamboo columns. In *13th International Conference on Non-conventional Materials and Technologies: Novel Construction Materials and Technologies for Sustainability*, pp. 51–62. Trans Tech Publications Ltd, Changsha, Hunan, China.
13. Eells, P., Pagliassotti, M., Brown, K., et al. (2013). Design of a rapidly deployable bamboo grid-shell structure. In *14th International Conference Non-conventional Materials and Technologies (IC-NOCMAT 2013)*, Trans Tech Publications Ltd, João Pessoa, Brazil.
14. Chen, X. H., & Wang, Z. (2005). Bamboo based panels for structural application. *World Bamboo and Rattan, 3*(3), 18–20 (in Chinese).
15. Xiao, Y., & Shan, B. (2013). *Glubam structures*. China Architecture & Building Press (in Chinese).
16. Wei, Y., Lv, Q. F., Zhang, Q. S., et al. (2009). Design and construction of modern bamboo anti-seismic living room. *Construction Technology, 38*(11), 52–54 (in Chinese).
17. Li, Y. S., Jiang, T. Y., Shan, W., et al. (2013). Quasi-static test on steel-bamboo composite beam-column exterior joints. *Engineering Mechanics, 4*, 241–248 (in Chinese).
18. Li, Y. S., & Zhang, J. L. (2016). State of the art of steel-bamboo composite and structural system. *Industrial Construction, 46*, 1–6 (in Chinese).
19. Shan, B., Zhou, Q., & Xiao, Y. (2009). Research and application of modern bamboo structure pedstrain bridge. *Journal of Hunan University: Natural Sciences, 36*(10), 29–34 (In Chinese).
20. Xiao, Y., Zhou, Q., & Shan, B. (2010). Design and construction of modern bamboo bridges. *Journal of Bridge Engineering, 15*(5), 533–541.
21. Frith, O. B., & Liu, K. W. (2013). Engineered modular bamboo transitional shelters for disaster relief: A case study from the 2008 Wenchuan earthquake, Sichuan Province, China. *Regional Development Dialogue, 34*(1), 114–131.
22. Lv, Q. F., Wei, Y., Zhang, Q. S., et al. (2008). Experimental study on mechanical properties of basic components for a new anti-seismic room with bamboo engineering materials. *Research & Application of Building Materials, 11*, 1–5 (in Chinese).
23. Wei, Y., Jiang, S. X., Lv, Q. F., et al. (2010). Experimental study on flexural performance of bamboo beams. *Building Structure, 40*(1), 88–91 (in Chinese).
24. Shan, B., Zhou, Q., & Xiao, Y. (2010). Research and construction of modern bamboo pedestrian bridges. *Building Structure, 40*(1), 92–96 (in Chinese).
25. Zhou, Q. (2013). *Experimental study and engineering application of glued laminated bamboo beams*. Hunan University (in Chinese).
26. Lv, X. H. (2011). *Experimental research and finite element analysis of Glubam columns under axial compression*. Hunan University (in Chinese).
27. Xiao, Y., Feng, L., Lv, X. H., et al. (2015). Experimental studies of glubam columns under axial loads. *Industrial Construction, 45*(4), 13–17 (in Chinese).
28. Varela, S., Correal, J., Yamin, L., et al. (2012). Cyclic performance of glued laminated guadua bamboo-sheathed shear walls. *Journal of Structural Engineering, 139*(11), 2028–2037.
29. Sinha, A., Way, D., & Mlasko, S. (2013). Structural performance of glued laminated bamboo beams. *Journal of Structural Engineering, 140*(1), 04013021.
30. Huang, D., Zhou, A., & Bian, Y. (2013). Experimental and analytical study on the nonlinear bending of parallel strand bamboo beams. *Construction and Building Materials, 44*, 585–592.
31. Zhou, A., & Bian, Y. (2014). Experimental study on the flexural performance of parallel strand bamboo beams. *The Scientific World Journal, 2014*(2), 181627.

32. Sharma, B., Gatóo, A., & Ramage, M. (2015). Effect of processing methods on the mechanical properties of engineered bamboo. *Construction and Building Materials, 83*, 95–101.
33. Sharma, B., Gatóo, A., Bock, M., et al. (2015). Engineered bamboo for structural applications. *Construction and Building Materials, 81*, 66–73.
34. Li, H. T., Su, J. W., Zhang, Q. S., et al. (2015). Experimental study on mechanical performance of side pressure laminated bamboo beam. *Journal of Building Structures, 36*(3), 121–126 (in Chinese).
35. Li, H. T., Su, J. W., Zhang, Q. S., et al. (2015). Mechanical performance of laminated bamboo column under axial compression. *Composites Part B: Engineering, 79*, 374–382.
36. Li, H. T., Chen, G., Zhang, Q. S., et al. (2016). Mechanical properties of laminated bamboo lumber column under radial eccentric compression. *Construction and Building Materials, 121*, 644–652.
37. Trayer, G. (1928). *Bearing strength of wood under steel aircraft bolts and washers and other factors influencing fitting design*. National Advisory committee for Aeronautics.
38. Fei, B. H., Zhang, D. S., Ren, H. Q., et al. (2008). The bearing capacity of the joint composed by laminated bamboo lumber. *Journal of Nanjing Forestry University (Natural Science Edition), 32*(3), 67–70 (in Chinese).
39. Yang, R. Z. (2009). *Research and application of glubam mechanical property and property of bolted joints*. Hunan University (in Chinese).
40. Yang, R. Z. (2013). *Research on material properties of Glubam and its application*. Hunan University (in Chinese).
41. Wang, Z. H., Liu, K. Z., & Ren, H. Q. (2010). Test on the bolts connection in bamboo laminated timber. In *2010 Academic Conference on Material Failure/Fracture across the Taiwan Strait*, Taiwan China (in Chinese).
42. Feng, L. (2015). *Bolt connection of modern bamboo-wood structures theoretical analysis and experimental study*. Hunan University (in Chinese).
43. Li, X. Z. (2013). *Research on bearing performance of bolt joint for recombinant bamboo*. Chinese Academy of Forestry (in Chinese).
44. Zhou, A. P., Huang, D. S., Tang, S. Y., et al. (2016). Experimental research on bearing capacity of the bolted PSB-steel-PSB joints. *Journal of Nanjing Tech University (Natural Science Edition), 38*(5), 34–39 (in Chinese).
45. Zhang, C. J., Lv, Q. F., & Cao, X. L. (2017). Experimental study for the key performance of the bamboo structure. *Building Structure, 47*(17), 1–7 (in Chinese).

Chapter 4
Standards

Standards are critical for the development of bamboo construction sector. This chapter will introduce relevant international standards issued by International Organisation for Standardisation (ISO) and those standards (including national standards, industrial standards, provincial standards, association standards) issued by different authorities in China.

4.1 International Standards

Since the early 2000s, INBAR has served as the liaison body of the Timber Structures Technical Committee (TC165) of the International Organisation for Standardisation (ISO) [1]. By 2004, the committee, led by an international expert group, developed two international Standards: ISO 22156:2004 *Bamboo—Structural design* and ISO 22157-1:2004 *Bamboo—Determination of physical and mechanical properties—Part 1: Requirements*; and a technical report: ISO TR22157-2:2004 *Bamboo—Determination of physical and mechanical properties—Part 2: Laboratory manual.* This work has had significant global impact, with a number of INBAR member countries, including India, Ecuador, Peru and Colombia, subsequently developing chapters on bamboo in their respective national building codes which referenced these ISO standards. Others countries, such as Jamaica, Vietnam, the Philippines, and the Netherlands, directly adopted ISO standards as national standards.

The ISO standards issued in 2004 were 'version zero' standards in many respects [2]. They identified needs and served as placeholders upon which revisions could be based. The 2004 ISO standards, in particular, lacked guidance on crucial issues such as connection design and strength grading. Additionally, structural engineered bamboo products, which were relatively new in the early 2000s, have gained commercial importance but are not covered by existing international standards [1]. It is worth mentioning that since the release of the first ISO bamboo standards in 2004, breakthroughs have been made in the development and revision of various national

K. W. Liu et al., *Contemporary Bamboo Architecture in China*,
https://doi.org/10.1007/978-981-16-8309-1_4

standards. Two of the most important standards examples are the Colombian *NSR-10 Colombian Standards for Seismic Resistant Design and Construction* which includes a chapter *Guadua Structures*; and ASTM D5456 *Standard Specification for Assessment of Structural Composite Lumber Products*, which included bamboo-wood composites for the first time in 2010.

Although this early work had a great impact, due to the lack of follow-up project support, INBAR was unable to continue to participate in the revision of ISO standards after 2004, and these standards were not maintained or revised. In 2013, with the support of core funds from the Chinese government, INBAR, together with partners from the University of Coventry in the United Kingdom and Colombia and Ecuador, launched a study on bamboo strength grading. The project was able to successfully obtain permission from ISO/TC165 to include a new Working Group 12 (WG12) on structural uses of bamboo. This working group has the mandate to amend existing ISO standards as well as to propose new standards. INBAR has served as the Convener for this working group since its inception in 2014.

Also in 2014, in order to strengthen its expert partnership and international influence on bamboo materials for structural uses, INBAR established the INBAR Bamboo Construction Task Force (INBAR TFC), whose members initially consists of 11 experts from eight countries. As of the end of 2021, the working group has 34 bamboo construction experts, including architects, structural engineers, materials scientists and related professional researchers representing 18 countries. By the joint efforts of INBAR and the INBAR TFC, WG12 of ISO TC165 began in 2014 to undertake revision of the original international standards and to develop new standards. As of the end of 2020, WG12 is responsible for the following international standards (Table 4.1), including a number of working draft (WD) or committee draft (CD) in preparation.

The revised Standard ISO 22157:2019 replaces the 2004 version, adding two test methods and revising most others with the intent of improving the utility and adoption of the methods as well as the standard.

In September 2018, ISO published its first new standard developed with the support of the INBAR TFC: ISO19624:2018 addresses mean and methods for structural grading of bamboo culms for construction. The standard provides a framework for any national or locally adopted grading process. It identifies criteria that should be considered in a visual grading protocol, and outlines the basis on which declared capacity or strength values are determined [1]. The new standard was the final deliverable of the INBAR-funded project "Strength grading of bamboo".

The completely revised ISO 22156, renamed *Bamboo structure—Bamboo culms—Structural design*, was published in June 2021. This revision permits both capacity and stress-driven approaches to design and provides means of design or acceptance of details, particularly connections [1]. The revised document builds upon the considerable research and development conducted in the last 20 years and better integrates ISO 22157, ISO 19624 and other modern ISO standards [1].

In addition to the work on full-culm bamboo, WG12 is currently developing four standards focusing on engineered bamboo products. DIS 23478: *Bamboo structures—Engineered bamboo products—Test methods for determination of physical*

Table 4.1 ISO TC165 standards for bamboo construction

Designation	Standard	Status
ISO 22156:2004	*Bamboo—Structural design*	Replaced by ISO 22156:2021
ISO 22157-1:2004	*Bamboo—Determination of physical and mechanical properties—part 1: Requirements*	Replaced by ISO 22157:2019
ISO/TR 22157-2:2004	*Bamboo—Determination of physical and mechanical properties—part 2: Laboratory manual*	Withdrawn
ISO 19624:2018	*Bamboo structures—Grading of bamboo culm—Basic principles and properties*	Active
ISO 22157:2019	*Bamboo structures—Determination of physical and mechanical properties of bamboo culms—Test methods*	Active
ISO 22156:2021	*Bamboo structures—Bamboo culms—Structural design*	Active
DIS 23478	*Bamboo structures—Engineered bamboo products—Test methods for determination of physical and mechanical properties*	TC 165 Committee draft
CD 5257	*Bamboo structures—Engineered bamboo products—Test methods for determination of physical and mechanical properties at small scale*	TC 165 Committee draft
NWIP 7567	*Bamboo structures—Glued laminated bamboo—Product specification*	TC 165 Working draft
Technical Specification	*Bamboo structures—Engineered bamboo products—Design of engineered bamboo structures*	Approved for development by TC 165

and mechanical properties focuses in test methods for glued laminated and scrimber engineered bamboo products intended for structural applications. Although standards have existed since the 1990s for non-structural engineered bamboo (such as flooring and other finishes), DIS 23478 represents the first materials test standard intended for structural engineered bamboo.

As DIS 23478 is developed, two associated Working drafts are also in preparation. CD5257: *Bamboo Structures—Engineered bamboo products—Test methods for determination of physical and mechanical properties at small scale* is aimed at establishing consistency in qualification and quality control of engineered bamboo products for structural applications, regardless of country of manufacture or country of end use. This standard will provide manufacturers, regulatory agencies, and end-users with a means to evaluate the material properties of engineered bamboo products intended for structural applications under dry use conditions. *NWIP 7567: Bamboo Structures—Glued laminated bamboo—Product specification* will provide a specification for glued laminated bamboo products for structural applications. It is anticipated that a second specification for scrimber products will also be initiated.

Additionally, in late 2019, WG12 proposed the development of *Bamboo structures—Engineered bamboo products—Design of engineered bamboo structures*, a standard for design of engineered bamboo structures. This standard will progress quickly but requires the completion of DIS 23478 and NWIP 7567 in order to have a basis on which to build. In order to address the significant gap in available design recommendations, WG12 is preparing a *Technical Specification* in parallel with the other activities. This Technical Specification will focus on design with existing engineered products and will evolve into a design *Standard* as the other documents are approved and adopted.

A critical role of INBAR TFC and WG12 will be continued maintenance and revision of these standards within the ISO TC 165 framework. Indeed, there are proposals for addition to ISO 22157:2019 [3] and ongoing activities that will ensure timely maintenance and updating of ISO 22156:2021.

In addition to TC 165, in 2016, ISO established the Technical Committee on Bamboo and Rattan (ISO/TC296), the Secretariat of which is located in the International Center for Bamboo and Rattan (ICBR), Beijing, China. This committee carries out standardization of bamboo- and rattan-derived materials and their applications, including terminology, classification, specifications, testing methods and quality requirements. The work of TC 296 excludes bamboo and rattan related standards associated with Civil structures. As a result, all standards developed by TC 165 will begin with the title "*Bamboo Structures*".

4.2 Chinese Bamboo Standards

By the end of 2020, China has issued 79 standards or codes relating to bamboo construction materials and bamboo structures, including:

- 15 National standards (Table 4.2);
- 50 Industry standards (Table 4.3), including 38 forestry industry standards, 5 construction industry standards, 3 Inspection and Quarantine industry standards, and 4 other standards, involving product terminology, mechanical processing, test methods, materials, energy consumption and inspection and quarantine;
- 10 Provincial standards (Table 4.4), including Hong Kong regional standards (2) and Taiwan regional standards (1);
- 4 Association Standards (Table 4.5).

Table 4.2 Current national standards related to bamboo construction in China

Designation	English and Chinese names of the standard
GB/T 15780-1995	Test methods for physical and mechanical properties of bamboo (竹材物理力学性质试验方法)
GB/T 2690-2000	Bamboo timber (毛竹材)
GB/T 13123-2003	Bamboo-mat plywood (竹编胶合板)
GB/T 21128-2007	Structural bamboo & wood composite board (结构用竹木复合板)
GB/T 21129-2007	Decorative bamboo veneered panel (竹单板饰面人造板)
GB/T 27649-2011	Bamboo and wood laminated composite flooring (竹木复合层积地板)
GB/T 26912-2011	Acceptance generality for production line of bamboo-wooden compound floor (竹木复合地板生产线验收通则)
GB/T 26536-2011	Bamboo strip (竹条)
GB/T 30364-2013	Bamboo scrimber flooring (重组竹地板)
GB/T 50920-2013	Code for design of culm-producing bamboo plantation engineering (用材竹林工程设计规范)
GB/T 20240-2017	Glued laminated bamboo flooring (竹集成材地板)
GB/T 36394-2018	Terminology of bamboo products (竹产品术语)
GB/T 36773-2018	Technical rule of phytosanitary treatment for bamboo products (竹制品检疫处理技术规程)
GB/T 37805-2019	Bamboo winding composite pipe (竹缠绕复合管)
GB/T 38071-2019	Structural laminated bamboo sliver curtain lumber (结构用竹篾层积材)

Table 4.3 Current industry standards related to bamboo construction in China

Designation	English and Chinese names of the standard	Classification
LY/T 1316-1999	Bamboo-processing machinery method of type designation (竹材加工机械型号编制方法)	Forestry industry standard
LY/T 1574-2000	Plybamboo for concrete-form (混凝土模板用竹材胶合板)	Forestry industry standard
LY/T 1072-2002	Laminated bamboo strips lumber (竹篾层积材)	Forestry industry standard
LY/T 1660-2006	Standard terminology for bamboo-based panel (竹材人造板术语)	Forestry industry standard
LY/T 1815-2009	Non-structural glued laminated bamboo (非结构用竹集成材)	Forestry industry standard
LY/T 1842-2009	Bamboo particleboard (竹材刨花板)	Forestry industry standard
LY/T 1998-2011	Split machine for bamboo strip (原竹劈条机)	Forestry industry standard
LY/T 1994-2011	Multi-spindle milling machine for bamboo strip (多轴竹条铣床)	Forestry industry standard
LY/T 2074-2012	Comprehensive Energy Consumption of ply bamboo (竹材胶合板生产综合能耗)	Forestry industry standard
LY/T 2222-2013	Sliced bamboo veneer (刨切竹单板)	Forestry industry standard
LY/T 2225-2013	Structural laminated bamboo sliver lumber (结构用竹篾层积材)	Forestry industry standard
LY/T 2395-2014	Comprehensive energy consumption of bamboo particleboard production (竹材刨花板生产综合能耗)	Forestry industry standard
LY/T 2396-2014	Comprehensive energy consumption of bamboo and wood composite board (竹木复合板生产综合能耗)	Forestry industry standard
LY/T 5004-2014	Code for design of bamboo plywood engineering (竹材胶合板工程设计规范)	Forestry industry standard
LY/T 2551-2015	Comprehensive energy consumption for bamboo floor production (竹地板生产综合能耗)	Forestry industry standard

(continued)

Table 4.3 (continued)

Designation	English and Chinese names of the standard	Classification
LY/T 2564-2015	Determination of physical and mechanical properties of bamboo culm (圆竹物理力学性能试验方法)	Forestry industry standard
LY/T 2565-2015	Bamboo plastic composite (竹塑复合材料)	Forestry industry standard
LY/T 2608-2016	Bamboo products classification (竹产品分类)	Forestry industry standard
LY/T 2609-2016	Rotary-cut bamboo veneer (旋切竹单板)	Forestry industry standard
LY/T 2614-2016	Indoor bamboo-based door (室内竹质门)	Forestry industry standard
LY/T 2711-2016	Bamboo laminated lumber for veneer (单板用竹集成材)	Forestry industry standard
LY/T 2712-2016	Bamboo veneer plywood (竹单板胶合板)	Forestry industry standard
LY/T 2713-2016	Bamboo decorative wood-based flooring (竹材饰面木质地板)	Forestry industry standard
LY/T 2905-2017	Bamboo Winding Composite Pipe (竹缠绕复合管)	Forestry industry standard
LY/T 3080-2018	Splitting machine for bamboo slice (竹片剖分机)	Forestry industry standard
LY/T 3164-2019	Comprehensive energy consumption of bamboo-wood composite production (竹木复合层积地板生产综合耗能)	Forestry industry standard
LY/T 3182-2020	Technical requirements for wood and bamboo products in forest ecotourism destinations (森林生态旅游地木 (竹) 材产品使用技术要求)	Forestry industry standard
LY/T 3193-2020	Terminology of bamboo-based materials for engineering (竹制工程材料术语)	Forestry industry standard
LY/T 3194-2020	Structural bamboo scrimber (结构用重组竹)	Forestry industry standard
LY/T 3195-2020	Quality of bamboo treated with preservative (防腐竹材的质量要求)	Forestry industry standard

(continued)

Table 4.3 (continued)

Designation	English and Chinese names of the standard	Classification
LY/T 3196-2020	Technical regulations for carbon accounting of bamboo forest (竹林碳计量规程)	Forestry industry standard
LY/T 3197-2020	Technical regulations for carbon accounting of bamboo products (竹材制品碳计量规程)	Forestry industry standard
LY/T 3201-2020	Flattened bamboo flooring (展平竹地板)	Forestry industry standard
LY/T 3202-2020	Bamboo winding utility tunnel (竹缠绕管廊)	Forestry industry standard
LY/T 2222-2020	Bamboo veneer (竹单板)	Forestry industry standard
LY/T 3204-2020	Flattened bamboo sheet (竹展平板)	Forestry industry standard
LY/T 3199-2020	Bamboo plastic composite profile reinforced with aluminum alloy (铝合金增强竹塑复合型材)	Forestry industry standard
LY/T 1926-2020	Test method for evaluating the antibacterial performance of wood (bamboo)-based panels and products (抗菌木(竹)质地板 抗菌性能检验方法与抗菌效果)	Forestry industry standard
JG/T 156-2004	Plybamboo form (竹胶合板模板)	Construction industry standard
JG/T 199-2007	Testing methods for physical and mechanical properties of bamboo used in building (建筑用竹材物理力学性能试验方法)	Construction industry standard
JGJ 254-2011	Technical code for safety of bamboo scaffold in construction (建筑施工用竹脚手架安全技术规范)	Construction industry standard
JG/T 428-2014	Composite plybamboo form with steel frame (钢框竹胶合板模板)	Construction industry standard
JG/T 537-2018	Bamboo-based composite for building and garden landscape engineering (建筑及园林景观工程用复合竹材)	Construction industry standard

(continued)

Table 4.3 (continued)

Designation	English and Chinese names of the standard	Classification
SN/T 1815-2006	Rules for the quarantine of bamboo products for import and export (进出境竹制品检疫规程)	Inspection and quarantine industry standard
SN/T 3275-2012	Rules of methyl bromide fumigation treatment for export bamboo products (出口竹制品溴甲烷熏蒸处理规程)	Inspection and quarantine industry standard
SN/T 4250-2015	Determination of sulphur dioxide in bamboo and wood products. Iron chromatography method (竹木制品中二氧化硫的测定 离子色谱法)	Inspection and quarantine industry standard
CCGF 407.4-2015	Implementation Code for Quality Supervision and Spot Inspection of Bamboo Flooring Products (竹地板产品质量监督抽查实施规范)	General Administration of Quality Supervision, Inspection and Quarantine of China
SB/T 11103-2014	Specification for grading of wood flooring enterprise (木地板企业等级划分规范)	Ministry of Commerce of China
TB/T 1781-2004	Specifications of uplifting tie plate underside rail for concrete sleeper (混凝土枕用轨下调高垫板技术条件)	Railway transportation
JC/T 2124-2012	Supporting plate made of plybamboo used for concrete block & concrete brick production (混凝土砌块(砖)生产用竹胶托板)	Ministry of Industry and Information Technology of China

Table 4.4 Provincial standards related to bamboo construction in China

Designation	English and Chinese names of the standard	Jurisdiction
DB43/T 556-2010	Standard for consumption quota of wood and bamboo raw materials for major forest products (主要林产品木竹原料消耗限额标准)	Hunan
DB43/517-2010	Formaldehyde emission limits of bamboo-wood plywood (竹·木胶合贴板甲醛释放限量)	Hunan
DB43/T 642-2011	Determination of pentachlorophenol in bamboo and wood products by gas chromatography (竹木制品中五氯苯酚含量的测定 气相色谱法)	Hunan
DB33/T 394-2009	Installation, inspection guidelines for wooden flooring (木质地板安装验收规范)	Zhejiang
DB33/T 952-2014	The quota & calculation method of energy consumption per unit products for bamboo scrimber flooring (重组竹地板单位产品能耗定额及计算方法)	Zhejiang
DB35/T 1145-2011	Bamboo scrimber board (竹重组板材)	Fujian
DB13/T 5089-2019	General specification for bamboo wood fiber integrated wallboard (竹木纤维集成墙板通用技术条件)	Hebei
2014	Safety rules for bamboo scaffolding practices (竹棚架工作安全守则)	Hong Kong
2006	Guidelines on the design and construction of bamboo scaffolds (竹棚架设计搭建指引)	Hong Kong
CNS 3219-2011	Preservative treated bamboo by pressure processes (加压注入防腐处理竹材)	Taiwan

Table 4.5 Current association standards related to bamboo construction in China

Designation	English and Chinese names of the standard	Issuing Association
CECS 434-2016	Technical specification for round bamboo structure building (圆竹结构建筑技术规范)	China Engineering Construction Association
CECS 470-2017	Technical specification for bamboo winding composite pipe (竹缠绕复合管道工程技术规程)	China Engineering Construction Association
T/ZZB 0079-2016	Bamboo scrimber flooring (重组竹地板)	Zhejiang-Made Brands Building and Promotion Association
T/CADBM 3-2018	Bamboo-Wood fiber integrated wall (竹木纤维集成墙面)	China Association of Decoration Building Materials

In addition to all the standards listed, the Building Environment Assessment Method (BEAM), jointly published by the Hong Kong Green Building Council (HKGBC) and the Building Environmental Assessment Association (BEAA), in Hong Kong, China, inscribed bamboo into the list of fast renewable building materials and encouraged designers to use such materials more frequently. Raw materials can replace high energy-consumption materials to reduce the impact of construction on the environment. Although not part of a formal standard, this will nonetheless promote the use of bamboo construction materials and the development of the bamboo construction sector.

With the exception of GB/T 15780-1995 *Test methods for physical and mechanical properties of bamboo*, the Hong Kong *Guidelines on the design and construction of bamboo scaffolds* and CECS 434-2016 *Technical specification for round bamboo structure building*, other Chinese standards are relevant to bamboo construction products and processing methods; they are not engineering construction standards, per se. Nonetheless, in order to promote bamboo construction, construction standards are required. In this regard, recent guidance is being developed at Shanghai Research Institute of Building Science (SRIBS) and Nanjing Forestry University (NFU) in the form of the following:

Technical specifications for engineered bamboo structures (SRIBS 2017); this document will inform the ISO technical specification on the same topic;
Technical specifications for construction and quality acceptance of engineered bamboo structures (SRIBS 2017);
Technical specifications for inspection of engineered Bamboo structures (SRIBS 2017); and,
Test methods of physical and mechanical properties of engineered bamboo (NFU 2018).

These new draft standards will have a positive impact on the development of the engineered bamboo structure sector in China and perhaps globally.

References

1. Liu, K. W., Trujillo, D., Harries, K., et al. (2020). INBAR construction task force—An explorative way for development in the bamboo construction sector. In Y. Xiao, Z. Li, & K. W. Liu (Eds.), *Modern Engineered Bamboo Structures, Sustainable Bamboo Building Materials Symposium of BARC 2018 & 3rd International Conference on Modern Bamboo Structures (3-ICBS-2018)*, pp. 51–60. June 25–27, 2018, Beijing, China: CRC Press, Taylor & Francis Group.
2. Harries, K., Alon, B. L., & Sharma, B. (2019). Chapter 4: Codes and standards development for nonconventional and vernacular materials in nonconventional and vernacular construction materials. In K. Harries & B. Sharma (Eds.). *Characterisation, properties and applications* (2nd ed.). Woodhead (Elsevier) Publishing Series in Civil and Structural Engineering.
3. Virgo, J., Moran, R., Harries, K., Garcia, J., & Platt, S. (2018). Flat ring flexure test for full-culm bamboo. *Materials Research Proceedings, 7*, 349–358.

Chapter 5
International Organizations, Research Institutions, and Production and Processing Enterprises in China

This chapter summarises organizations, institutions and relatively large-scale production and processing enterprises in China, which are committed to promoting the development of a bamboo construction industry.

5.1 International Organization

The International Bamboo and Rattan Organisation (INBAR) is the first intergovernmental development organization that headquartered in Beijing, China. It is currently made up of 48 Member States. Since its establishment in 1997, INBAR has been dedicating to promoting environmentally sustainable development using bamboo and rattan. The Global Bamboo Construction Programme (GBCP) of INBAR, established in 2006, is aimed at promoting bamboo as an environmentally-friendly substitute for other building materials in order to provide solutions to housing shortages, climate change mitigation and adaptation, and livelihood development. GBCP has cooperated with public and private sectors in more than 20 countries to carry out bamboo construction-related scientific research, project demonstration, standard development and policy promotion.

In order to promote further collaborations between international research institutes and commercial companies interested in structural uses of bamboo, the INBAR Bamboo Construction Task Force (INBAR TFC) was established in 2014. At present, there are 34 senior bamboo construction experts from 18 countries in the group, aiming to serve as the world's premier information and knowledge repository on structural uses of bamboo. Since its inception, the INBAR TFC has contributed many achievements including: standardization work for structural uses of bamboo; high quality peer-reviewed publications; international projects; consultancy services; international conferences; and capacity-building for the bamboo construction professional [1].

K. W. Liu et al., *Contemporary Bamboo Architecture in China*,
https://doi.org/10.1007/978-981-16-8309-1_5

5.2 Research Institutions

In order to obtain a 'snapshot' of institutions and organisations conducting research and development in bamboo construction, a bibliometric analysis was performed as described in the following section.

5.2.1 *Bibliographic Analysis Methodology*

Two similar searches were conducted. Both were conducted 2 July 2021 and are intentionally more refined than that presented in Chap. 3. The first search, of the China National Knowledge Infrastructure (CNKI), focused on Chinese-language publications and yielded 682 results. The second, of the Web of Science Core Collection (WSCC), was limited to English-language publications and yielded 530 results. A summary of the search parameters used in each case are given in Table 5.1.

5.2.2 *Findings*

Table 5.1 shows the year of publication of the 682 and 530 results reported in CNKI and WSCC, respectively. The essentially exponential increase in publication since 2000 is evident. The continued dominance of Chinese contributions to the research portfolio is also clear from the WSCC data shown. With the exception of the USA and UK, most research output has its origins in regions where bamboo construction is part of the vernacular and is endemic.

Beyond national origin, it is informative to identify the institutional affiliations of this work. Figure 5.1 shows the top institutional affiliation identified in the CNKI and WSCC searches. The dominance of three Chinese institutions—Nanjing Forestry University, Hunan University and Southeast University—is apparent. It is noted that a number of papers will, of course, be affiliated with multiple institutions. Furthermore, many of the affiliations having ten or fewer papers, often represent the work of a single principal investigator.

A second interesting data set that can be obtained from the bibliographic analysis is the dominant sources of funding for the published works. The published sources of funding obtained from the WSCC analysis are shown in Fig. 5.2. Care should be taken interpreting this data; it shows the number of times a funding agency is acknowledged and may not be reflective of the extent of funding. For example, although EPSRC (UK) is acknowledged multiple times, the funding from multiple projects has been relatively modest; whereas the National Science Foundation (US) has made only a single, albeit larger, award.

Table 5.1 Parameters for bibliometric analysis

Database	China National Knowledge Infrastructure (CNKI)	Web of Science Collection Core (WSCC)
Language	Chinese only	English only
Categories	Architecture and Engineering	Engineering Civil Construction Building Technology Architecture
Document types	Journals and Conference Proceedings Theses and dissertations	Journals and Conference Proceedings
Search terms	Bamboo architecture Bamboo structure Bamboo-based engineering Glue-laminated bamboo Bamboo scrimber	Bamboo [and] construction Bamboo [and] building Bamboo [and] structure Engineered bamboo Laminated bamboo Bamboo scrimber
Search range	Title, abstract and key words	Title, abstract and key words
Results	682	530
Distribution by year of publication	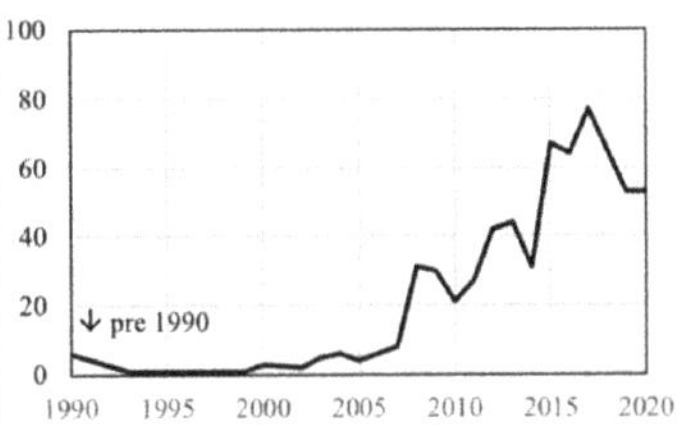	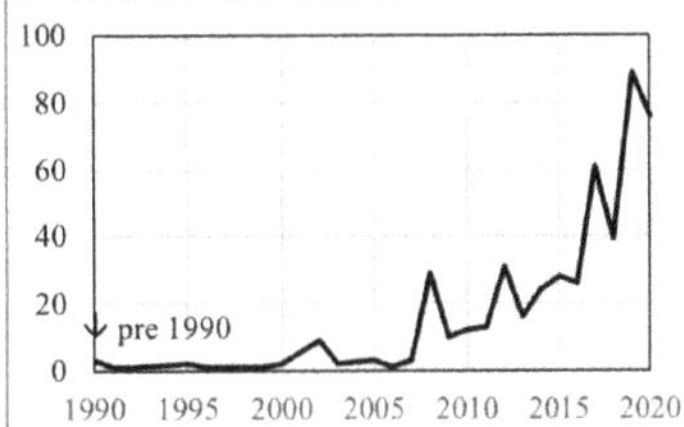
Distribution of English language publications by country of origin		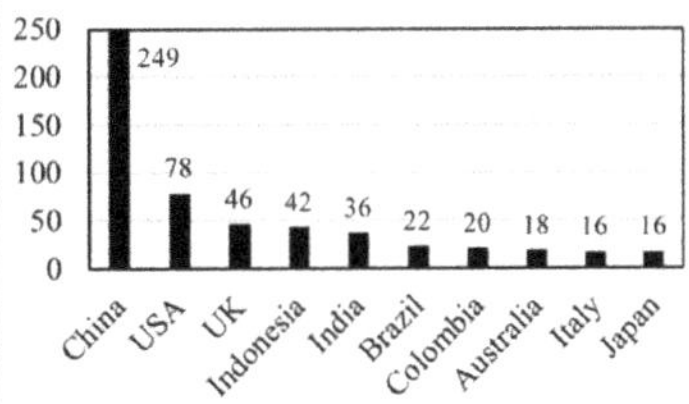 n = 543 due to international co-authorships

5.2.2.1 Chinese Research Institutions

There remain relatively few institutions with bamboo construction-related research programs. There are even fewer universities providing bamboo construction-related courses. The broad focus areas of Chinese institutions having such programs is summarised in Table 5.2. The dominant interest in engineered bamboo products is apparent.

Strengthening the capacity of building of professionals is critical to promoting the development of a Chinese bamboo construction industry; this is one of strategies of *Research on Development Strategy and Key Technologies of Bamboo Construction Sector in China towards 2035*, a project funded by the Chinese Academy of

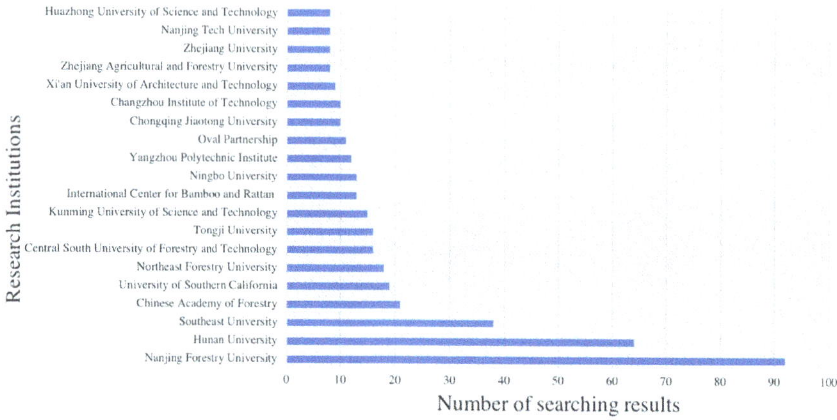

(a) Chinese language papers from CNKI search

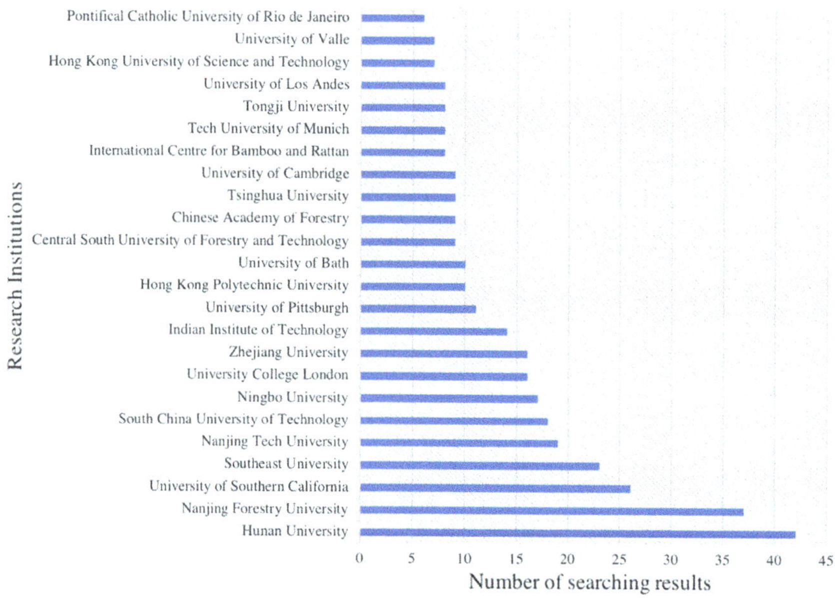

(b) English language papers from WSCC search

Fig. 5.1 Institutional sources of papers

Engineering. Several suggestions for future research and applications in this field are identified in this report as follows [2]:

1. Develop environmentally-friendly green bamboo construction materials with high-durability and good fire-resistant performance.
2. Establish a relatively complete standard system for bamboo construction.

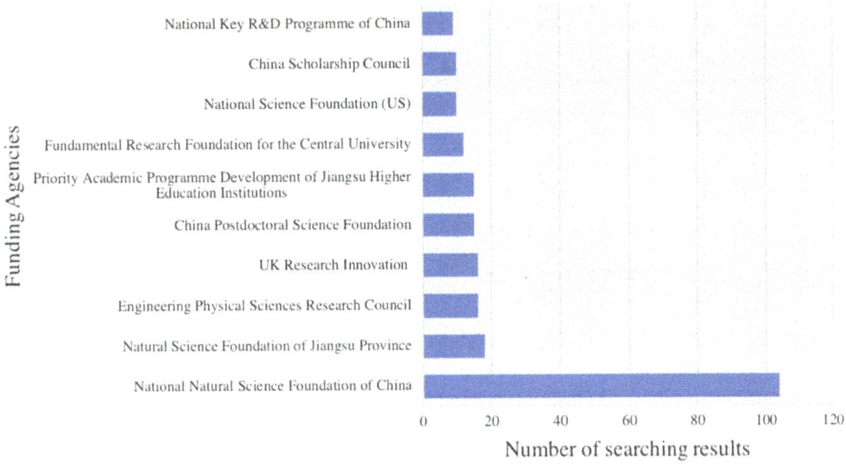

Fig. 5.2 Funding agencies identified in WSCC search

3. Develop a series of innovative prefabricated bamboo construction systems which could resist various disasters.
4. Promote large-scale bamboo construction demonstration projects with different types of structures for multiple applications in China.

5.3 Production and Processing Enterprises

In order to develop a list of the main production and processing enterprises of bamboo construction materials in China, the authors invited the China Bamboo Industry Association to provide information. Based on the data provided, the following 25 major enterprises (Table 5.3) were selected. These include manufacturers of bamboo construction materials, adhesives and specialised machinery for bamboo product manufacturing enterprises. Most product manufacturers include high-value bamboo-flooring in the production portfolios. Table 5.3 is intended to be a snapshot of bamboo construction-related industry in China; it is not all-inclusive.

Table 5.2 Major institutions in China with bamboo construction-related research programs

Research institution (shaded entries are top sources of English-language publications as shown in Figure 5.1b)	Location	Research area									
		Full-culm bamboo	Engineered bamboo	Bamboo scrimber	FRP reinforced bamboo	Bamboo-plastic composites	Bamboo fibre reinforced concrete	Bamboo reinforced rammed earth	Bamboo-steel composite structures	Bamboo scaffolding	Bamboo construction for landscape
Hunan University	Hunan		✓								
Nanjing Forestry University	Jiangsu	✓	✓								
Southeast University	Jiangsu		✓								
Nanjing Tech University	Jiangsu		✓								
South China University of Technology	Guangdong		✓								
Ningbo University	Zhejiang		✓						✓		
Zhejiang University	Zhejiang		✓								
Hong Kong Polytechnic University	Hong Kong									✓	
Central South University of Forestry and Technology	Hunan		✓								
Chinese Academy of Forestry	Beijing		✓								
Tsinghua University	Beijing	✓	✓		✓						
International Centre for Bamboo and Rattan	Beijing	✓	✓								
Tongji University	Shanghai	✓	✓								
Hong Kong University of Science and Technology	Hong Kong						✓				
Northeast Forestry University	Heilongjiang		✓						✓		✓
Kunming University of Science and Technology	Yunnan	✓						✓			
Yangzhou Polytechnic Institute	Jiangsu			✓							
Chongqing Jiaotong University	Chongqing	✓	✓								
Changzhou Institute of Technology	Jiangsu			✓							
Xi'an University of Architecture and Technology	Shanxi										✓
Zhejiang Agriculture and Forestry University	Zhejiang										✓
Beijing Forestry University	Beijing	✓	✓								✓
Shanghai Research Institute of Building Sciences	Shanghai		✓								
Shanghai Jiaotong University	Shanghai				✓		✓				
Fujian Agriculture and Forestry University	Fujian										
Hebei Agricultural University	Hebei										✓
Guizhou University	Guizhou										✓

Table 5.3 Major production and processing enterprises of bamboo construction-related products in China

Name	Location	Bamboo construction-related products							
		Bamboo flooring	Structural glue laminated bamboo	Structural bamboo scrimber	Bamboo wallboard or grille	Bamboo-wood composite materials	Bamboo filament wound composite products	Decorative bamboo materials	Structural full-culm bamboo
Hangzhou Dasso Technology Co., Ltd.	Zhejiang	✓	✓		✓			✓	
Zhejiang YOYU Bamboo Industry Co., Ltd.		✓			✓				
Hangzhou Zhengtian Industrial Co., Ltd.		✓							
Zhejiang Tianzhen Bamboo & Wood Development Co., Ltd.		✓							
Zhejiang Tenglong Bamboo Industry Group., Ltd.		✓							
Anji Zhujing Bamboo Industry Technology Co.,Ltd.									✓
Zhejiang Xinzhou Bamboo-Based Composites Technology Co., Ltd.							✓		
Hangzhou Bamboo Technology Co., Ltd.									✓
Ganzhou Sentai Bamboo & Wood Co., Ltd.	Jiangxi	✓	✓						✓
Jiangxi Kangda Bamboo Products Group Co., Ltd.		✓	✓					✓	
Jiangxi Feiyu Bamboo Industry Group Co., Ltd.		✓	✓					✓	
Jiangxi Kangtilong Bamboo Industry Co., Ltd		✓							
Jiangxi Songtao Bamboo Industry Co., Ltd		✓							
Fujian Jianou Huayu Bamboo Industry Co., Ltd.	Fujian	✓							
Fujian He Qi Chang Bamboo Product Co., Ltd.						✓			
Fujian Youzhu Technology Co., Ltd.		✓	✓						
Anhui Longhua Bamboo Products Co., Ltd.	Anhui	✓				✓			
Anhui Hongyu Bamboo & Wood Product Co., Ltd.		✓							
Hubei Chu Feng Zhu Yun Technology Co., Ltd.	Hubei							✓	
Hubei Juning Bamboo Science and Technology Co., Ltd.		✓		✓					
Hongyazhuyuan Science and Technology Company.	Sichuan	✓		✓					
Hunan Taohuajiang Bamboo Technology Co., Ltd.	Hunan		✓	✓					
Dynea (Guangdong) Co., Ltd.	Guangdong	adhesive for engineered bamboo products							
Qingdao Guosen Machinery Co., Ltd.	Shandong	hot press machinery for bamboo veneer and multi-layer bamboo flooring; cold/hot press production line for bamboo scrimber							
HOMAG China Golden Field Ltd.	Jiangsu	hot press machinery for bamboo-based panel							

Source China Bamboo Industry Association

References

1. Liu, K. W., Trujillo, D., Harries, K., et al. (2020). INBAR construction task force—An explorative way for development in the bamboo construction sector. In Y. Xiao, Z. Li, Liu, & K. W. (Eds.), *Modern Engineered Bamboo Structures, Sustainable Bamboo Building Materials Symposium of BARC 2018 & 3rd International Conference on Modern Bamboo Structures (3-ICBS-2018),* pp. 51–60. June 25–27, 2018, Beijing, China: CRC Press, Taylor & Francis Group.
2. Jiang, W. A., Liu, K. W., Zhang, X., et al. (2021). Trends and path for the development of bamboo structural engineering towards 2035 in China. *China Civil Engineering Journal, 54*(10), 125–132.

Chapter 6
Case Studies

This chapter introduces examples of commercially-built bamboo structures built in China in recent years. The chapter also presents number of non-commercial examples illustrate other promising application directions for bamboo construction. Based on information provided by China's major manufacturers of bamboo construction products, the authors contacted relevant architects or designers to collect first-hand information on the projects, including their context and photos of completed works. The information presented in most cases was obtained directly from architects or designers.

Through this curation the authors attempt to illustrate general information on the projects and to demonstrate the characteristics of different bamboo applications and the functional attributes of different bamboo products. The authors believe that when each architect or designer of those projects chose bamboo—still a non-mainstream construction material—they must have clear design intentions. This places these projects in the unique architectural context of modern Chinese bamboo architecture and demonstrates the spiritual and cultural attributes represented by bamboo in modern Chinese architecture and, indeed, society.

The chapter is arranged in the following sections:

6.1 Bamboo as decorative materials
6.2 Bamboo as structural materials
6.3 Landscapes
6.4 Rural construction
6.5 Transportation facilities
6.6 Water pipes and urban municipal tunnels.

Through the introduction of typical cases, the authors hope to arouse people's recognition of natural materials, traditional cultures and the concept of sustainable development.

K. W. Liu et al., *Contemporary Bamboo Architecture in China*,
https://doi.org/10.1007/978-981-16-8309-1_6

6.1 Bamboo as Decorative Materials

Engineered bamboo applications:

Fuzhou Strait Culture and Art Centre (2018)

Ilan and Assaf Ramon International Airport, Israel (2019)

Fuyang Gongwang Art Museum (2016)

MUJI Hotel Beijing (2018)

Hang Seng University of Hong Kong (2016)

Calypso Mediterranean Restaurant, Shangri-La Hotel in Jing'an District, Shanghai (2014)

Old Bailey Restaurant, Tai Kwun of Hong Kong (2018)

Seaside Library (2015)

Seaside Church (2015)

Bamboo and Rattan Pavilion of the 10th China Flower EXPO (2021)

Nanjing Aohong Resort Hotel (2018)

Changbai Mountain Squirrel (Bicycle) Station (2015).

Full-culm bamboo applications:

Tangxing No. 5 Residential House (2018)

China Buoy Town Fishing Center (2017)

Linxiang Global Tourism Service Center (2018)

Gudu Wharf in Yunnan Sea (2016)

Kunqu Opera Society (2017).

Fuzhou Strait Culture and Art Center

Location: Fuzhou, Fujian, China

Completed: 2018

Architects: Pekka Salminen (Finland), Xu Zongwu

The Fuzhou Strait Culture and Art Center covers an area of 150,000 m^2, with a total cost of 2.7 billion yuan (approximately 400 million USD). The project is composed of five buildings, including a multi-functional theatre, opera house, concert hall, art museum and film center, as well as a central cultural hall. It looks like a "jasmine flower" and is known as the "Sydney Opera House" in the East. The building uses a large number of ceramic elements and engineered bamboo, which endow it with distinct Chinese elements, support a self-cleaning function, and meet required acoustic standards.

Application of bamboo products: The application of bamboo in the whole project was adopted, including bamboo floor, sound absorbing bamboo wallboard, fireproof bamboo wallboard, fireproof bamboo grille and outdoor anti-corrosive bamboo materials, etc.

Photo source: Figs. 6.1, 6.2, 6.3, 6.4 and 6.5 are provided by PES-Architects Ltd.

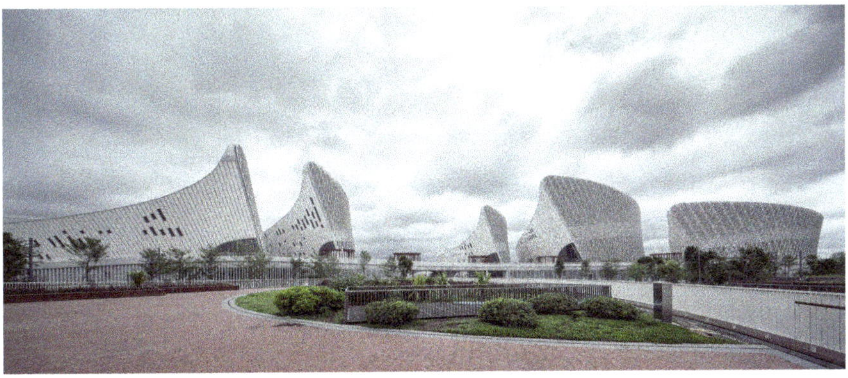

Fig. 6.1 Five jasmine petals of Fuzhou Strait Culture and Art Center (*photo by Zhang Yong*)

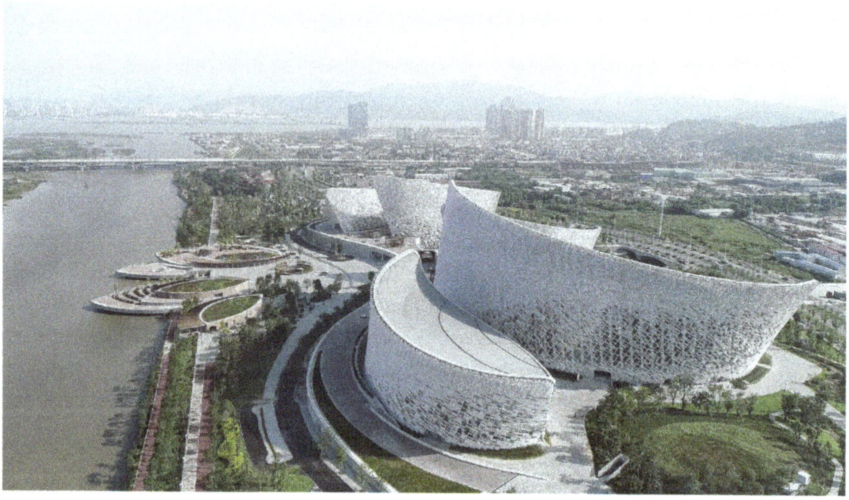

Fig. 6.2 Aerial view of Fuzhou Strait Culture and Art Center (*photo by Marc Goodwin*)

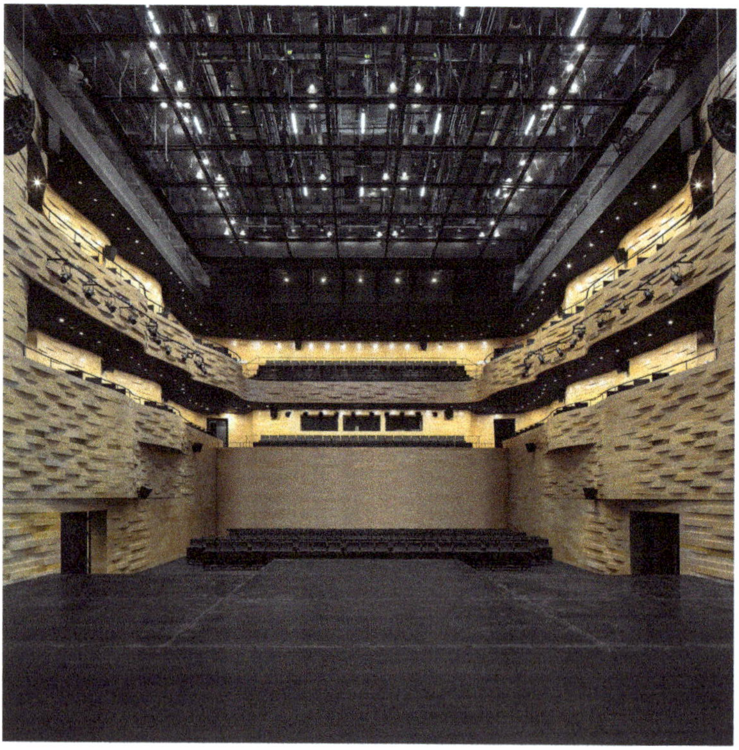

Fig. 6.3 Multifunctional hall (*photo by Marc Goodwin*)

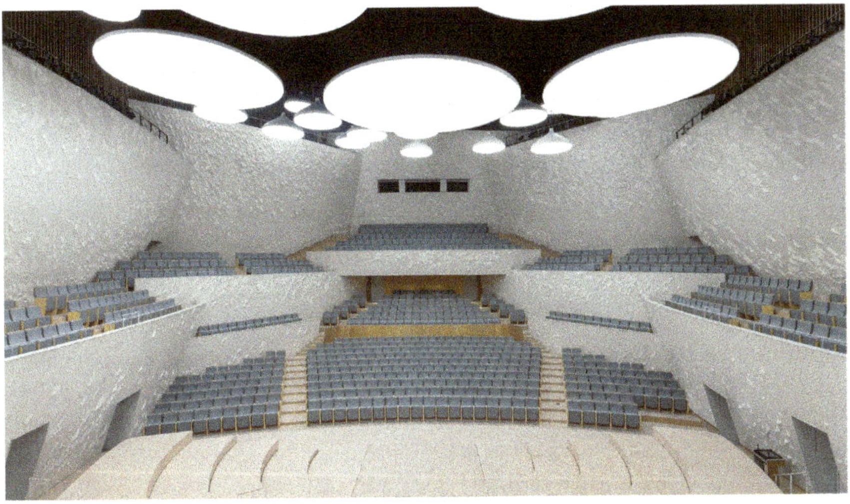

Fig. 6.4 Concert hall (*photo by Marc Goodwin*)

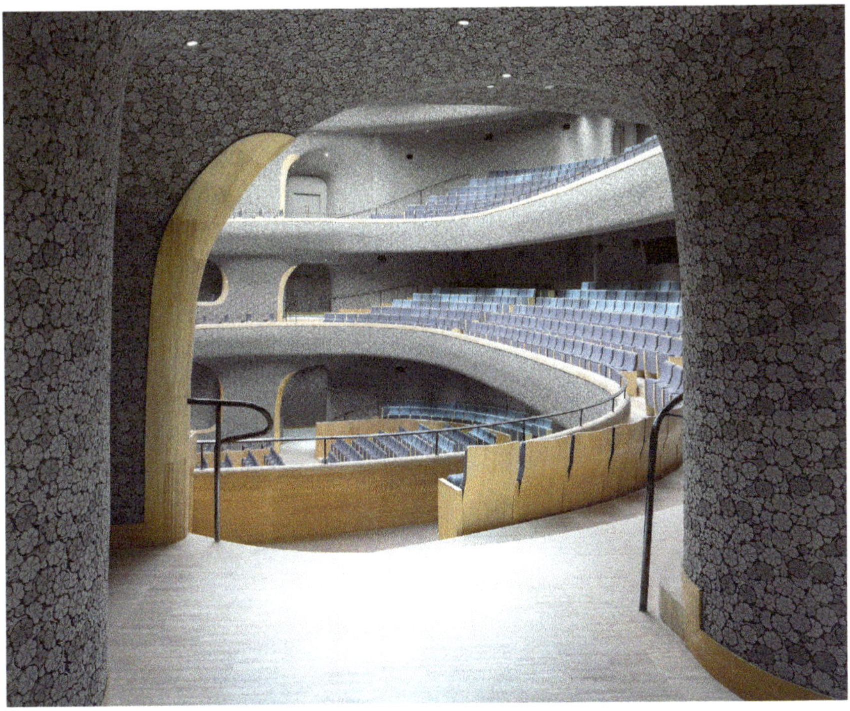

Fig. 6.5 Entrance of Opera hall (*photo by Marc Goodwin*)

Ilan and Assaf Ramon International Airport, Israel

Location: Eilat, Israel

Completed: 2019

Architects: Amir Mann/Ami Shinar Architects & Planners Ltd. (Israel)

Ilan and Assaf Ramon International Airport is Israel's second new international airport, located in the southern coastal tourist city, Eilat. In this unique desert environment, the designer created a sculpture-like airport in a minimalist style, with a geometric shape similar to weathered rocks. The interior combines a variety of styles.

Application of bamboo products: The airport takes "maximizing the use of natural materials" as a guiding principle. The ceiling and wall of the airport are made of glued laminated bamboo specially processed to meet the EU B1 fire-protection standard. Breaking through the constraints of traditional natural materials, glued laminated bamboo 6 m in size are used, skillfully integrating the airport into its surroundings.

Photo source: Figs. 6.6, 6.7 and 6.8 are provided by Ganzhou Sentai Bamboo & Wood Co., Ltd. Fig. 6.6 is a rendering by Amir Mann/Ami Shinar Architects & Planners Ltd and Green Sky Ltd.

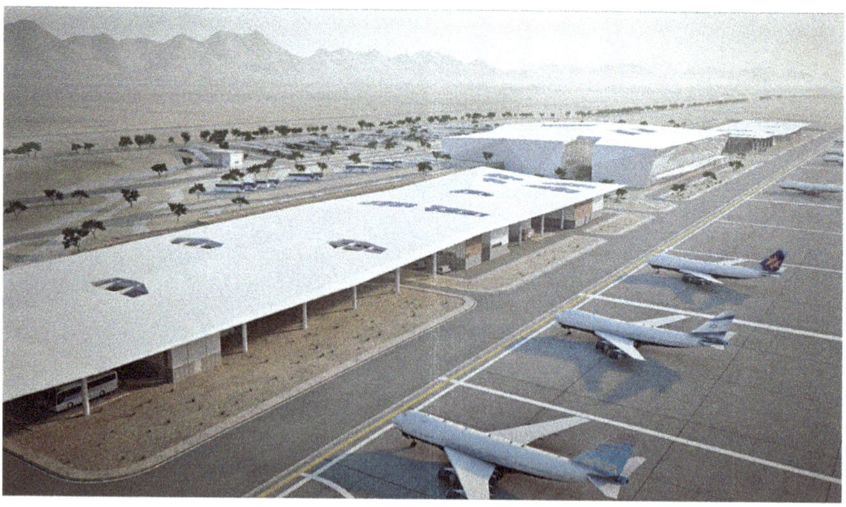

Fig. 6.6 Aerial view of Ilan and Assaf Ramon International Airport

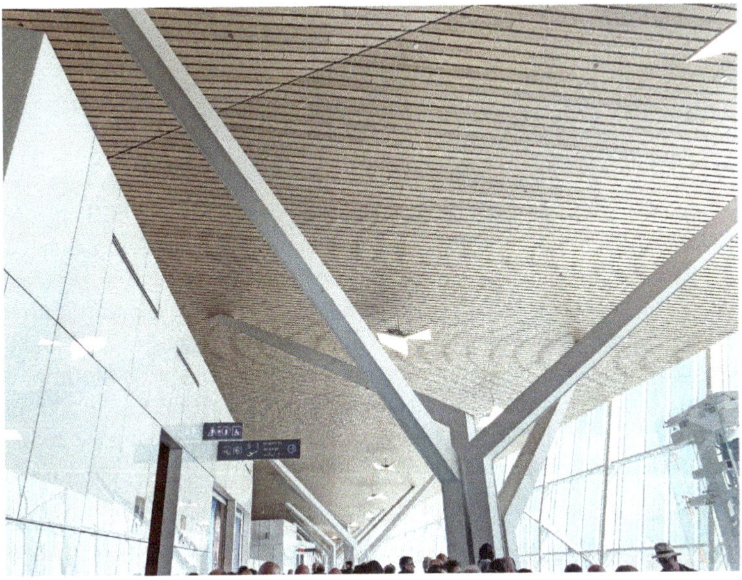

Fig. 6.7 Arrival hall (*photo by Xiong Zhenhua*)

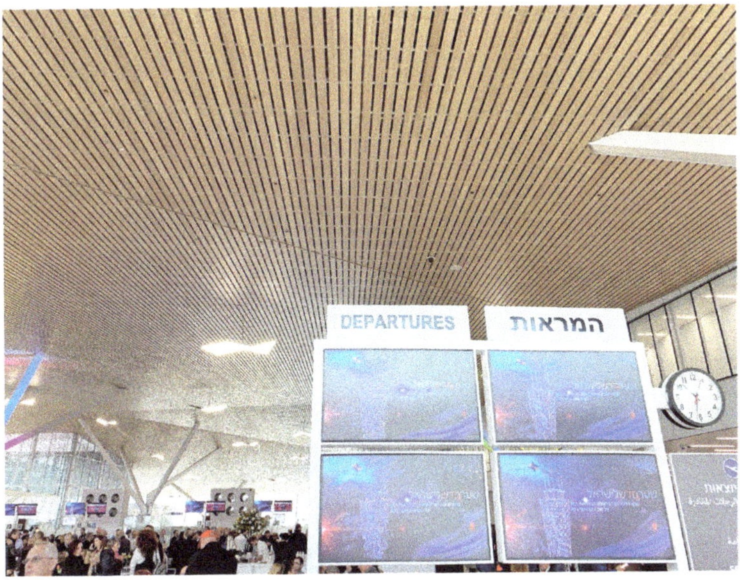

Fig. 6.8 Departure hall (*photo by Xiong Zhenhua*)

Fuyang Gongwang Art Museum

Location: Hangzhou, Zhejiang province, China

Completed: 2016

Architects: Wang Shu

Fuyang Gongwang Art Museum is located in the Fuyang District of Hangzhou City, west of Dongwu Cultural Park. It is one of three museums in the Fuchun Cultural Complex. The Fuchun Cultural Complex is designed by Wang Shu, winner of the 2012 Pritzker Architecture Prize, who took a famous Chinese painting "*Dwelling in the Fuchun Mountains*" as the background and theme of the construction. Combining this with the urban landscape, Wang arranged the buildings in the form of a main mountain, a secondary mountain and distant mountain, forming "livable image". By extracting the local architectural language of Fuyang, reflecting the connotation of Fuchun Mountain and River landscape and regional cultures, the museum is a constructed version of *Dwelling in the Fuchun Mountains* [1].

Application of bamboo products: Bamboo braided fence, bamboo grid ceiling, bamboo decorative wall ceiling and heavy bamboo railings.

Photo source: Fig. 6.9 is provided by Hunan Taohuajiang Bamboo Science and Technology Co., Ltd.; Figs. 6.10 and 6.11 are provided by Hangzhou Dasso Technology Co., Ltd.

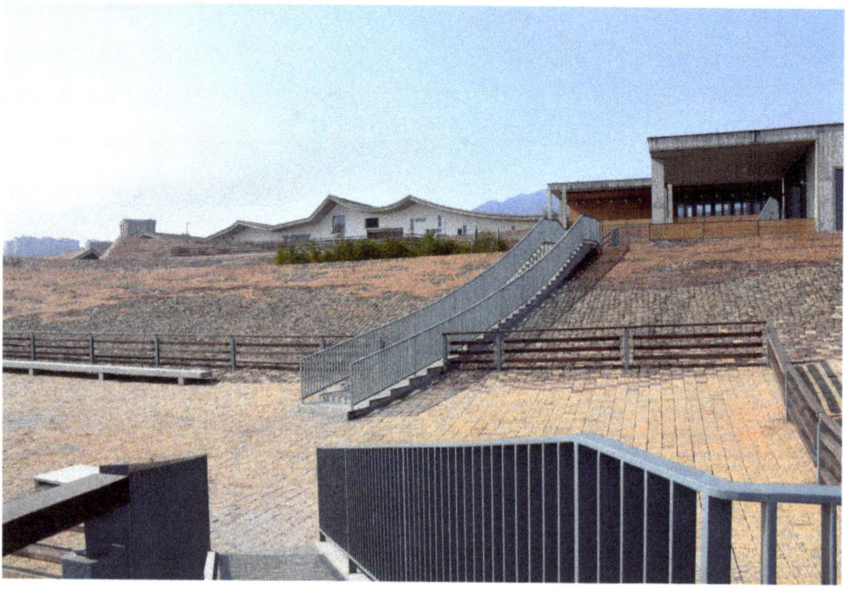

Fig. 6.9 Exterior view of Fuyang Gongwang Art Museum

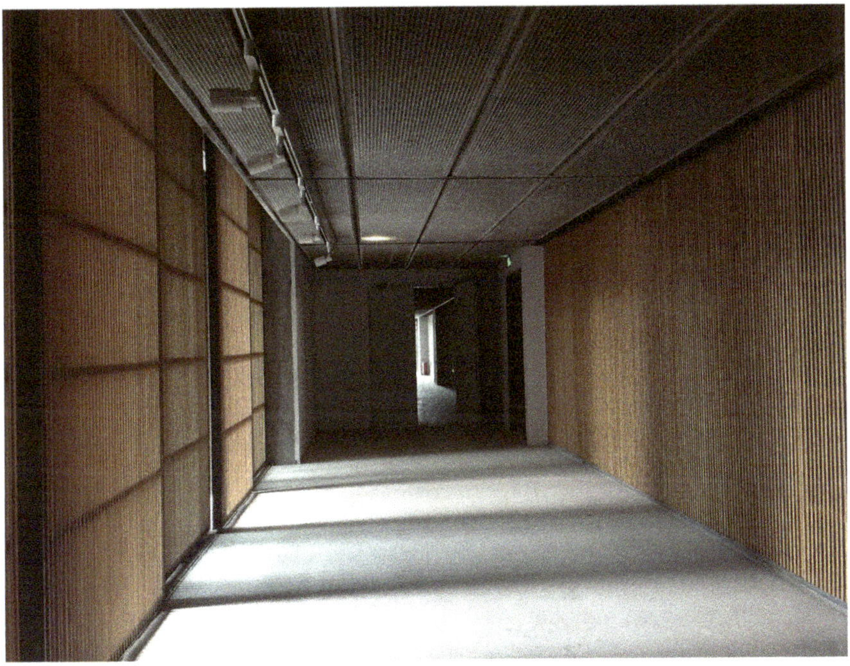

Fig. 6.10 Bamboo walls

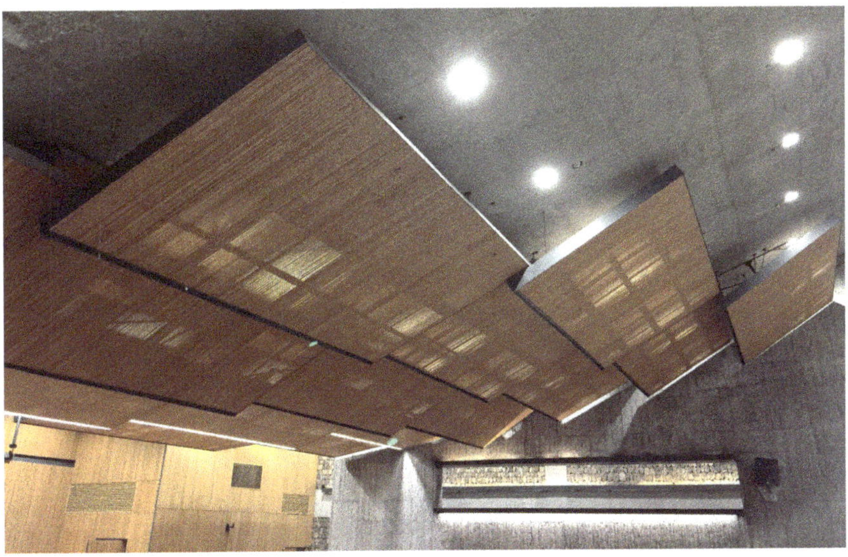

Fig. 6.11 Suspended bamboo celling

MUJI Hotel Beijing

Location: Beijing, China

Completed: 2018

Architects: UDS Co., Ltd. (Japan)

As a modern metropolis with rapid development, Beijing still retains its flavour of traditional old Beijing in hutongs; although the original charm of local culture is gradually disappearing in people's live. A Japanese company, MUJI, opened a hotel in Beijing to convey the concept of "returning to the origin of things". The hotel is located in the "Beijing Fun", Qianmen district of Beijing, 450 m from Tiananmen Square. In line with the principle of "anti-luxury, anti-crudeness", the designer used bamboo, which has been popular among Chinese people since ancient times, and transformed green bricks, obtained when surrounding old streets were reconstructed, into a building treasure that passes on the flavour of old Beijing.

Application of bamboo products: The modern bamboo material, bamboo scrimber was widely used in public spaces of the hotel. Old bricks and beams from nearby streets were used to create a new space in the style of historical and cultural blocks. Functionally, from the front desk, the book bar, to the 4-storey high MUJI display wall, the building is unified by using bamboo scrimber. This reflects the diversity of the application of bamboo scrimber, and represents the concept of the hotel brand.

Photo source: Figs. 6.12, 6.13, 6.14, 6.15, 6.16 and 6.17 are provided by UDS Co., Ltd.

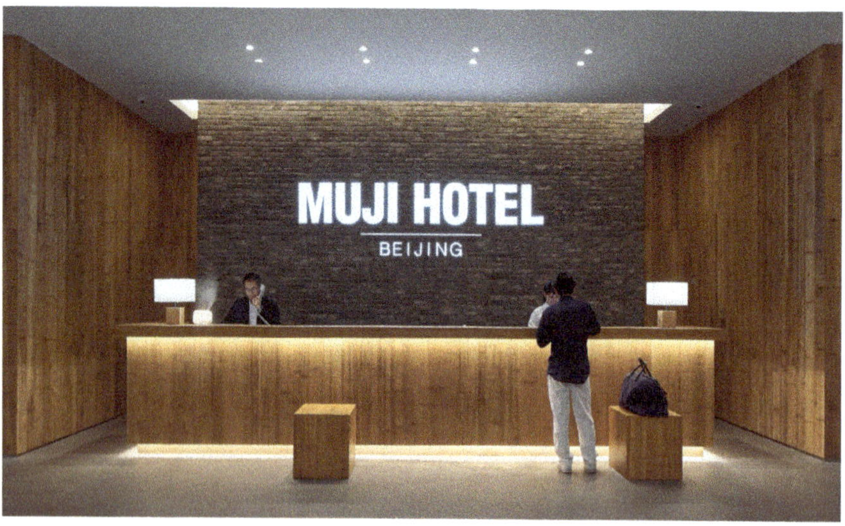

Fig. 6.12 Front desk

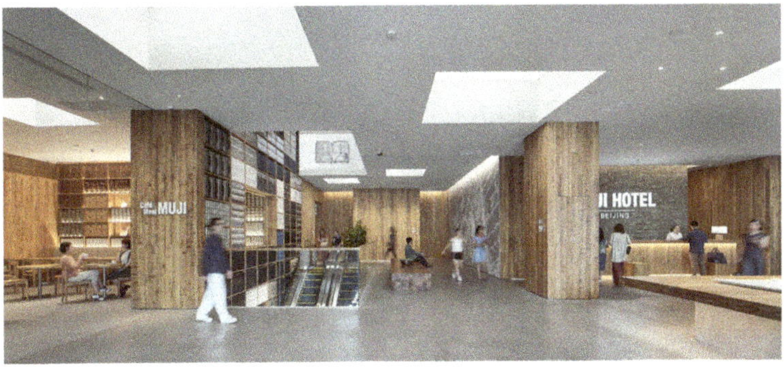

Fig. 6.13 Lobby

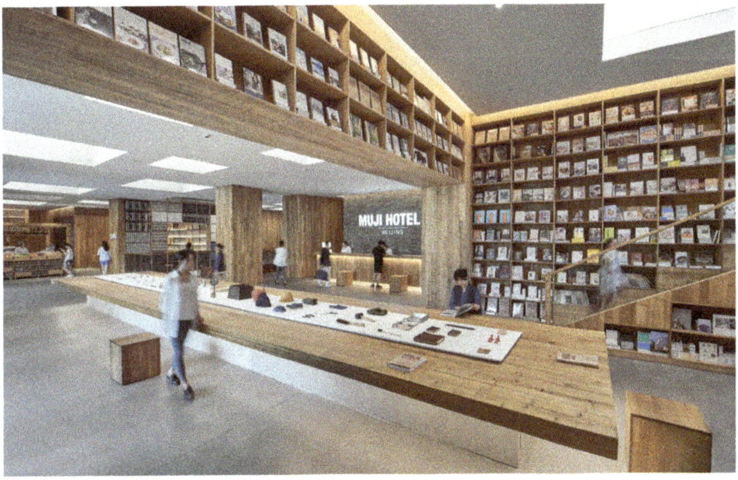

Fig. 6.14 Book lounge

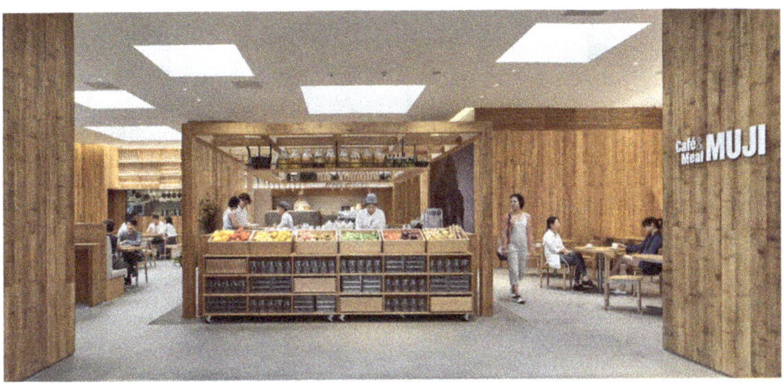

Fig. 6.15 Café & Meal

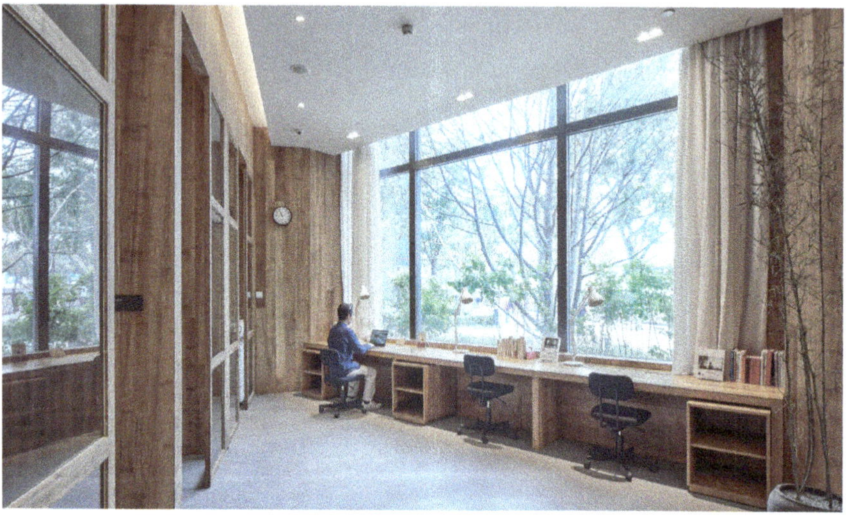

Fig. 6.16 Business Centre

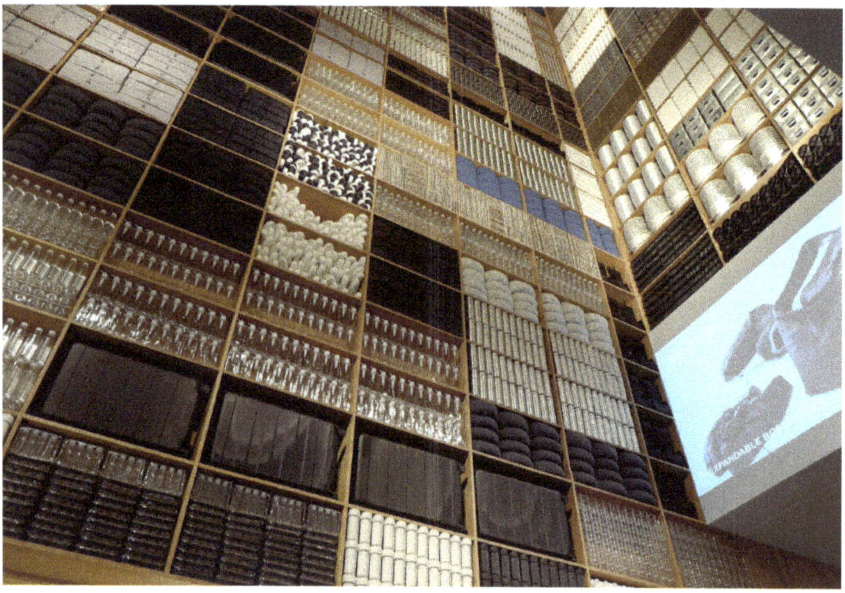

Fig. 6.17 Display wall

Hang Seng University of Hong Kong

Location: Shatian, Hongkong, China

Completed: 2016

Architects: Wong & Ouyang (HK) Ltd.

The Hang Seng University of Hong Kong is the first university in Hong Kong that widely uses bamboo products and bamboo landscape in campus construction. Based on the concept of "environmental education", advanced architectural service technology has been actively introduced into campus design, creating an elegant, fresh and natural campus environment. The design and construction process advocate environmental protection, in particular using a large number of environmental protection materials. The design has been certified as "BEAM Plus Platinum Grade of Green Environmental Assessment" issued by the Hong Kong Green Building Council.

Application of bamboo products: All decorations and furniture in the campus are made of custom-made bamboo materials. These include bamboo wallboard, bamboo sound absorption board, bamboo furniture in the office, bamboo desks, bamboo playground floor, bamboo outdoor floor, and bamboo furniture in student dormitories, etc. The whole school seems to be a museum for bamboo, setting a very good model for the application of environmental protection concepts.

Photo source: Figs. 6.18, 6.19, 6.20, 6.21 and 6.22 are provided by Ganzhou Sentai Bamboo & Wood Co., Ltd.

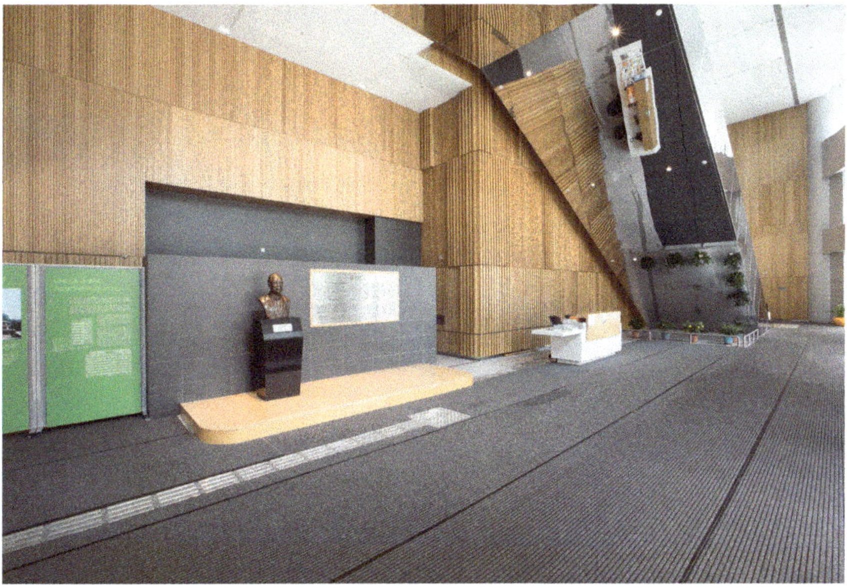

Fig. 6.18 Lobby of the teaching building

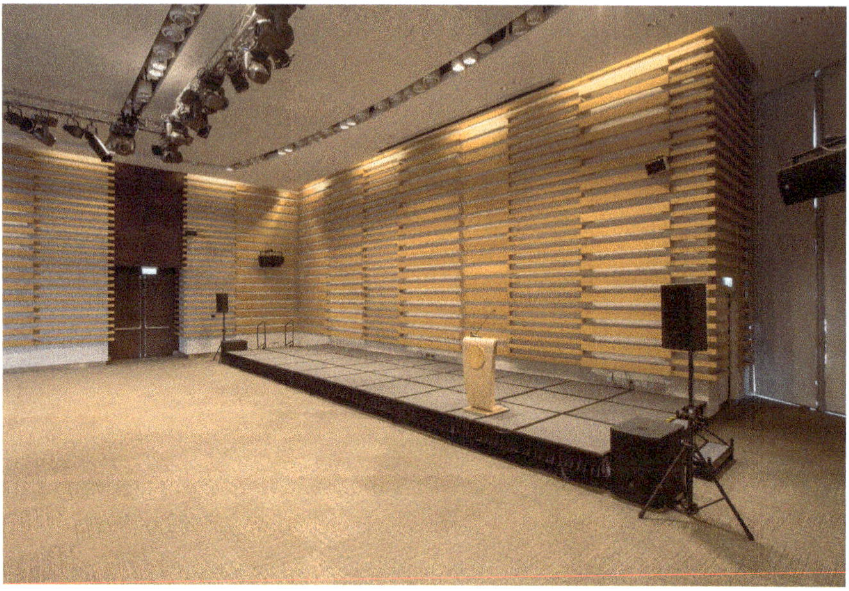

Fig. 6.19 Multi-function activity room

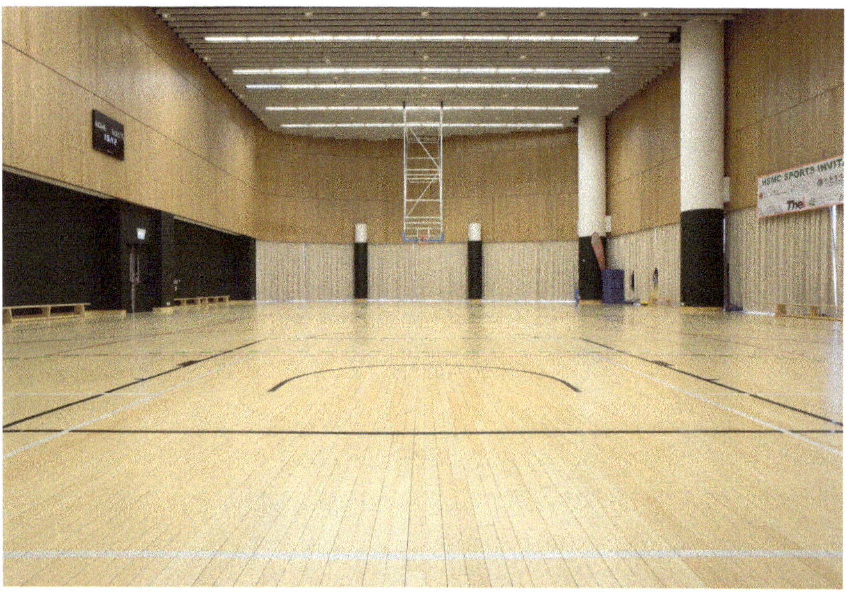

Fig. 6.20 Gymnasium

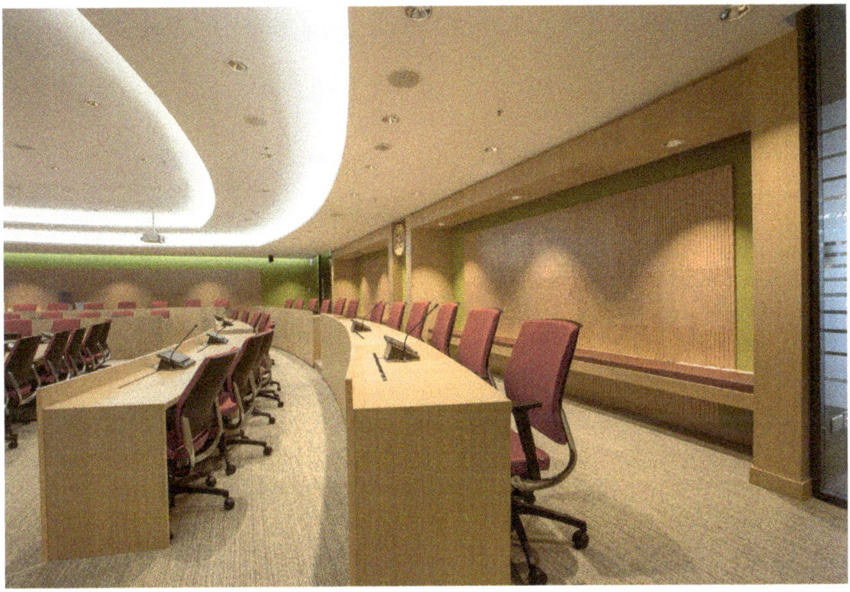

Fig. 6.21 Meeting room

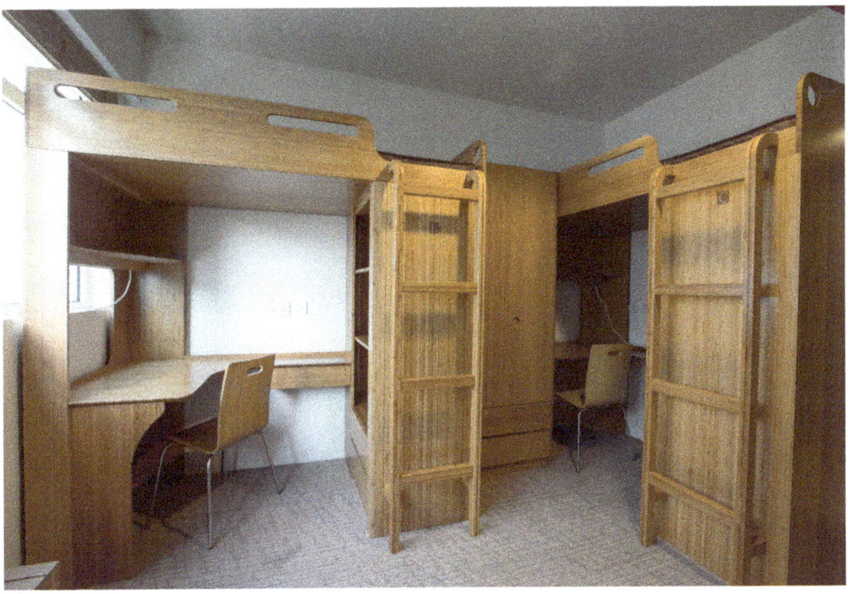

Fig. 6.22 Student dormitory

Calypso Mediterranean Restaurant, Shangri-La Hotel in Jing'an District, Shanghai

Location: Shanghai, China

Completed: 2014

Architects: Shigeru Ban (Japan)

The Calypso Mediterranean Restaurant, located in the promenade galleria of Shangri-La Hotel was designed by 2014 Pritzker Architecture Award winner Shigeru Ban. The theme of the restaurant is bamboo, using various forms of bamboo woven decorations, including roof shading and floor partitions. Indoor and outdoor wallboards are wrapped with bamboo, forming a bamboo house. Combined with the permeability effect of a large amount of glass, it is particularly unique in the prosperous city.

Application of bamboo products: Bamboo woven wallboard, bamboo woven screen, indoor and outdoor bamboo wallboard.

Photo source: Figs. 6.23, 6.24, 6.25, 6.26 and 6.27 provided by Ganzhou Sentai Bamboo & Wood Co., Ltd.

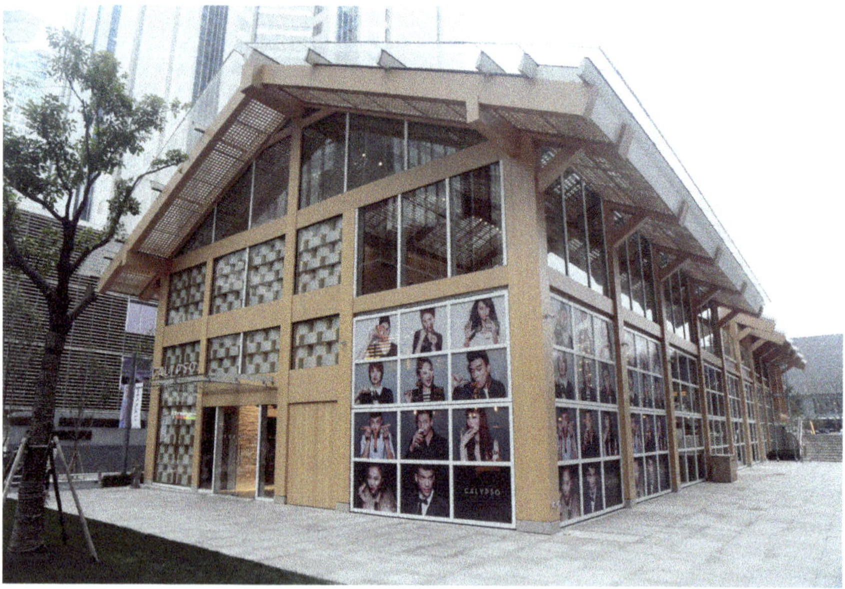

Fig. 6.23 Exterior facades of Calypso Mediterranean Restaurant

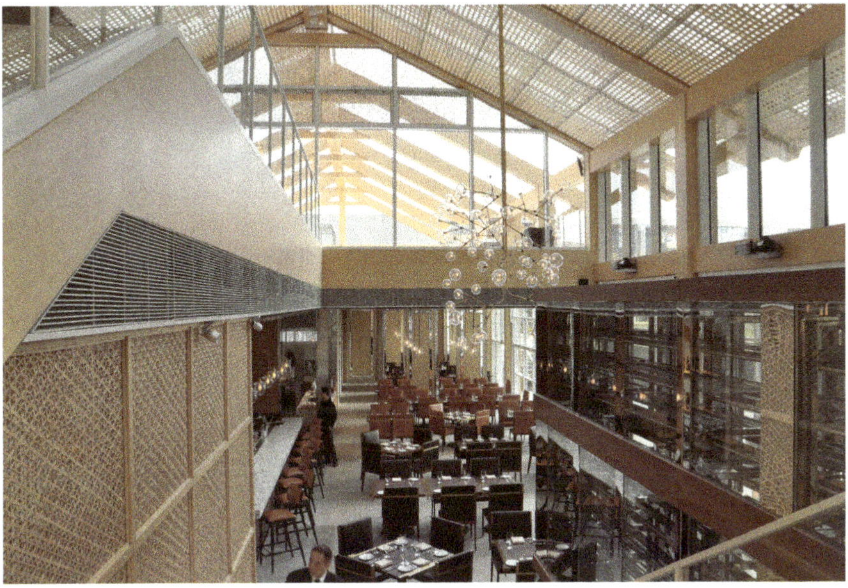

Fig. 6.24 Interior space of Calypso Mediterranean Restaurant

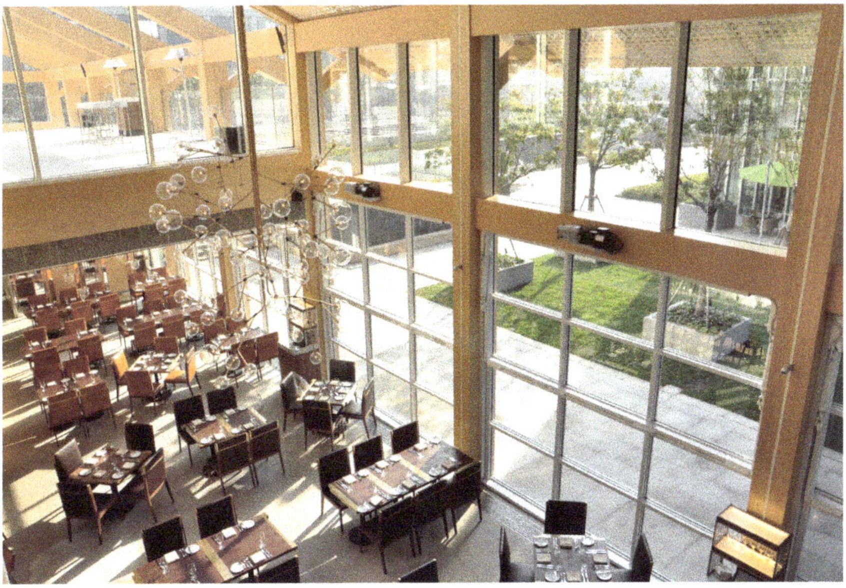

Fig. 6.25 Interior space of Calypso Mediterranean Restaurant

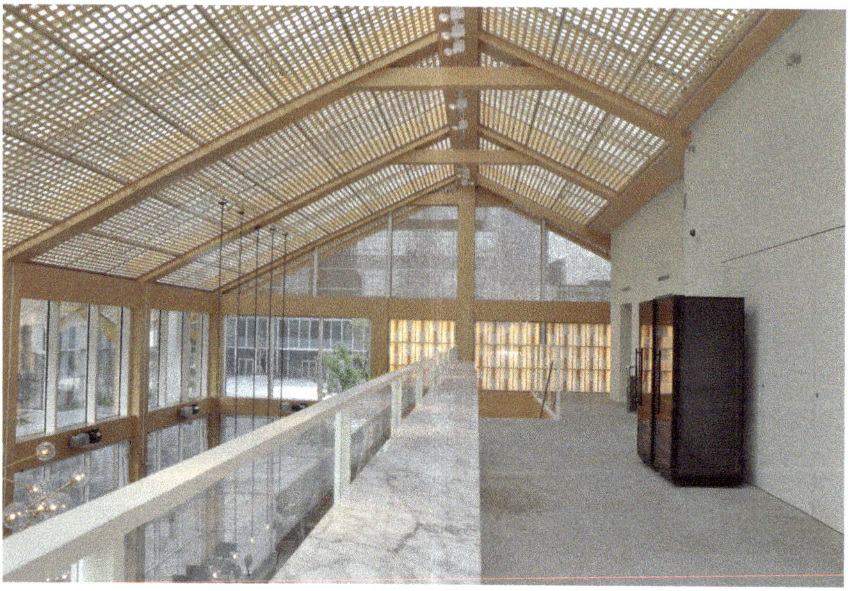

Fig. 6.26 Second floor and roof

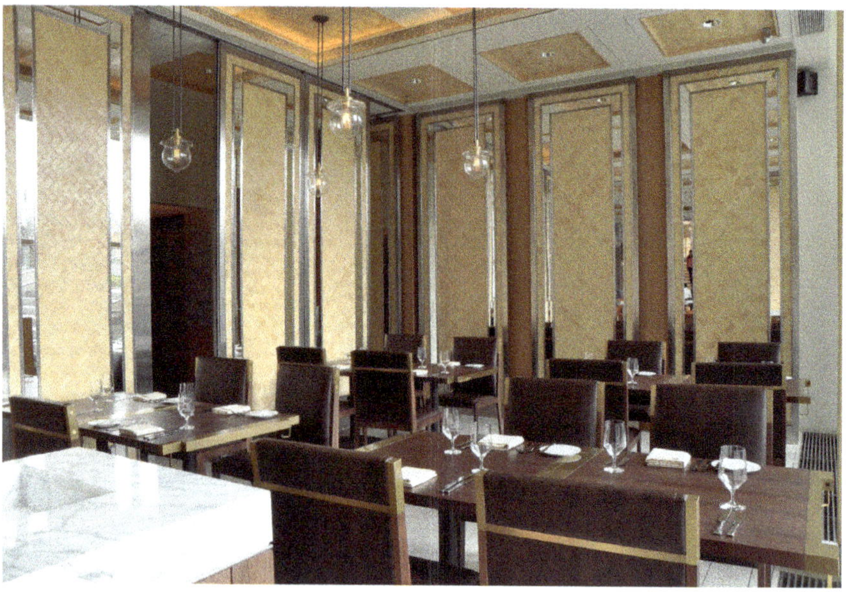

Fig. 6.27 Bamboo weaving for wall and ceiling decoration

Old Bailey Restaurant, Tai Kwun of Hong Kong

Location: Central District, Hong Kong, China

Completed: 2018

Architects: Herzog & de Meuron (Switzerland)

Located in the central part of Hong Kong Island, the Pavilion, formerly known as the Central Police Station, Central Magistrate and Victoria Prison, is one of the most important historical complexes in Hong Kong. The project was designed by 2001 Pritzker Architecture Award winners Jacques Herzog and Pierre de Meuron. The Hong Kong Jockey Club was responsible for project renovation. It took 10 years to complete with the highest level of building renovation standards.

Application of bamboo products: Old Bailey Restaurant is decorated in a retro style, with bamboo as the dominant material, combined with wire drawing, deep caramelised technology, and the original design of rhombic tenon and mortise structure. Bespoke bamboo furniture does not use any hardware. It is innovative in inheritance and presents a stunning retro decorative effect in a unique modern design manner.

Photo source: Figs. 6.28, 6.29, 6.30, 6.31 and 6.32 are provided by Ganzhou Sentai Bamboo & Wood Co., Ltd.

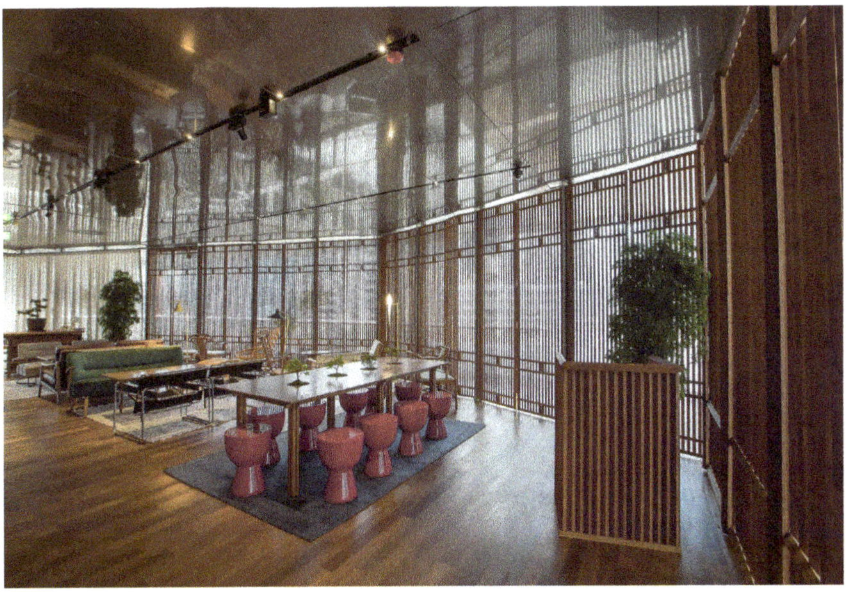

Fig. 6.28 Entrance space

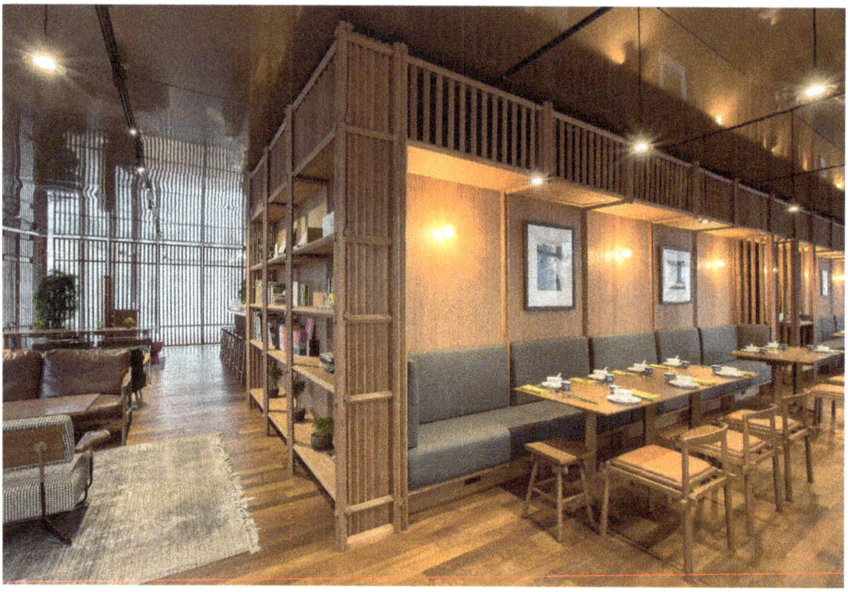

Fig. 6.29 Interior space near bamboo bookshelf

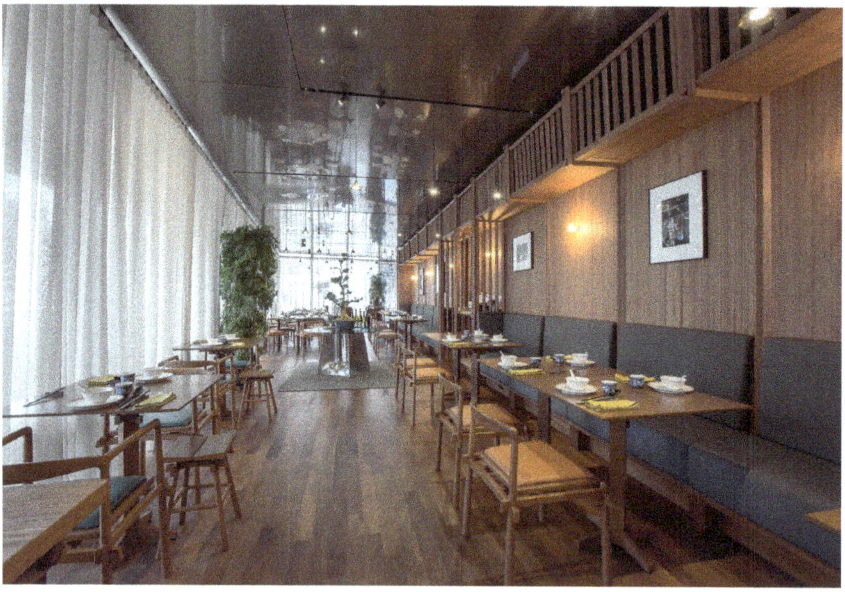

Fig. 6.30 Interior space of Old Bailey Restaurant

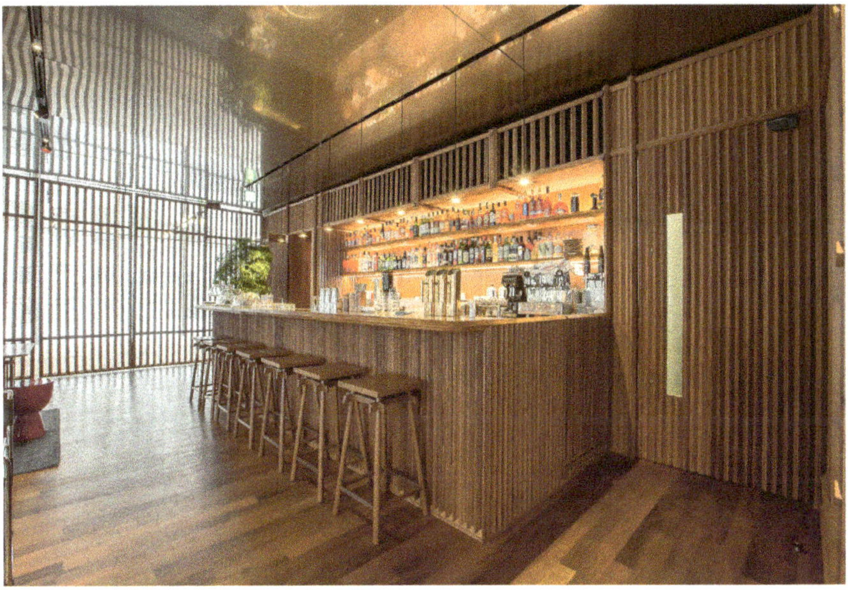

Fig. 6.31 Bar

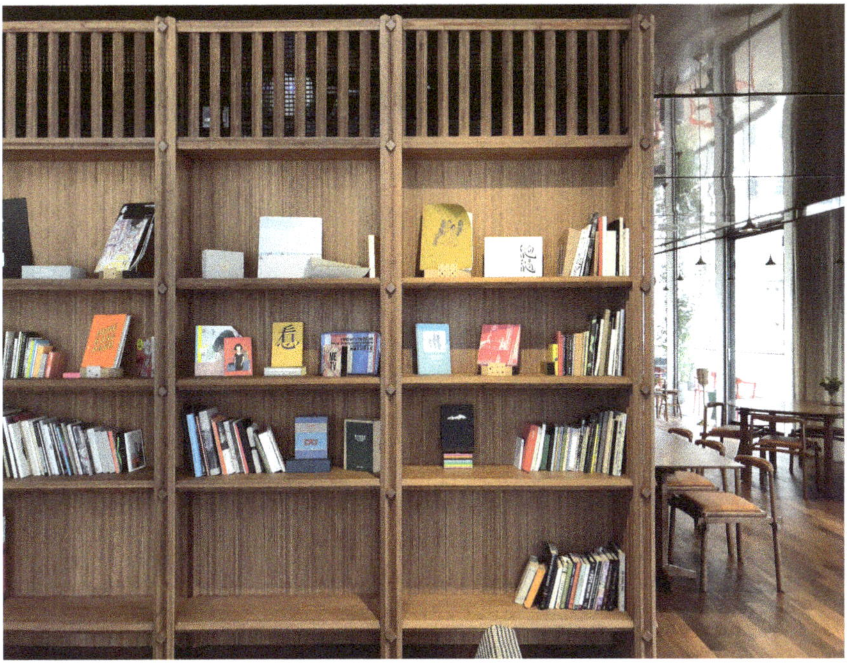

Fig. 6.32 Bamboo bookshelf

Seaside Library

Location: Qinhuang Island, Hebei, China

Completed: 2015

Architects: Dong Gong

The main idea of this design is to explore the coexistence among the boundaries of space, the activities of the body, changes of the light atmosphere, circulation of air, and the ocean view. The east side of library faces the sea. It serves the community residents in the west residential areas and is open to the public. The library consists of a main reading space, a meditation space, an activity room and a small water bar rest space. According to the different functional requirements of each space, architects define the specific relationship between space and sea as well as the way light and wind enter the room. If the house is split along the north–south direction, it can be more clearly perceived that this group of spaces interpret the specific relationship between each space and the sea, and the elements of the series of relationships are precisely the human body's wandering and memory in space.

Application of bamboo products: Furniture (bookcases, chairs and desktops), door and window frames, floor.

Photo source: Figs. 6.33, 6.34 and 6.35 are provided by Vector Architects.

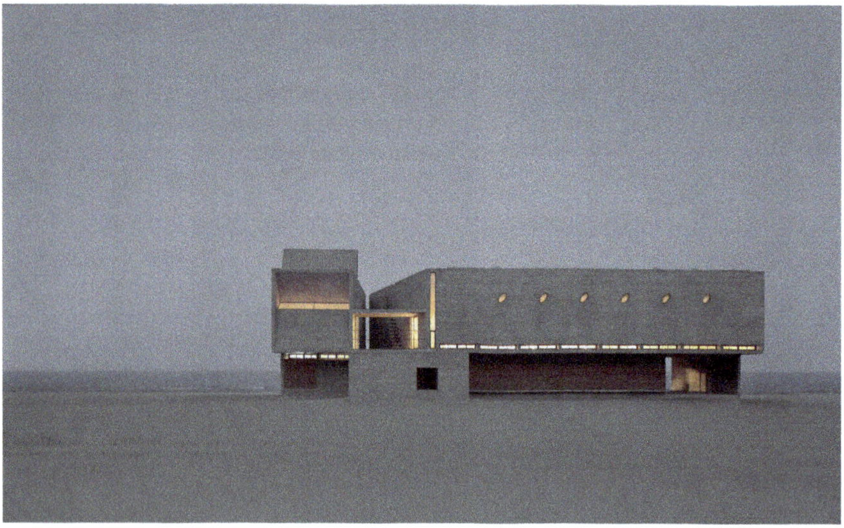

Fig. 6.33 West elevation (*photo by Su Shengliang*)

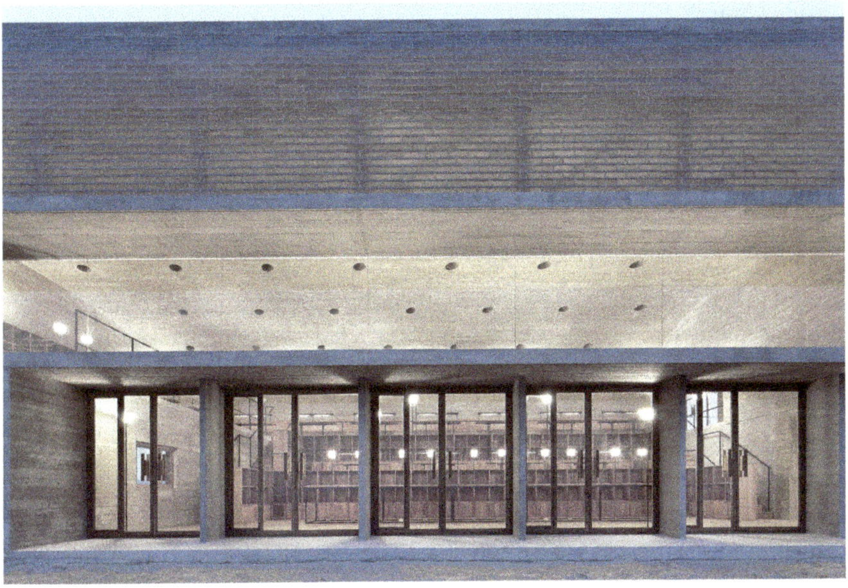

Fig. 6.34 Reading area viewed from the beach (*photo by Xia Zhi*)

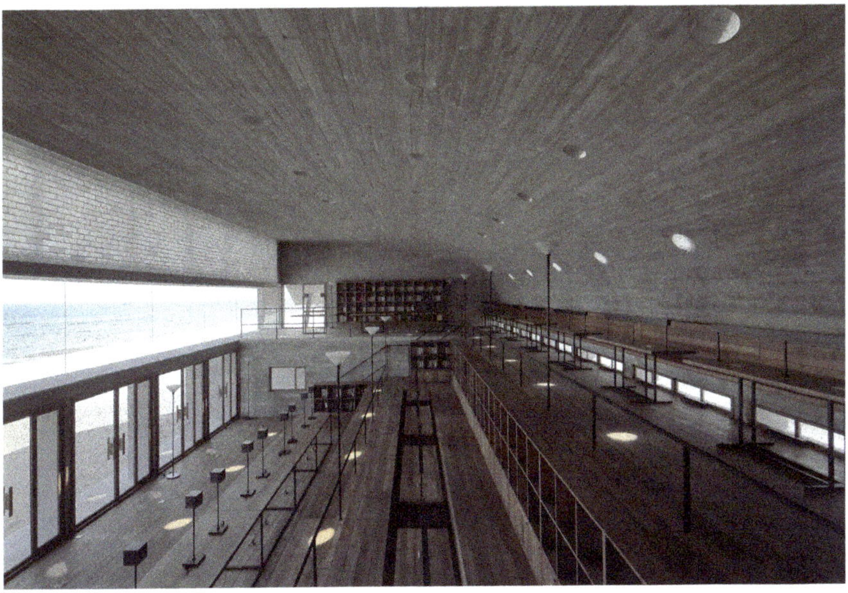

Fig. 6.35 Reading area (*photo by Su Shengliang*)

Seaside Church

Location: Qinhuang Island, Hebei, China

Completed: 2015

Architects: Dong Gong

Imagine that this church is a ship that once drifted on the sea a long time ago. As time goes by, the seawater fades away, leaving behind a house that is still floating over the beach. People climb up the stairs, going through the gate, bypassing the shadow wall, and finally entering a church facing the sea. Because of the elevated space, people don't have the view of the beach and people on the beach, only the vast sea beyond. Together with the seaside library 300 m away, the two form a group of spiritual places where people can calm down, feel nature and feel themselves.

Application of bamboo products: Furniture (bookcases, chairs and desktops), door and window frames, floor.

Photo source: Figs. 6.36 and 6.37 are provided by Vector Architects.

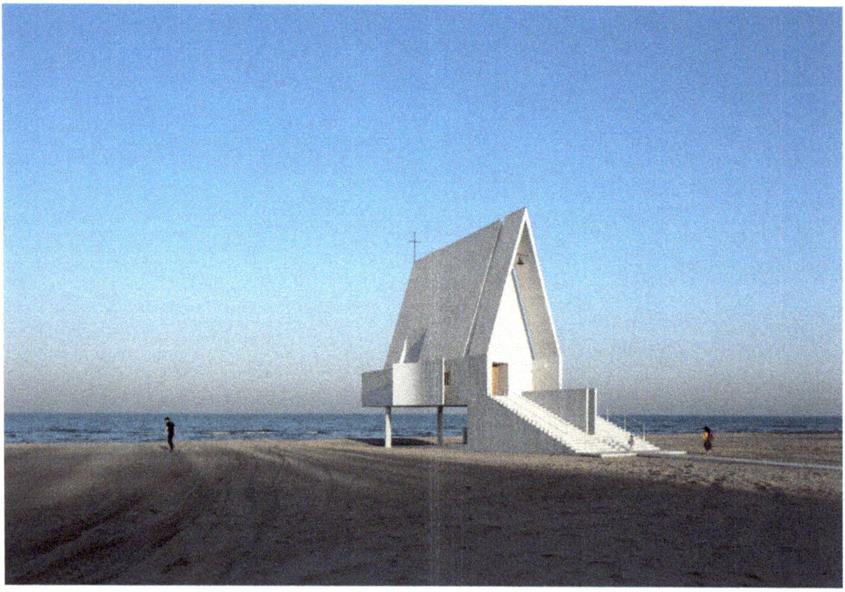

Fig. 6.36 Northwest view (*photo by Chen Hao*)

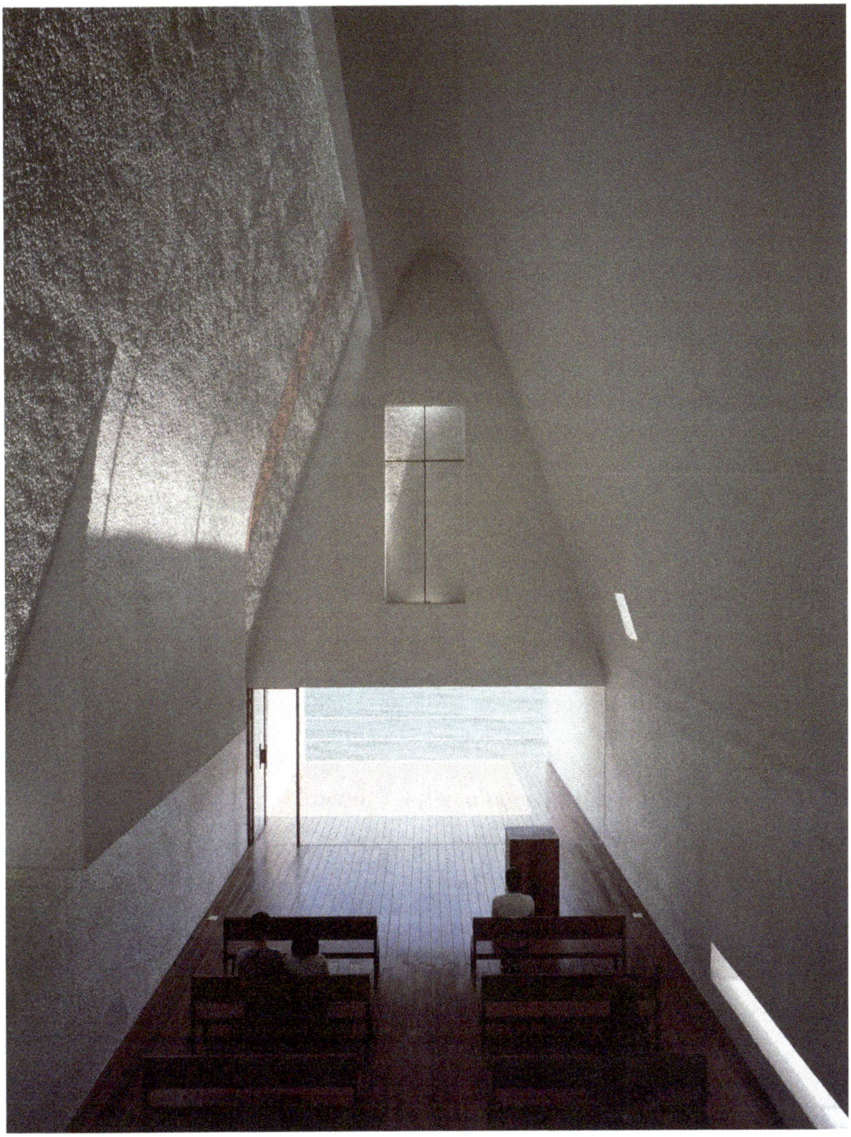

Fig. 6.37 View from Organ Loft to Auditorium (*photo by Chen Hao*)

Bamboo and Rattan Pavilion of the 10th China Flower EXPO

Location: Chongming, Shanghai, China

Completed: 2021

Architects: Shen Di, Du Jin et al.

The Bamboo and Rattan Pavilion is one of three permanent pavilions of 10th China Flower Expo, located in the Bamboo and Rattan Garden on the west side of the main axis of the Expo. The design stems from the relationship between architecture and landscape, introducing the functional space underground and the upper part a landscape sculpture resembling a craft floating on the water. The pavilion forms like a "Bamboo Cocoon" with a partial openness, taking the meaning of "breaking cocoon as a butterfly". Using the construction logic of bamboo crafts, the cocoon was constructed by bamboo sheets layered and woven between steel cable nets, synthesising external form, functional space and construction method. Through weaving, the form and the space were constructed uniformly, and traditional architectural elements such as wall and roof were transformed into a continuous interface. In this way, the pavilion was made to be artistically expressive, like a bamboo craft, and presents a unique space where light and shadow flow, and change endlessly.

Application of bamboo products: Bamboo scrimber was selected as the main material for weaving. Nearly 10,000 bamboo scrimber units constitute the "warp and weft" of the woven mesh surface. In addition, bamboo scrimber was also used in doors and windows in the exhibition hall, as well as overhead platforms and guide boards in outdoor courtyard. The uses of bamboo scrimber as a decorative material and furniture in the permanent building were fully explored in this case.

Photo source: Figs. 6.38, 6.40 and 6.42 are provided by International Centre for Bamboo and Rattan; Figs. 6.39, 6.41 and 6.43 are provided by Arcplus Group PLC.

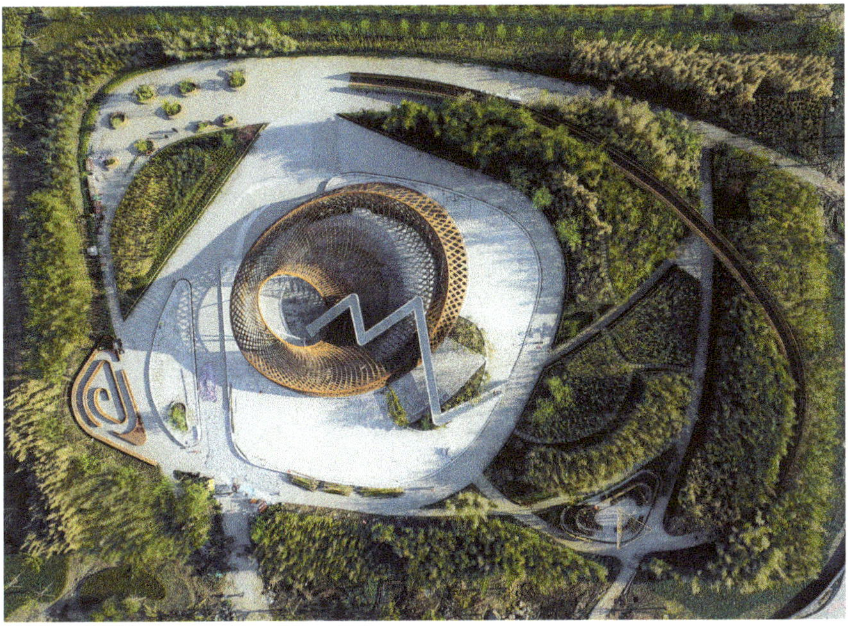

Fig. 6.38 Aerial view of Bamboo and Rattan Garden

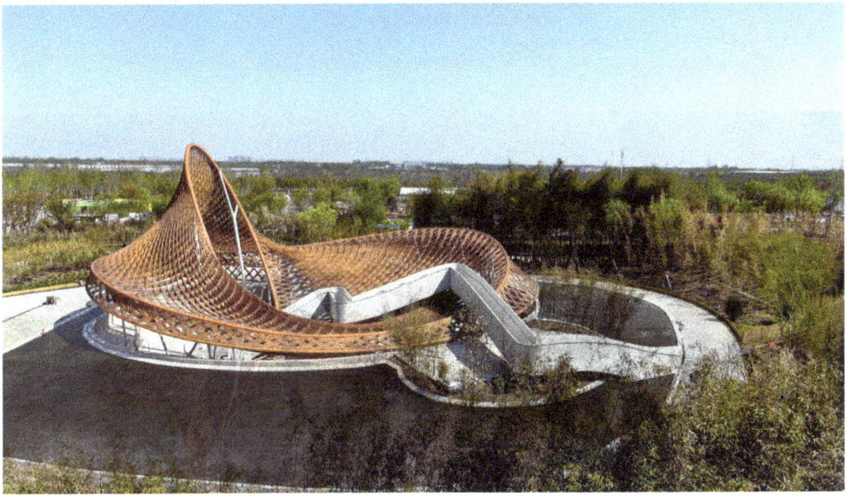

Fig. 6.39 South view of Bamboo and Rattan Pavilion (*photo by Wang Cheng*)

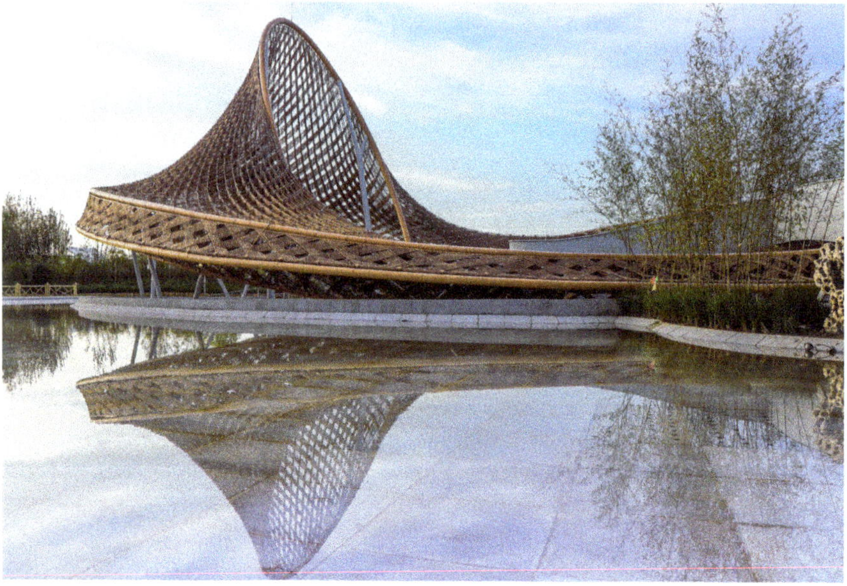

Fig. 6.40 Bamboo and Rattan Pavilion "floating" on the water

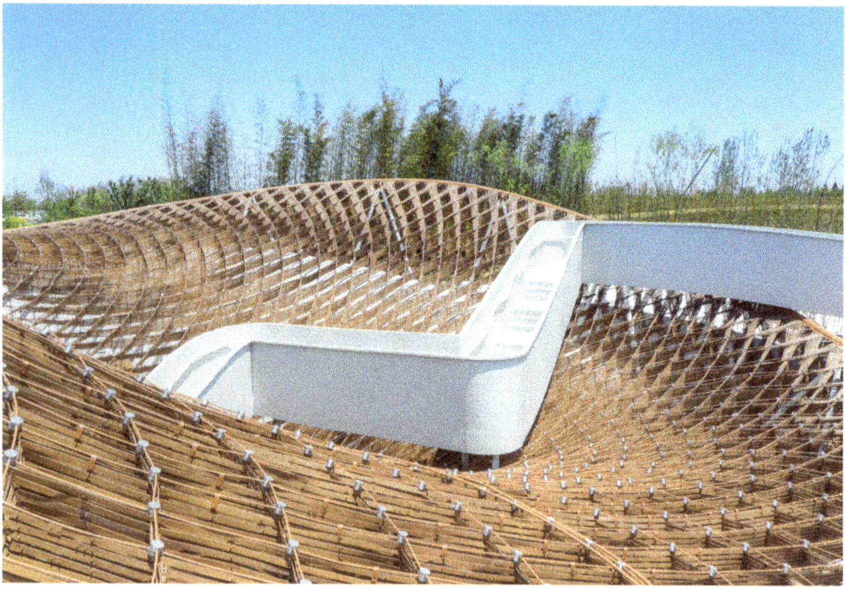

Fig. 6.41 Bridge over the woven skin

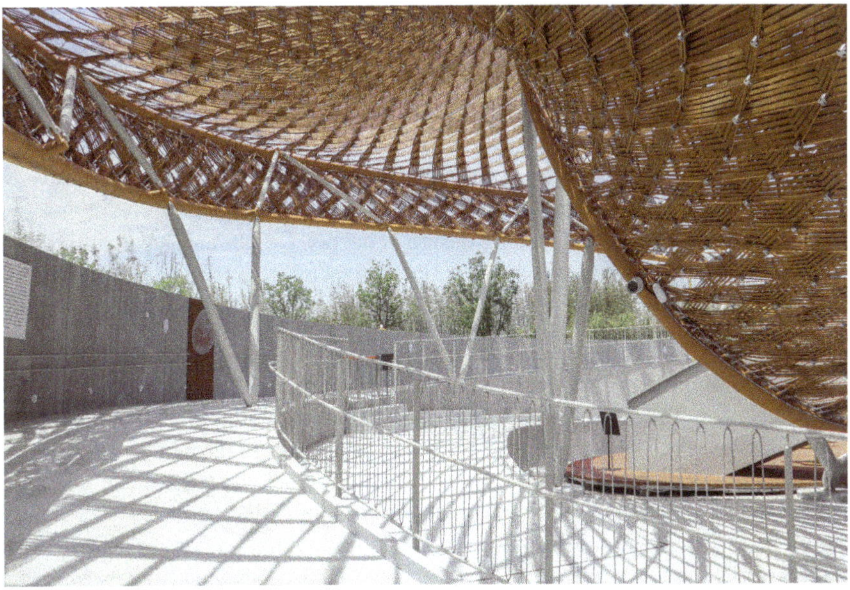

Fig. 6.42 Sunken courtyard

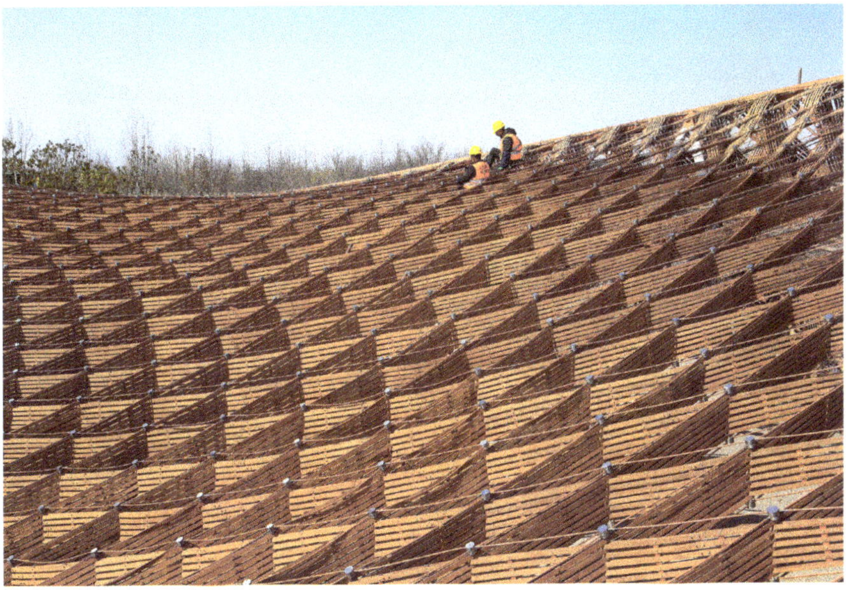

Fig. 6.43 Texture of bamboo weaving

Nanjing Aohong Resort Hotel

Location: Nanjing, Jiangsu, China

Completed: 2018

Architects: Tang Chun Sang

Nanjing Aohong Resort Hotel regards private island, courtyard and experiencing nature at zero distance as its leading design concepts. Combining local culture and customs, the hotel uses bamboo as the main element of interior design, and incorporates comfortable and low-key design style, aimed at creating a landmark spa holiday high-end hotel in Nanjing.

Application of bamboo products: The exterior facade uses bamboo scrimber decorative materials. These possess not only toughness and hardness, but also have the characteristics of ecological and environmental protection. In the men's bathing area, the combination of bamboo scrimber and modern stone creates a clean and pure space, which is simple but with modern characteristics. Villa rooms are divided into two floors in which the first is quiet and elegant by taking bamboo scrimber as an element while the second floor is integrated with landscape elements, reconnecting with nature.

Photo source: Figs. 6.44 and 6.45 are provided by Dome Design Consultant.

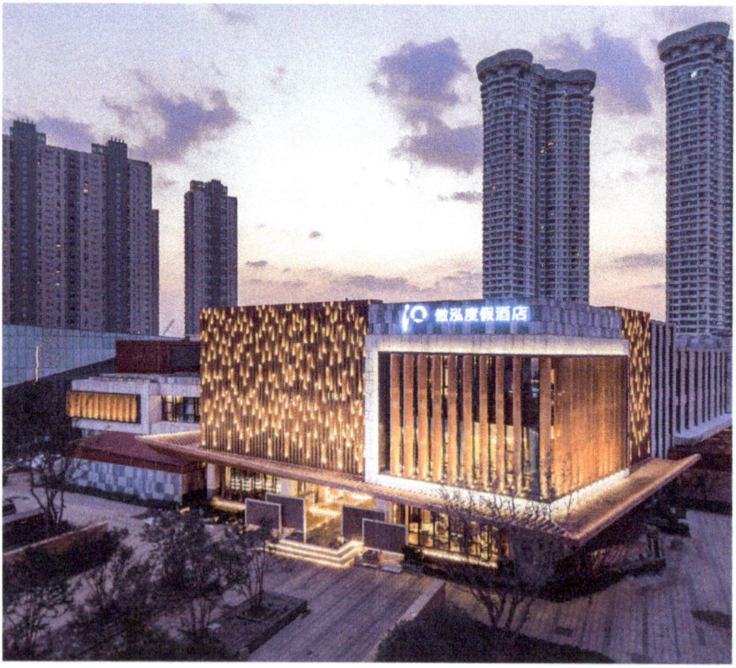

Fig. 6.44 Exterior facades of Nanjing Aohong Resort Hotel

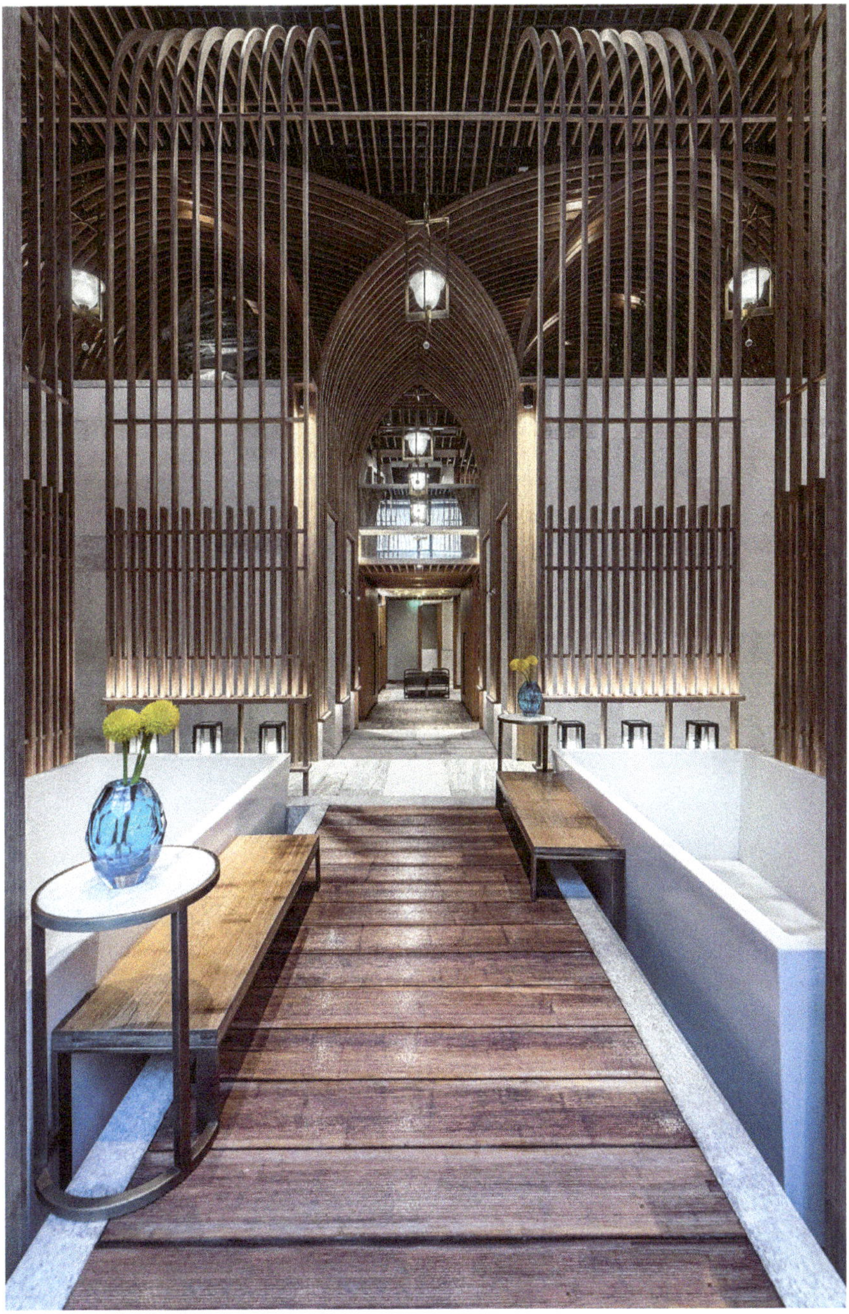

Fig. 6.45 Men's bathing area

Changbai Mountain Squirrel (Bicycle) Station

Location: Changbai Mountain, Jilin, China

Completed: 2015

Architects: Xu Xiaodong

Entrusted by Changbai Mountain Management Committee, the architect made overall planning and design for the green slow lane along Changbai Mountain. Squirrel Station is the starting point of the green slow lane around Changbai Mountain. The Station has the functions of bicycle cultural experience and dissemination, Changbai Mountain scenery exhibitions, the organization and publicity of bicycle events, urban residents' leisure activities and so on.

Application of bamboo products: The project is located between the open space of pine forest and a small town. How to deal with the relationship between the building and forest is one of the main challenges in this project. The project adds a layer of bamboo scrimber grid curtain wall to the circular building. The grey space formed by the bamboo scrimber grid greatly weakens the sense of conflict between the closed building and the naturally permeable pine forest, and makes the relationship between the building and its environment harmonious.

Photo source: Figs. 6.46, 6.47 and 6.48 are provided by BES Architectural Design Co., Ltd.

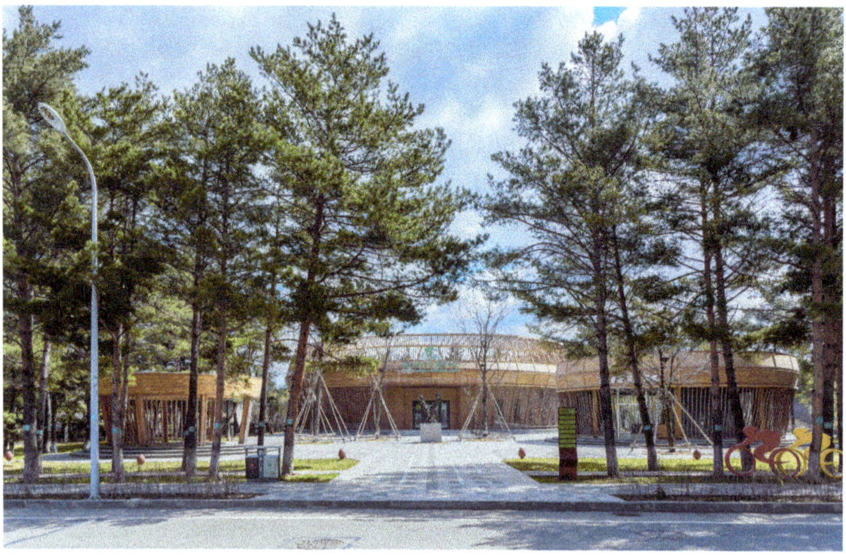

Fig. 6.46 Green slow lane along Changbai Mountain

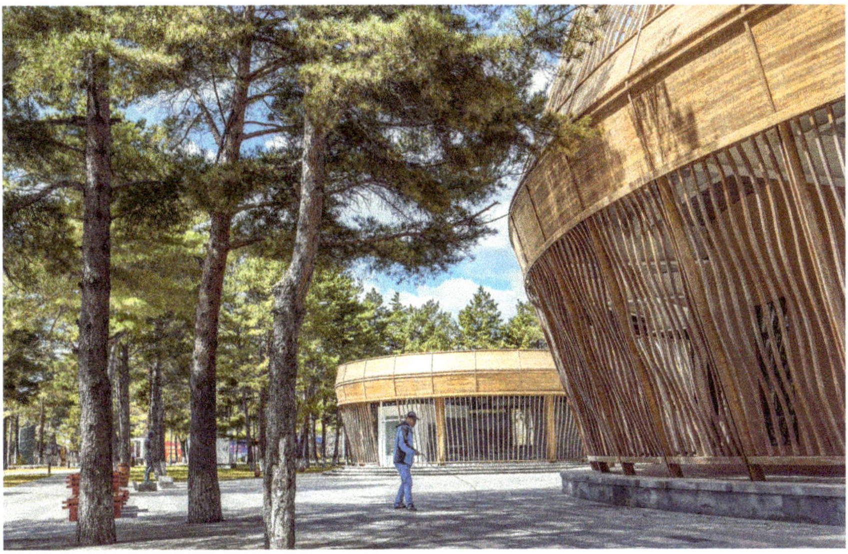

Fig. 6.47 Open space of pine forest around the station

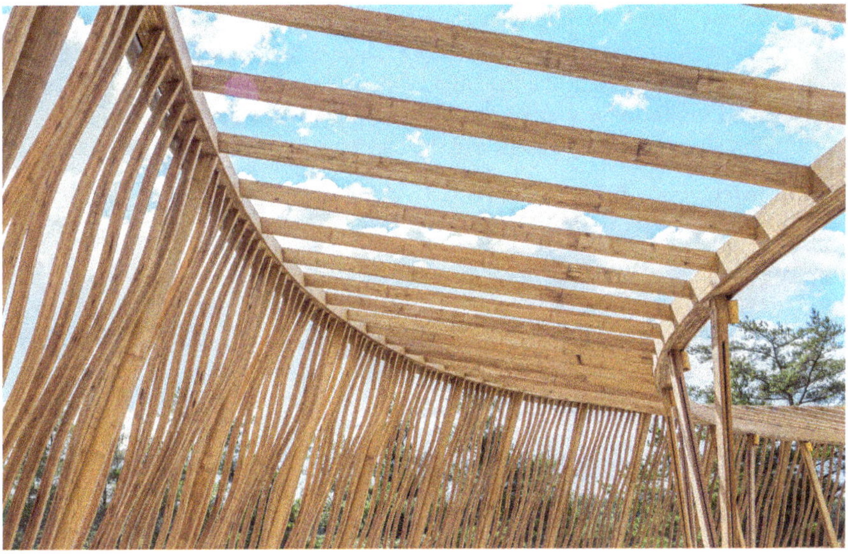

Fig. 6.48 Detail of the structure

Tangxing No. 5 Residential House

Location: Xi'an, Shanxi, China

Completed: 2018

Architects: Liu Saiwen

Based on the concept of "close to nature" and the Zen thought of "combination of movement and quiet", this five-storey villa takes a large number of original ecological bamboo materials into its interior decoration. Because of nearly 6 m high first storey having an area of more than 500 m^2, the space scale is too large to live in. Through the separation and scattering of space, the designer makes the original dull and empty become rich and pleasant. The decorative materials are mainly made from full-culm bamboo and white cement with hemp knife to wipe the wall. The floor is made mostly of local natural stone and log.

Application of bamboo products: On the first floor, a hollow circular window is set at the entrance using a traditional Chinese framing technique, and a tea drinking area is encircled by bamboo culms; on the second floor there is a Buddhist hall comprising a semi-circular luminous area composed of 726 full-culm bamboo barrels on the top and divergent full-culm bamboo arranged to demonstrate the meaning of "Temple under sun shine". The third floor is the activity space, a place for tea tasting is specially set up in the middle of the pool. The top is a dome, which hides the blue LED light band imitating the sky. At night, the LED tube lamp in the center of the dome is like moonlight falling from the dome 12 m above. It shows an artistic conception of "the bright moon shining between pines and the clear spring stone flowing upward".

Photo source: Figs. 6.49, 6.50 and 6.51 are provided by U-Find International Design & Consulting Co., Ltd.

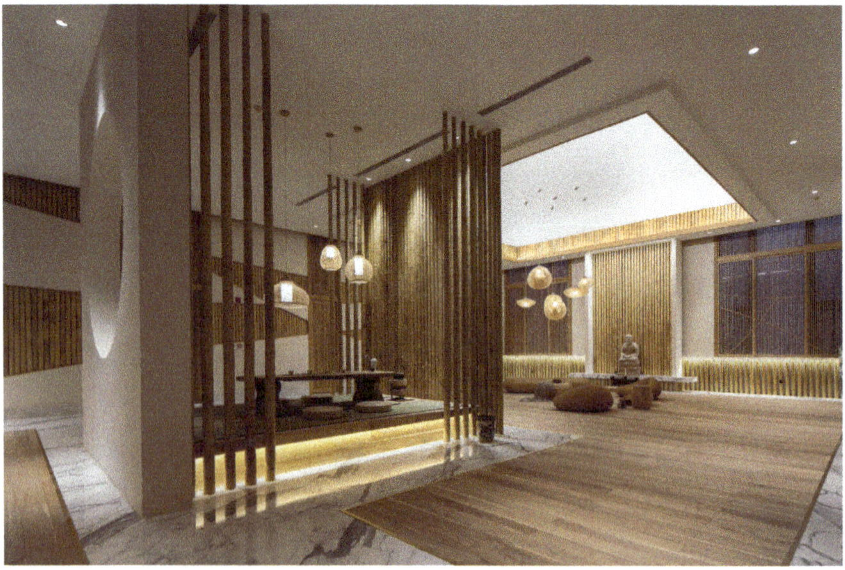

Fig. 6.49 Living room on the first floor

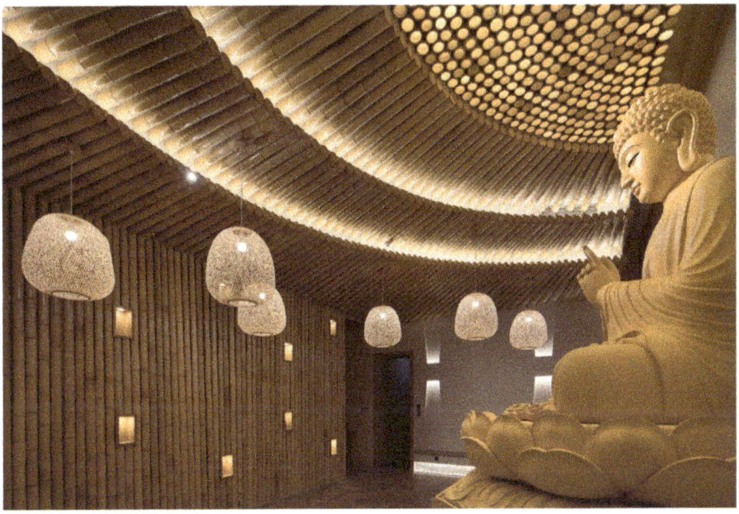

Fig. 6.50 Buddhist hall on the second floor

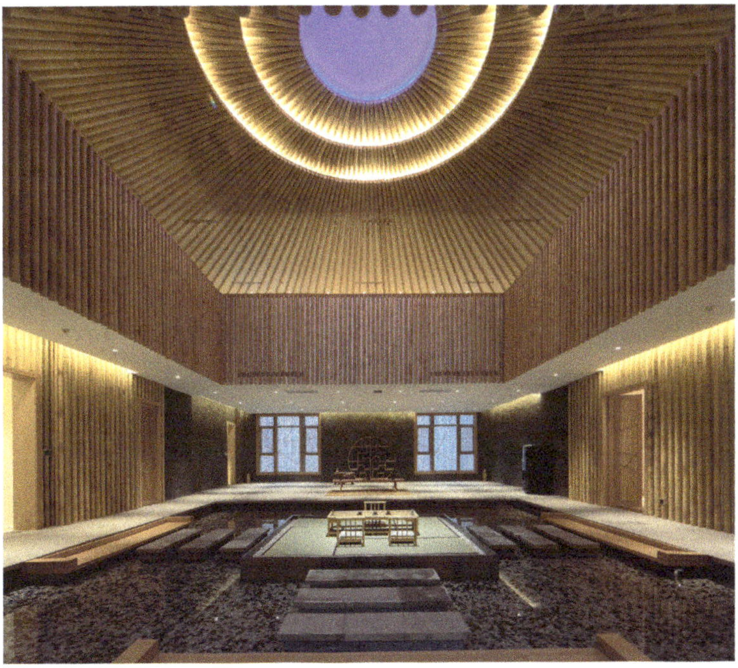

Fig. 6.51 Living room on the third floor

China Buoy Town Fishing Center

Location: Linxiang, Hunan, China

Completed: 2017

Architects: Guo Ming

Based on local resource endowment and historical tradition, the bamboo art industry, buoy industry and cultural tourism industry are skilfully combined to realise a feasible and sustainable development path which is in line with the local people's conditions. The architectural style, whose shape is the image of two fish, uses bionic architectural techniques. Local buoys are abundant for fishing, accounting for 80% of the country's buoy industry. In ancient times, people also used bamboo poles to fish. Therefore, the architect combined the two and hoped to use bamboo materials and a unified design method to continue the design of the whole space. The building is used as the service center for the CAA International Fishing Competition. The design of interior space makes good use of natural lighting, which makes the concrete space of bamboo formwork and interior full-culm bamboo decoration reflect each other. Bamboo elements run through the whole design.

Application of bamboo products: Full-culm bamboo as indoor and outdoor decorative materials.

Photo source: Figs. 6.52, 6.53 and 6.54 are provided by Sharing Design Group Co., Ltd.

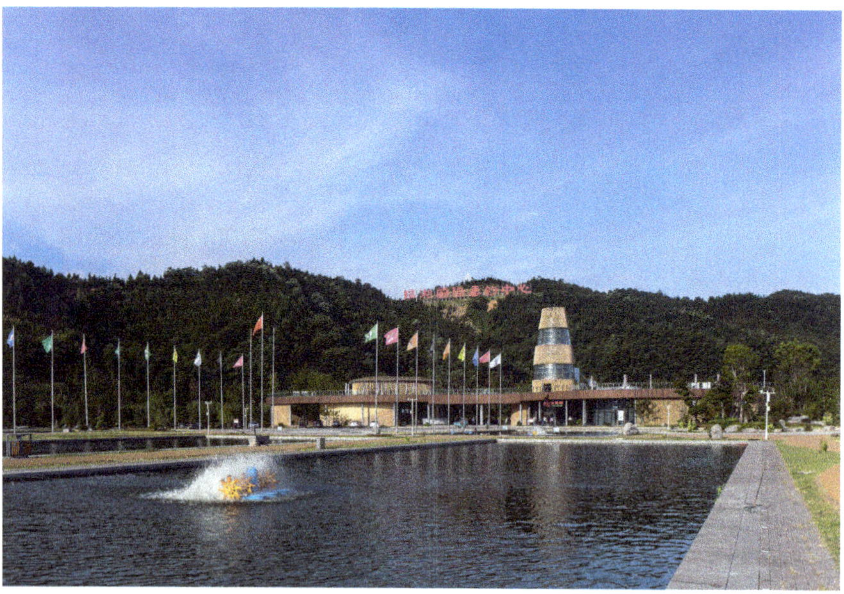

Fig. 6.52 Surroundings of China Buoy Town Fishing Center

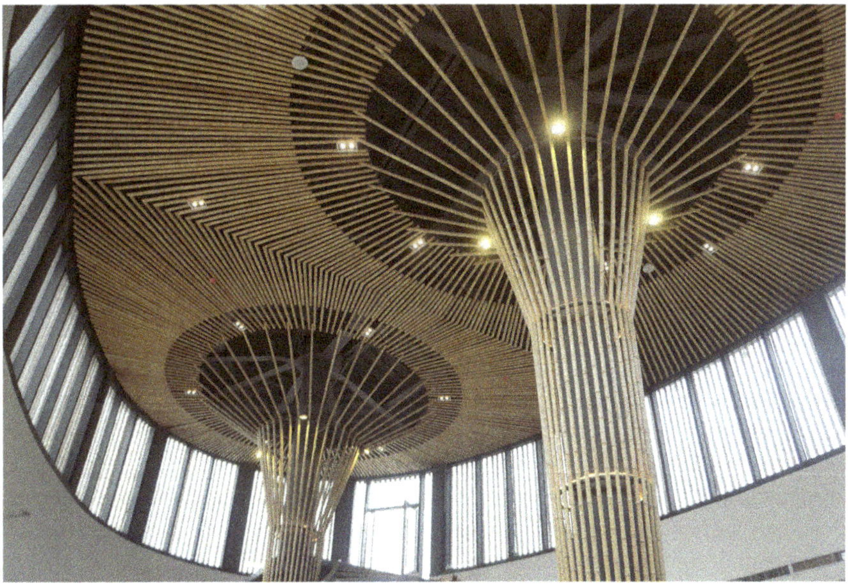

Fig. 6.53 Interior space with bamboo decoration

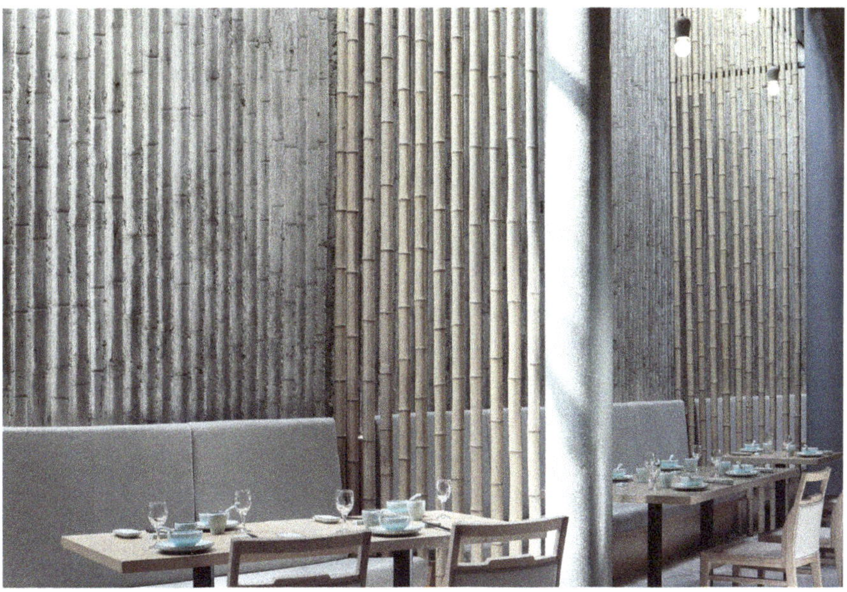

Fig. 6.54 Bamboo partitions and concrete wall formed with bamboo formwork

Linxiang Global Tourism Service Center

Location: Linxiang, Hunan, China

Completed: 2018

Architects: Guo Ming

There are many bamboo forests in the area, and the buoy industry as its pillar industry is closely related to fishing tackle and bamboo poles. Using the bionic architectural technique, the architect designed three main buildings with the appearance of "bamboo leaves" and a landscape tower with the appearance of a "bamboo shoot". The three large "bamboo leaf" main buildings are arranged in the form of product characters, and are used as a tourist reception center, administrative offices and a tourist food center. The design of the "bamboo shoot" landscape tower comes from the concept of "bamboo shoot after rain". At the bottom, natural rocks with warm colours are used to create a scene where the bamboo shoot grow in the soil and break through the soil after rain. The top of the tower is equipped with lighting equipment. Most materials of the tower body are full-culm bamboo with glass elements inserted. The overall design is clear and natural.

Application of bamboo products: Full-culm bamboo as indoor and outdoor decorative materials.

Photo source: Figs. 6.55 and 6.56 are provided by Sharing Design Group Co., Ltd.

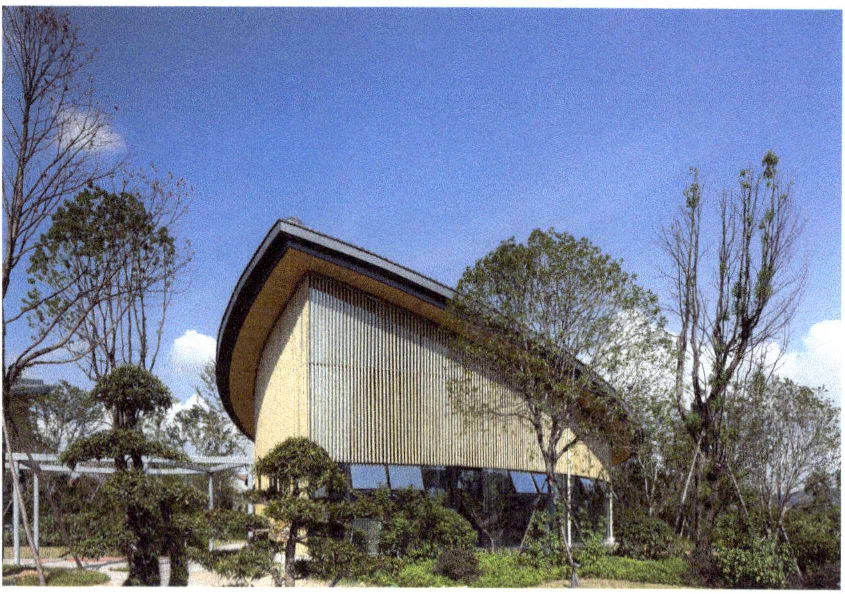

Fig. 6.55 "Bamboo leaf" building

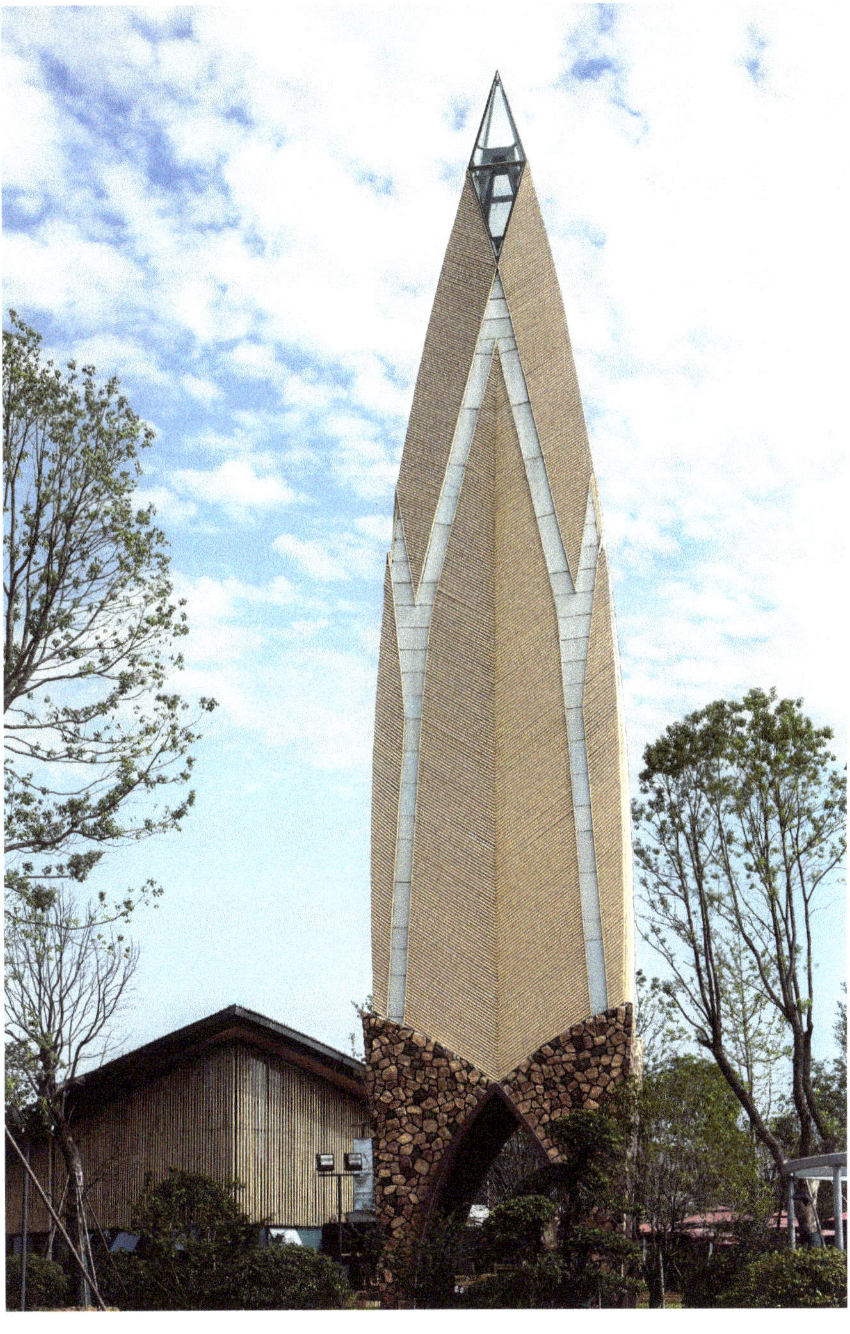

Fig. 6.56 "Bamboo shoot" landscape tower

Gudu Wharf in Yunnan Sea

Location: Kunming, Yunnan, China

Completed: 2016

Architects: Ren Lizhi

Gudu Wharf in Yunnan Sea is located in the famous cultural tourist city of the ancient Dian Kingdom. It is located in the north of Dianchi Lake and is surrounded by ecological wetland parks on three sides. Because of the long history and culture of the ancient Dian Kingdom and the beautiful wetland scenery of Dianchi Lake, the use of bamboo with local characteristics has endowed the wharf with unique architectural charm and site characteristics. In line with the principle of "in-place" design, architects adopt low-tech design means, excavate and inherit the traditional construction technology and technology of in-place materials, advocate the original technical aesthetics, and achieve the creative concept of "rooted in the environment, integrated with nature".

Application of bamboo products: The circular bamboo pillars resist bending and are seamless stitched resulting a unique aspect of "multiple V-shaped" members. The circular bamboo pillars stand in the center of the corridor, making the two sides of the circular corridor up to 500 m without any shelter, creating a shocking visual effect. The full-culm bamboo of the ceiling is curved and applied with a reed mat on the top, which suits the scenery of the lake and the mountain, and constructs a harmonious relationship between human and architecture, architecture and nature, and between human and nature.

Photo source: Figs. 6.57, 6.58 and 6.59 are provided by Tongji Architectural Design (Group) Co., Ltd.

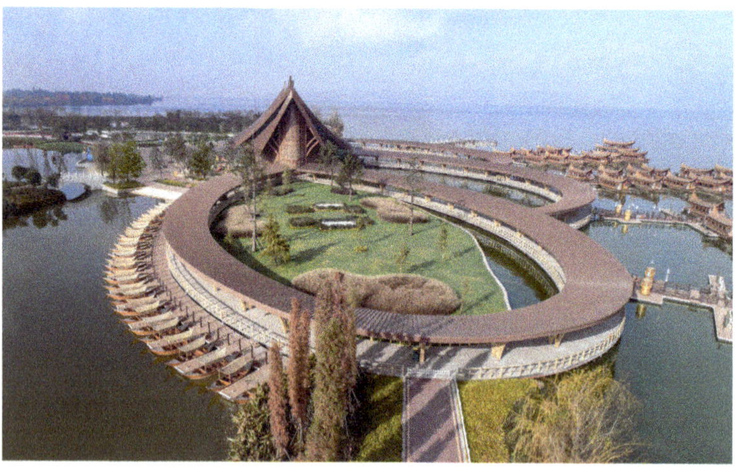

Fig. 6.57 West façade

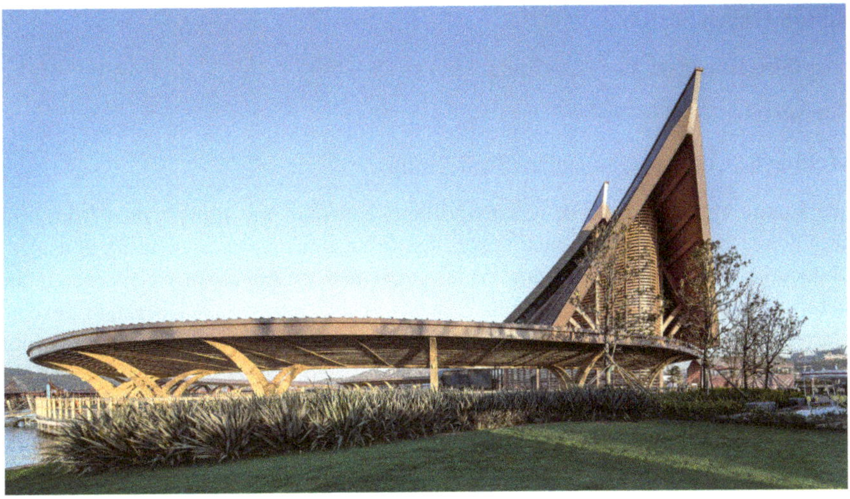

Fig. 6.58 Aerial view from the east

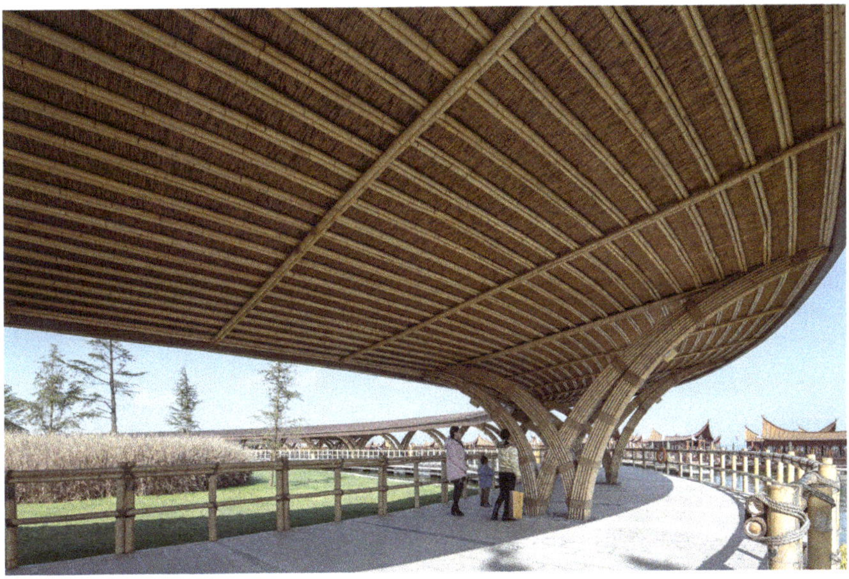

Fig. 6.59 The corridor

Kunqu Opera Society

Location: Kunshan, Jiangsu, China

Completed: 2017

Architects: Cui Kai, Guo Haian, Shen Yiting

The Kunqu Opera Society in Xibang Village, Kunshan was transformed from four rural courtyards. It was designed according to the conception of "reading house" of 24 good places of Yushan Caotang, which is the most prominent private garden in the history of Kunshan at the end of Yuan Dynasty. In the Gathering of Literati in Yushan, a large number of stories about literati and bamboo were illustrated. Beautiful poems like "there is a bamboo patch in front of my house, and lotus behind" were written. A large number of bamboo materials are used in the architectural design of the Kunqu Opera Society to show the character and sentiment of men and women of literature and writing who "could not live without bamboo".

Application of bamboo products: Bamboo curtain wall: using bamboo spacing to express the music rhythm of Kunqu Opera—*Peony Pavilion*. Bamboo lamp: the designer innovatively uses the bamboo cavity to make line slots. Bamboo doors and windows: bamboo elements are used in the design of doors and windows, bamboo shavings-wrapped doors, bamboo pavilion with small diameter full-culm bamboo windows for shading. Bamboo pavilion: a full-culm bamboo structure was used for creating the pavilion and the stage. Bamboo ceiling: a changeable roof space was treated with full-culm bamboo. Bamboo purlins, floor and wallboards, bamboo curtain and bamboo furniture, highlighting bamboo as the preferred material for home decoration.

Photo source: Figs. 6.60, 6.61, 6.62 and 6.63 are provided by China Architecture Design & Research Group.

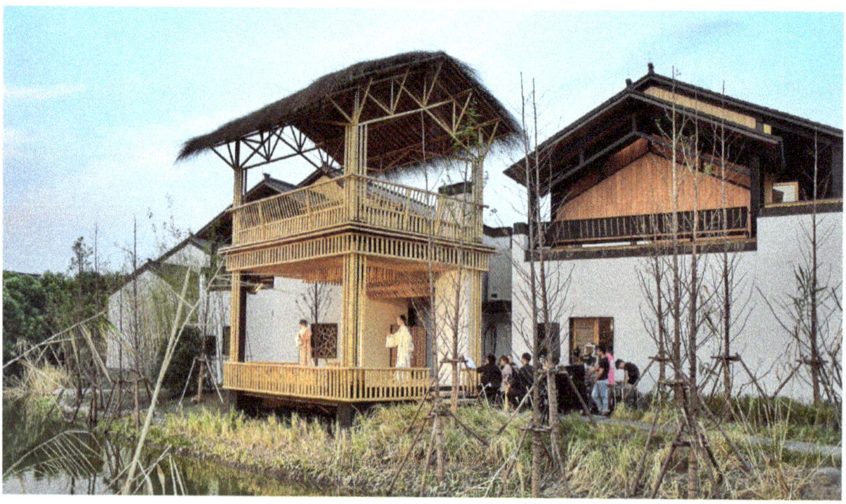

Fig. 6.60 Watching a play on the bridge (*photo by Jiang Yanzhi*)

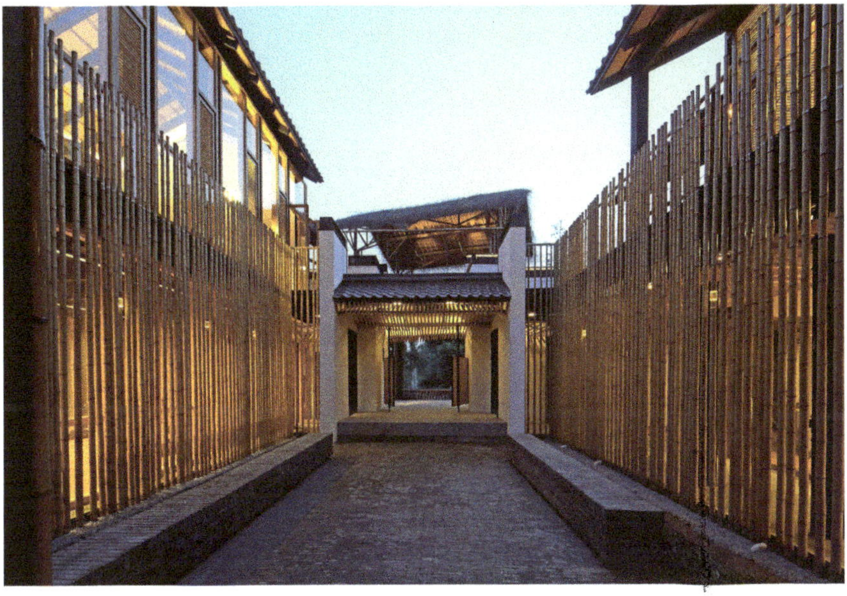

Fig. 6.61 Teaching backcourt (*photo by Zhang Guangyuan*)

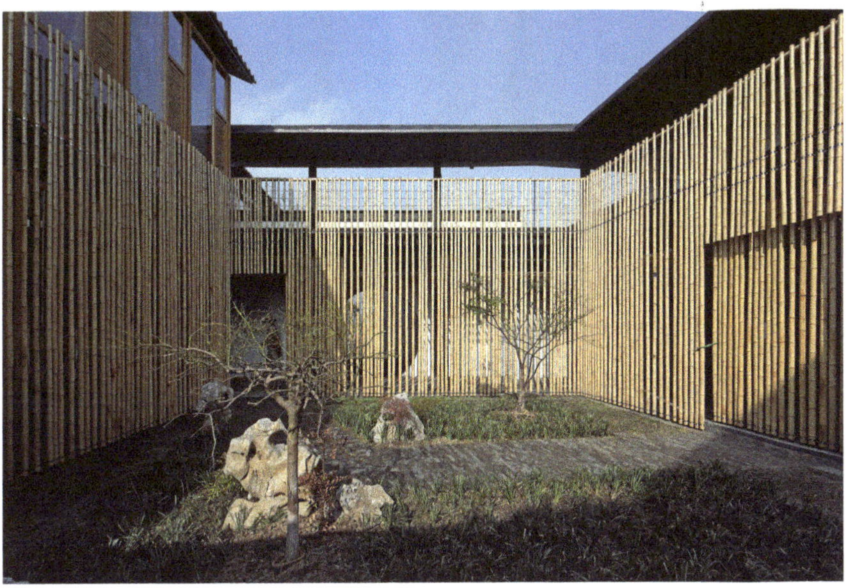

Fig. 6.62 Orchid courtyard (*photo by Zhang Guangyuan*)

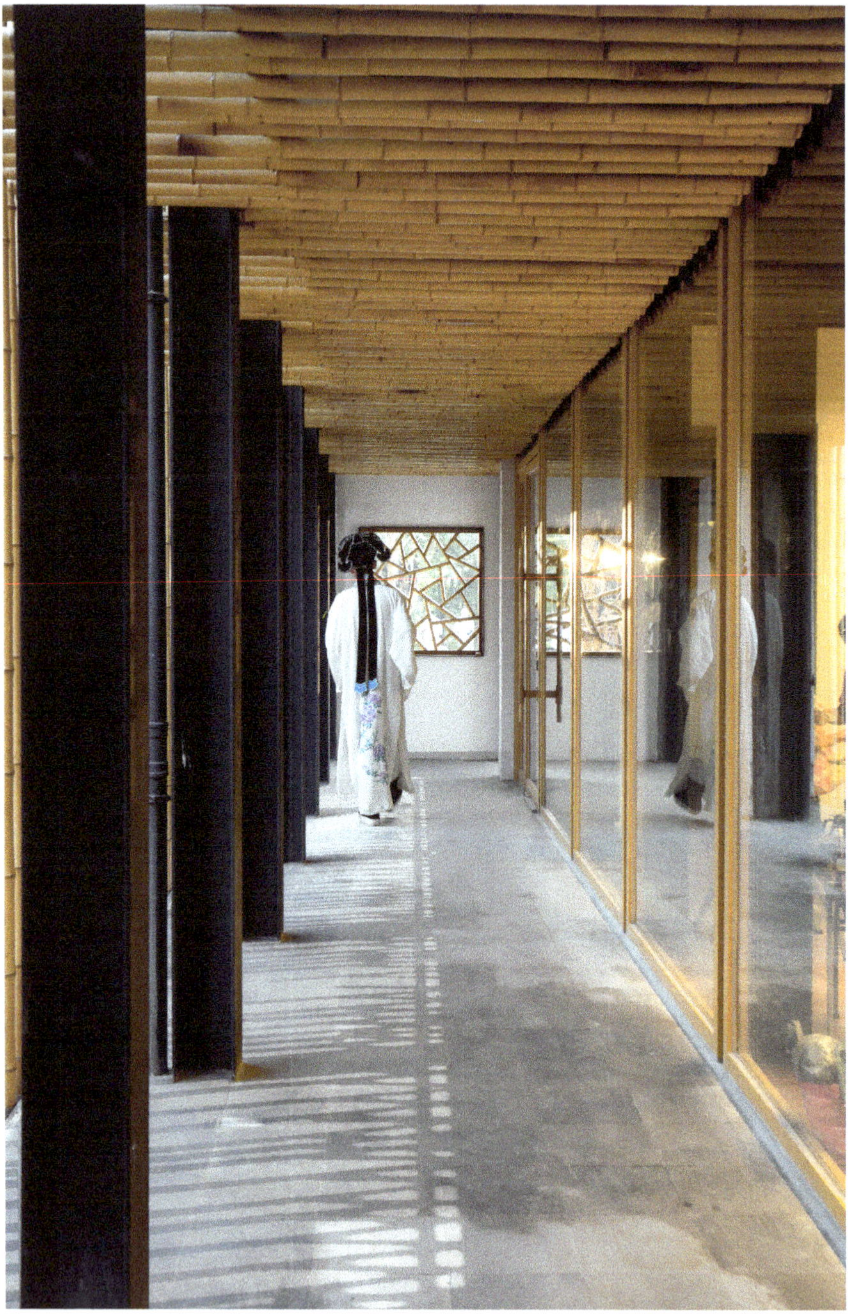

Fig. 6.63 Step on the shadow to seek song (*photo by Jiang Yanzhi*)

6.2 Bamboo as Structural Materials

Full-culm bamboo structures:

Fish-shaped Pavilion of INBAR Garden at 2021 Yangzhou International Horticultural Expo (2021)

INBAR Garden at 2019 Beijing Expo (2019)

Tea Market of Zhuguanlong Township (2021)

Shangcun Village Lounge (2018)

Bamboo Exhibition Corridor of Ancient Kiln in Jinshan Mountain (2017)

Tourist Service Centre on Nantian Island in Xiangshan, Ningbo (2020)

Baili Teahouse (2020)

Reception Centre of Jiufeng Village (2019)

Liu Yun: Full-culm Bamboo Architectures in Huangjueya Old Street, Nan'an District of Chongqing (2020)

Wing of the Wind (2020)

Gap of the Cloud (2020)

Baizhishan Tourist Reception Center (2019)

Yibin International Bamboo Products Trading Centre (2021).

Engineered bamboo structures:

Waterside Bamboo Pavilion of INBAR Garden at 2021 Yangzhou International Horticultural Expo (2021)

Qishe Courtyard Renovation (2020)

Office Building of Ganzhou Sentai Bamboo & Wood Co., Ltd. (2020)

Tourist Service Room for Tangshan Mine camp (2019)

Zhaojun Museum (2018)

Youth Hostel at Niubeishan Mountain (2014)

Site of Chu River and Han Border World King of Chinese Chess Tournament: Contest Pavilion of World King of Chinese Chess (2018)

Zen Centre at Baoguosi Buddhism Temple in Lezhi (2016)

Sanhe Canteen (2016)

Villas in Saudi Arabia (2016)

Caishan Urban Vegetable Bar for Parents and Children (2015)

Beijing Youth Camp International (2013)

Swire Properties Bamboo Community Center (2017)

No.37 Luanqing Hutong—“Quad of Gourd” (2017)

Shuangxi Academy (2018)

Qing Bridge—Public Facility (2017)

Rest Station of Nanjing Qixia Mountain Scenic Area (2018)

Courtyard N4, Lixiang Village (2019)

Bamboo Chalets in the West Historic Block of Gongchen Bridge, Hangzhou Section, Grand Canal, China (2017).

Fish-shaped Pavilion of INBAR Garden at 2021 Yangzhou International Horticultural Expo (2021)

Location: Yangzhou, Jinagsu, China

Architects: Song Yehao, Chen Xiaojuan, Xie Dan et al.

Completed: 2021

The INBAR Garden is located in the centre of the International Horticultural Exposition 2021 (Yangzhou Expo 2021). It has a total area of about 3300 m^2. The Garden includes two low-carbon, environmentally friendly, bamboo pavilions built in different styles, which integrate modern and traditional elements: Fish-Shaped Bamboo Pavilion and Waterside Bamboo Pavilion (described later in this chapter).

The Fish-Shaped Bamboo Pavilion is located in the north of the INBAR Garden, covering an area of about 280 m^2. It is symmetrical along the centre of the ridge, resembling a fish with an arched back. A series of bamboo arches of different heights and spans change with the shape of the roof ridge, which shape the characteristic indoor space with great rhythm. The main entrance, at the head of the fish is larger to the west, and the scale of the secondary entrance at the tail towards the East is slightly smaller. Considering that the air in Yangzhou is humid in spring and summer during the Yangzhou Expo 2021, the pavilion is transparent from east to west to permit air through-flow. Ceiling fans are set indoors to enhance air flow and improve the comfort of users by energy saving. The middle of the pavilion, a "lighting belt" combines the highest roof ridge, and a section of full-height windows yielding rich light and shadow changes to the room.

Application of bamboo products: After determining the concept, architects introduced a digital model and determined construction method and logistics together with the structural engineer. The key components of different levels were generated and the structural model was locked after structural calculation. In collaboration with the manufacturer, the section sizes and erection sequences of the primary and secondary structure were determined. The full-culm bamboo structure includes main columns, secondary columns and purlins. The enclosure from inner to outer was arranged as a decorative bamboo curtain, oriented strand boards, waterproof roll material and bamboo tiles. Diamond-shaped woven bamboo strips and endurance sunshine boards are set at the lighting belt. In order to ensure reliable waterproofing, double layer waterproof roll was laid, and bamboo tiles were laminated to each other. The entire structure is in harmony with indoor bamboo furniture, bamboo umbrella lamps, etc., which fully shows the unique charm of bamboo.

Photo source: Figs. 6.64, 6.65, 6.66, 6.67 and 6.68 are provided by SUP Atelier of Architectural Design and Research Institute of Tsinghua University.

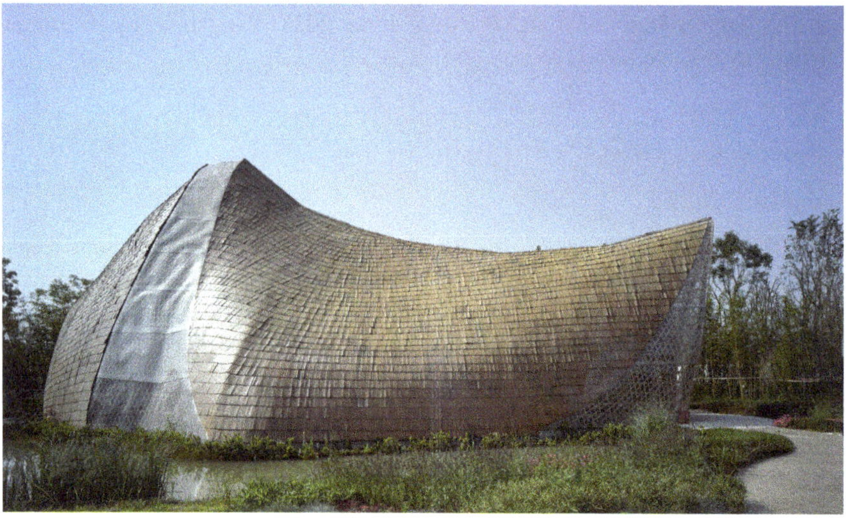

Fig. 6.64 Fish-shaped pavilion (*photo by Chu Yingnan*)

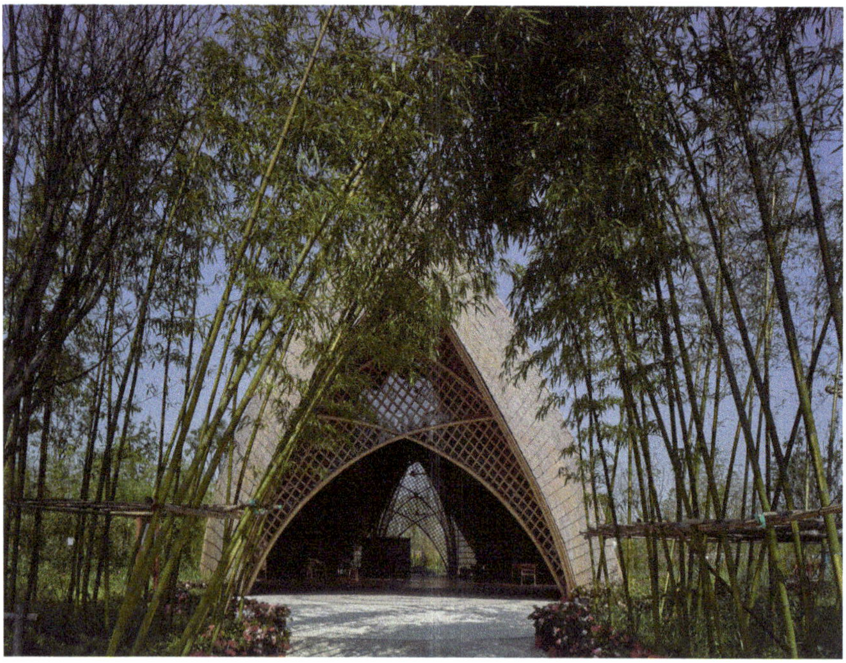

Fig. 6.65 Main entrance of fish-shaped pavilion (*photo by Chu Yingnan*)

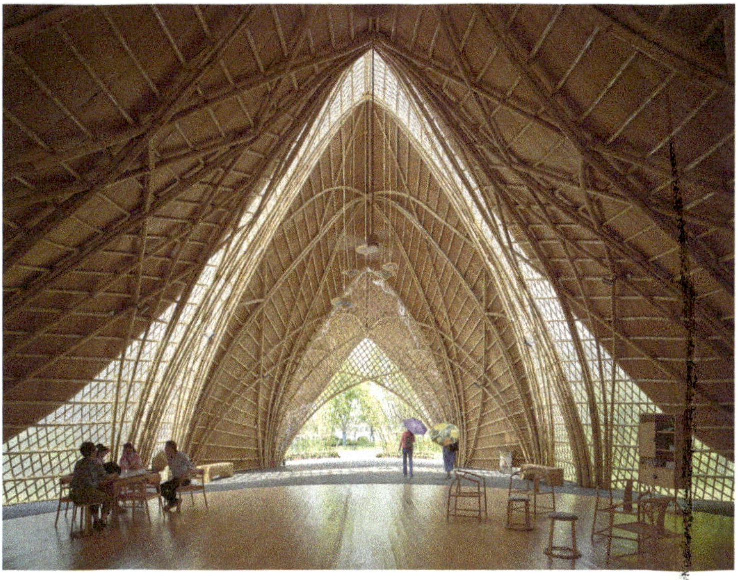

Fig. 6.66 Interior of function hall (*photo by Chu Yingnan*)

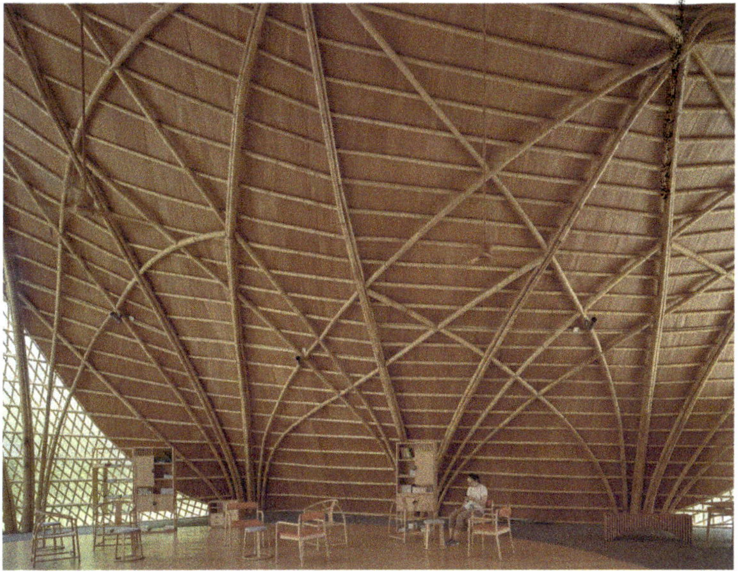

Fig. 6.67 Woven full-culm bamboo structure (*photo by Chu Yingnan*)

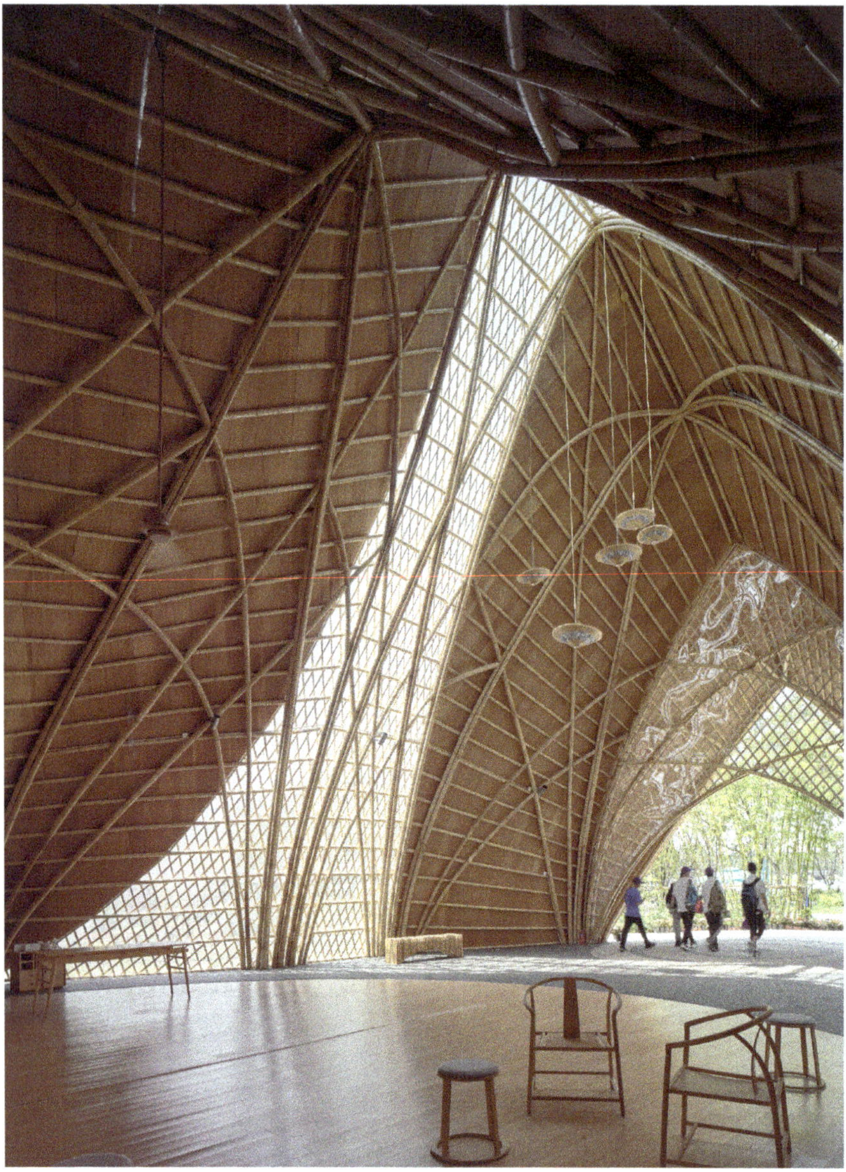

Fig. 6.68 Daylight into the hall (*photo by Chu Yingnan*)

INBAR Garden at 2019 Beijing Expo

Location: Beijing, China

Completed: 2019

Architects: Mauricio Cardenas Laverde (Italy), Wang Xuesong, Wang Zhaoxia et al.

The INBAR Garden with the theme of "Creative Bamboo and Rattan, Five Continents' Scenery" is located in the east of the main axis of 2019 Beijing Expo. The area of the exhibition hall is about 1600 m^2, and the outdoor greening landscape area is about 2000 m^2 (including the green roof of the pavilion). Through the integrated design of the bamboo garden of the exhibition hall, the architect narrates the traditional "exhibition hall in the bamboo garden" in modern design language. The space of the pavilion is carefully hidden under the green garden supported by nine bamboo arches. The green space of the garden spreads to the roof along the arch feet and gradually disappears. Architecture, bamboo structure and bamboo landscape are all in one, forming a vivid eye on the earth, which is known as the "Bamboo Eye" of 2019 Beijing Expo.

Application of bamboo products: The main structure of the exhibition hall is made of more than 5000 *P. edulis* (moso) bamboo culms with diameters of 8–10 cm. The single arch span is 32 m. It is the largest full-culm bamboo structure in northern China. The nine bamboo arches supporting the exhibition hall are designed in the form of variable cross-section truss arches. This combines the advantages of both arch and truss structure, making the large-span bamboo structure look very light and graceful.

Photo source: Figs. 6.69, 6.70, 6.71, 6.72 and 6.73 are provided by INBAR.

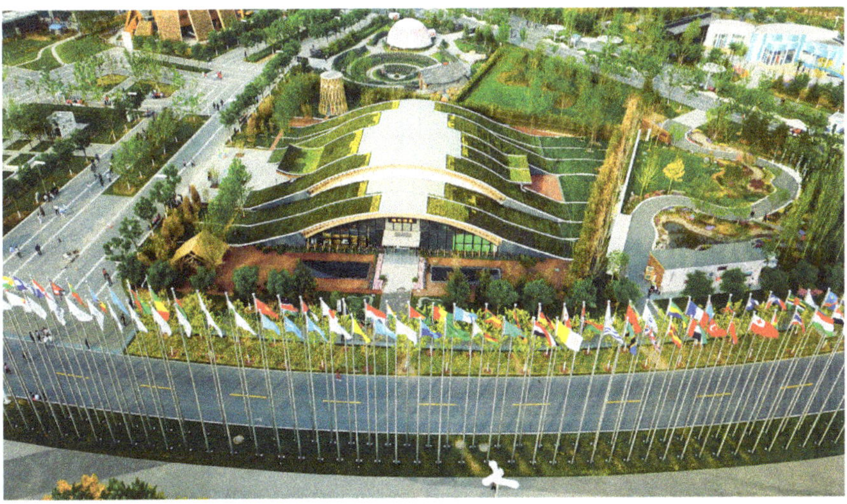

Fig. 6.69 Aerial view of INBAR Garden at 2019 Beijing Expo

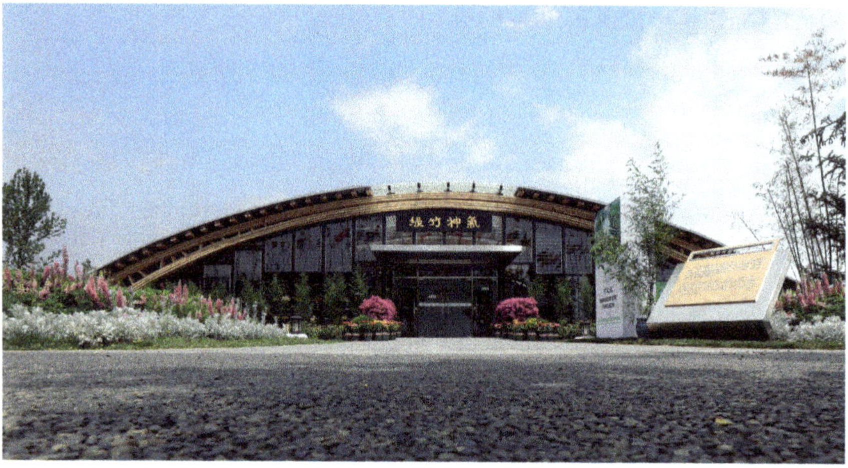

Fig. 6.70 West facade of INBAR Pavilion (*photo by Wang Xudong*)

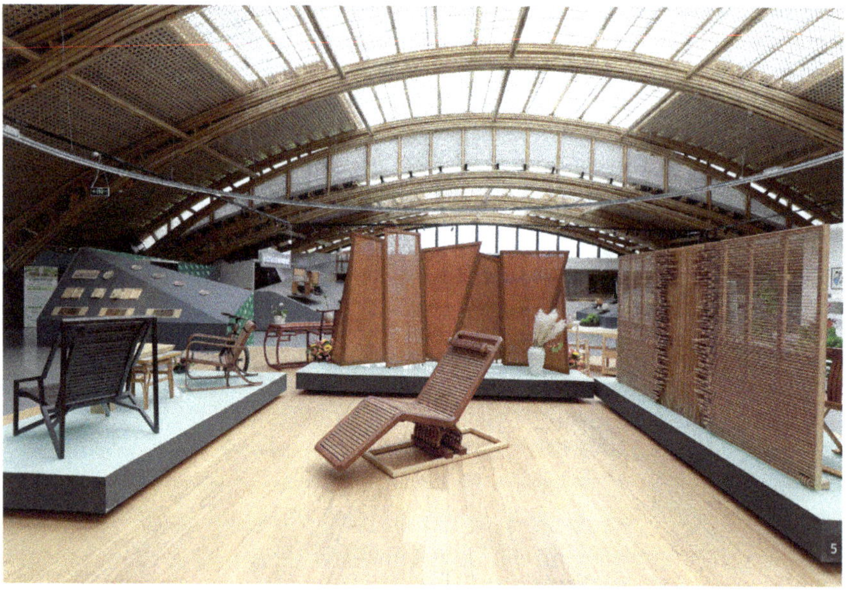

Fig. 6.71 Exhibition at INBAR Pavilion (*photo by Wang Xudong*)

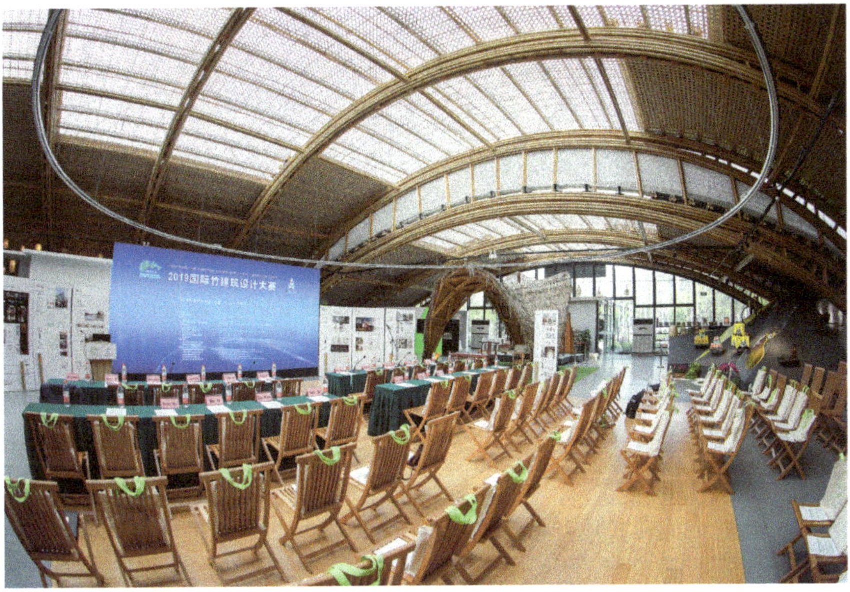

Fig. 6.72 2019 International Bamboo construction competition held at INBAR Pavilion

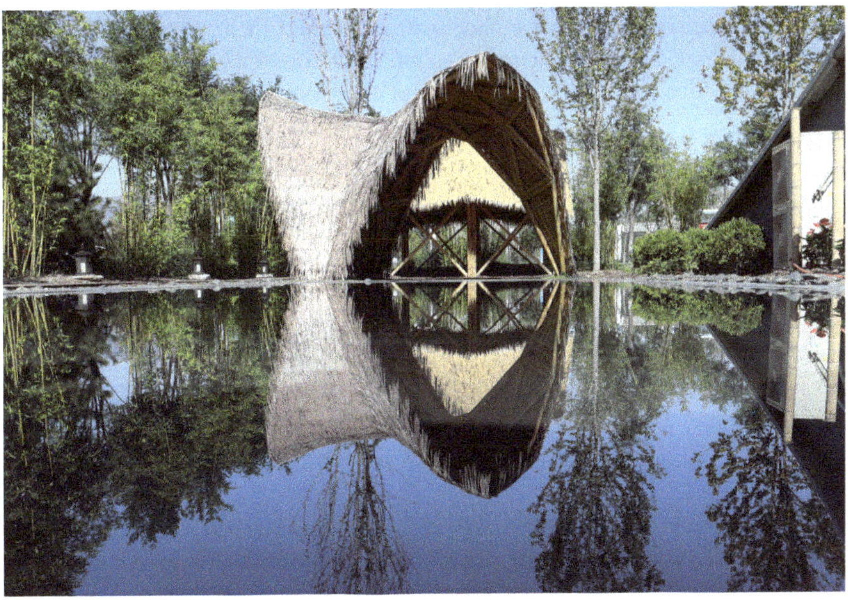

Fig. 6.73 One of winning works of 2019 International Bamboo Construction Competition at INBAR Garden (*photo by Wang Xudong*)

Tea Market of Zhuguanlong Township

Location: Ningde, Fujian, China

Completed: 2021

Architects: Song Yehao, Sun Jingfen, Zhu Yingnan et al.

The Tea Market of Zhuguanlong Township is located in Shouning County, Ningde City, Fujian Province. It is a place of tea trading for villagers. Initially, architects and local government discussed the possibility of building a multi-functional large-span space. Because of the influence of the seasons, the building is not only used for tea trading, but also serves several other functions: a farm trading market, a place for villagers' rest and children's sports activities, or a centralized activity site for village special days, such as beggars' banquet, marriages or funerals. The biggest challenge for the project is to use full-culm bamboo to realize a large-span space. After studying the construction technology of traditional Chinese Hongqiao (Woven Timber Arch Bridge), architects decided to build a roof comprised of woven full-culm bamboo arches. A load test was carried out before construction to ensure its structural reliability. The construction team arranged full-culm bamboo structural members of fixed length and regular combination, and set up four bamboo arches. Each bamboo arch contains 12 rows of full-culm components, and in this manner, a woven bamboo arch structure with an 18 m span was constructed. The project also uses local traditional rammed earth technology to build enclosure walls, which are perfectly integrated with a concrete structure and terrazzo ground, forming a sense of nature, and also a reflection of the concept of Chinese architecture.

Application of bamboo products: Four woven full-culm bamboo arches with 18 m span were built.

Photo source: Figs. 6.74, 6.75 and 6.76 are provided by SUP Atelier of Architectural Design and Research Institute of Tsinghua University.

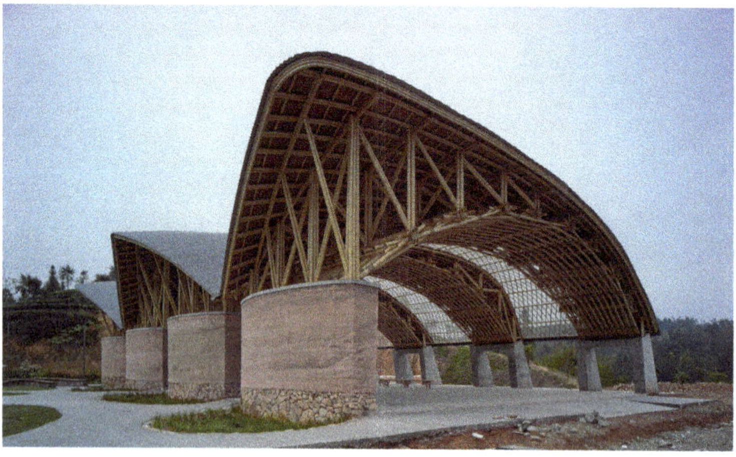

Fig. 6.74 Northwest view (*photo by Chu Yingnan*)

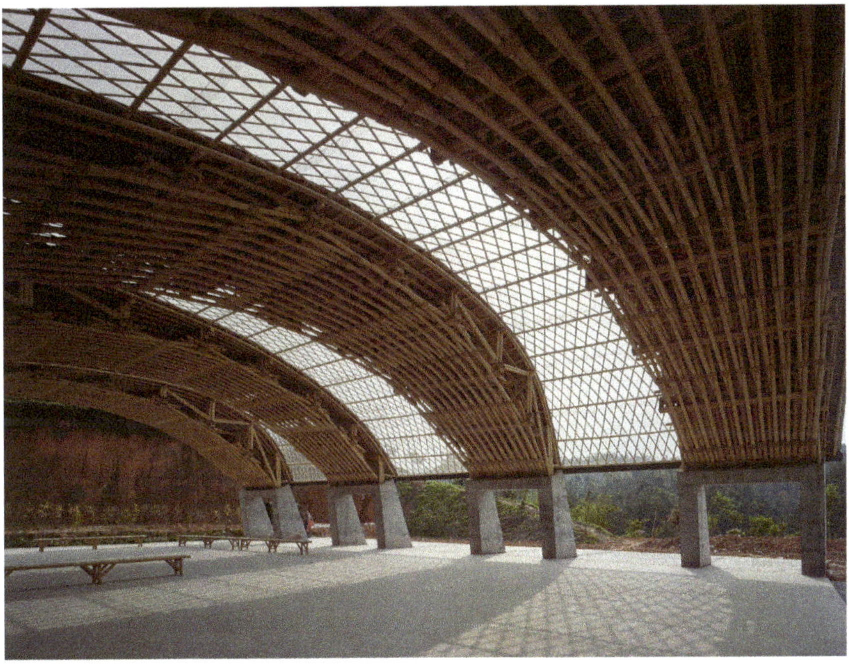

Fig. 6.75 Full-culm bamboo arch bridge structure (*photo by Chu Yingnan*)

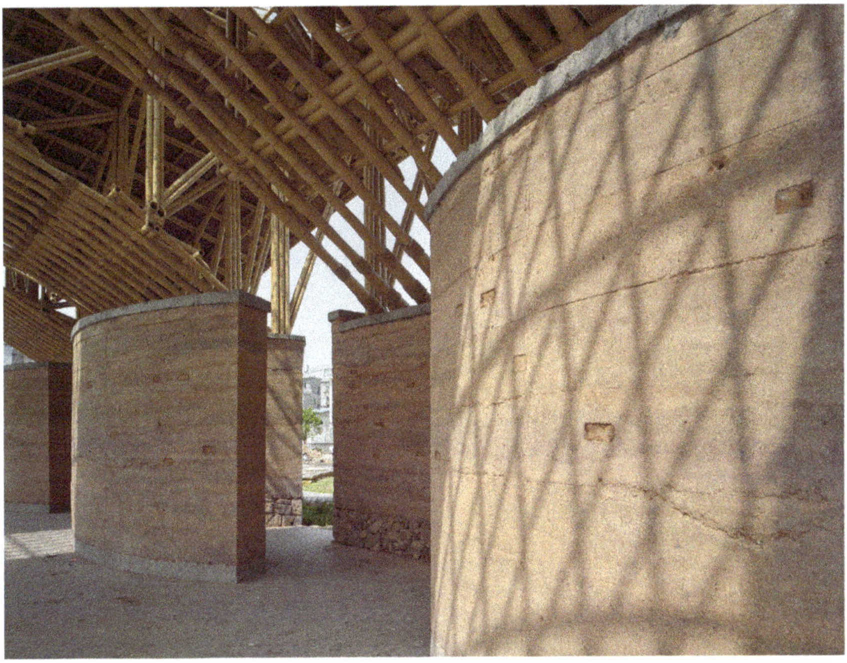

Fig. 6.76 Detail of rammed earth wall (*photo by Chu Yingnan*)

Shangcun Village Lounge

Location: Jixi, Anhui, China

Completed: 2018

Architects: Song Yehao, Sun Jingfen, Chen Xioajuan et al.

The team, composed of China Academy of Urban Planning and Design, Tsinghua University and Peking University, tries to explore a sustainable development path for the protection of traditional villages in Shangcun. The team selected Gaojia Old House as the villagers' public living room, and gradually carried out the work of renovating the village's human settlement and environment, promoting the development of industry, protecting the traditional style and features, and restoring the dwellings. A long-term mechanism for the protection and development of Shangcun Village will be gradually established, so as to guide Shangcun Village to achieve sustainable development in the future.

Application of bamboo products: The architectural design team selected bamboo as the main material of the building, skillfully designing six "bamboo umbrellas" and three groups of "awnings", to form a bamboo awning shared by villagers and tourists. The combination of bamboo structural system and traditional old houses is not only the collision of old and new materials with old and new technology, but also the superposition of open space represented by the bamboo umbrella units and enclosed space surrounded by walls. It is an attempt to construct easy-to-disassemble units and upgrade residential waste sites with collage.

Photo source: Figs. 6.77, 6.78 and 6.79 are provided by SUP Atelier.

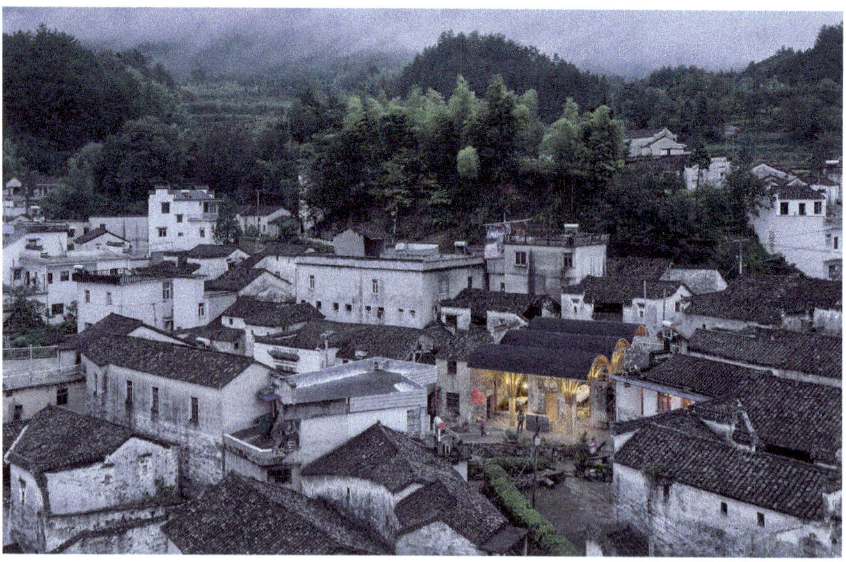

Fig. 6.77 Aerial view at nightfall (*photo by Xia Zhi*)

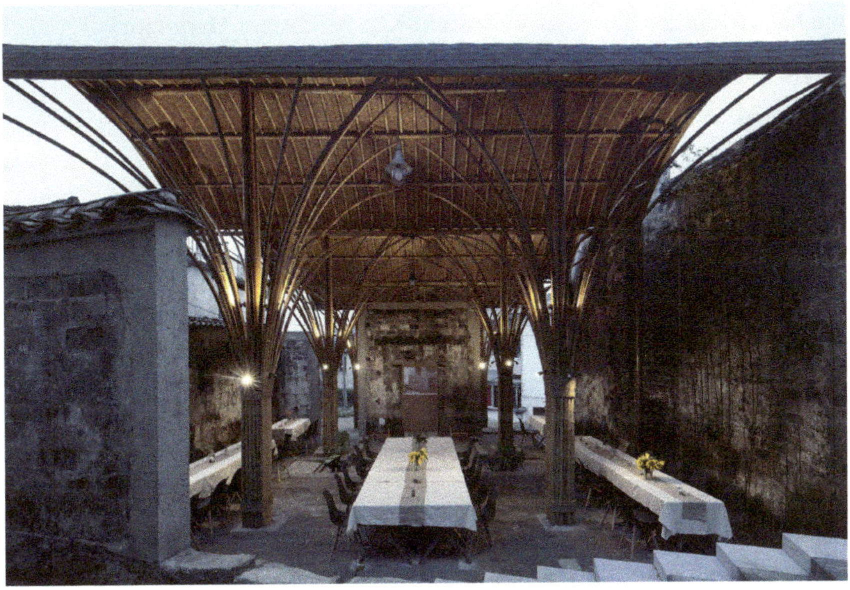

Fig. 6.78 View from South back street (*photo by Xia Zhi*)

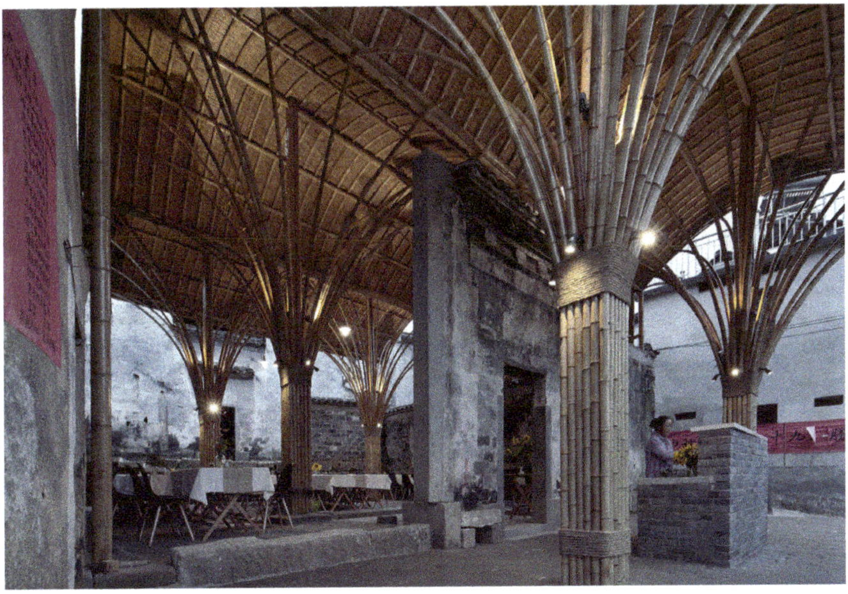

Fig. 6.79 Northeast view under eaves (*photo by Xia Zhi*)

Bamboo Exhibition Corridor of Ancient Kiln in Jinshan Mountain

Location: Shang Yu, Zhejiang, China

Completed: 2017

Architects: Zhong Dekun, Peng Xiaosong, Guo Ziyi

The site of Ancient Kiln in Jinshan Mountain consists of three dragon kilns shown to date from the Eastern Han, Three Kingdoms and Western Jin Dynasties, respectively. They witnessed the transformation from primitive celadon to mature celadon. The kilns are situated between valleys, with streams flowing through, facing a reservoir and lake in the South and with a beautiful natural environment all around. The building conforms to the terrain. The building system is a portal frame, which is composed of full-culm bamboo composite columns and a full-culm bamboo composite truss. The roof is also covered with bamboo tube tiles. The architecture is fully integrated into the environment, people can see through the building, feeling the lush forests and quiet valleys.

Application of bamboo products: Local full-culm is the main structural material of the building.

Photo source: Figs. 6.80, 6.81 and 6.82 are provided by Anji Zhujing Bamboo Industry Technology Co., Ltd.

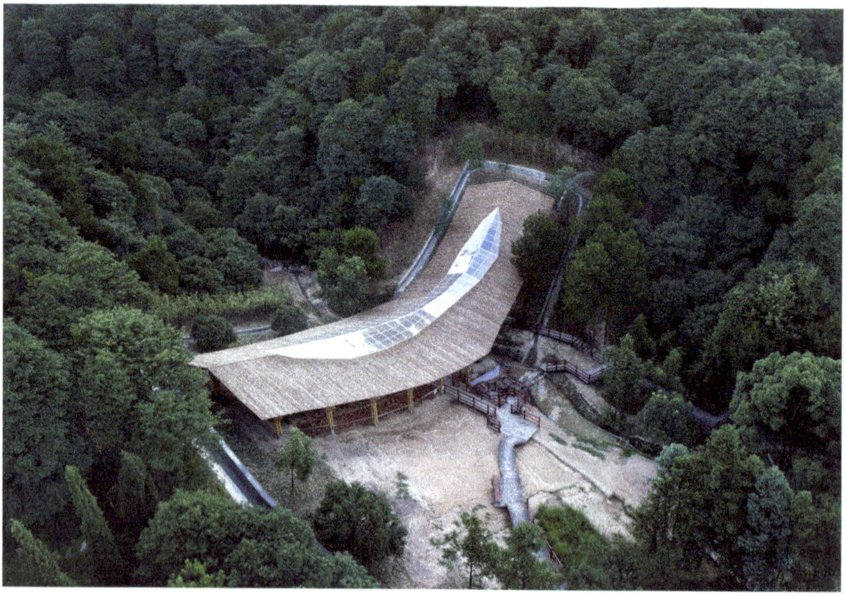

Fig. 6.80 Aerial view of bamboo exhibition corridor

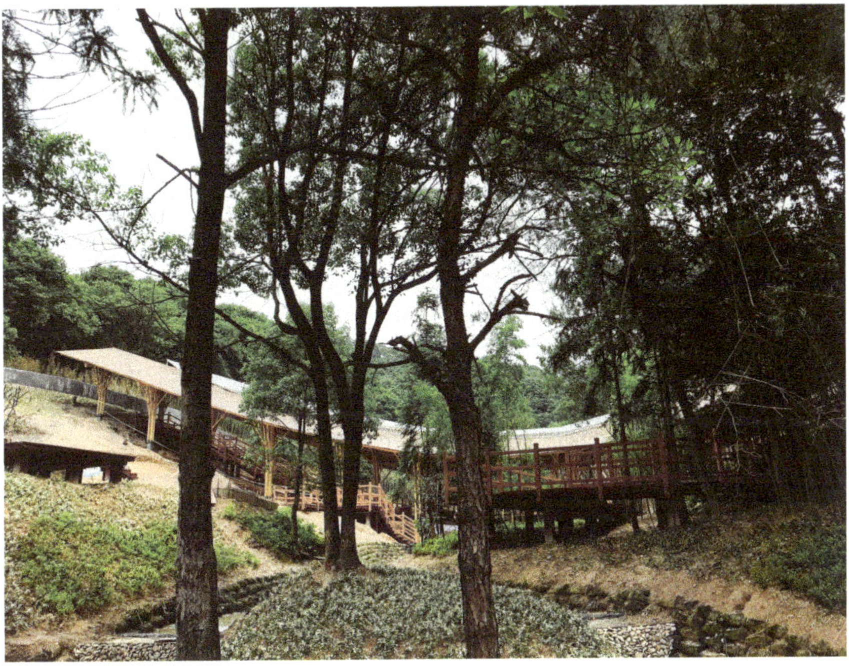

Fig. 6.81 Bamboo exhibition corridor through the woods

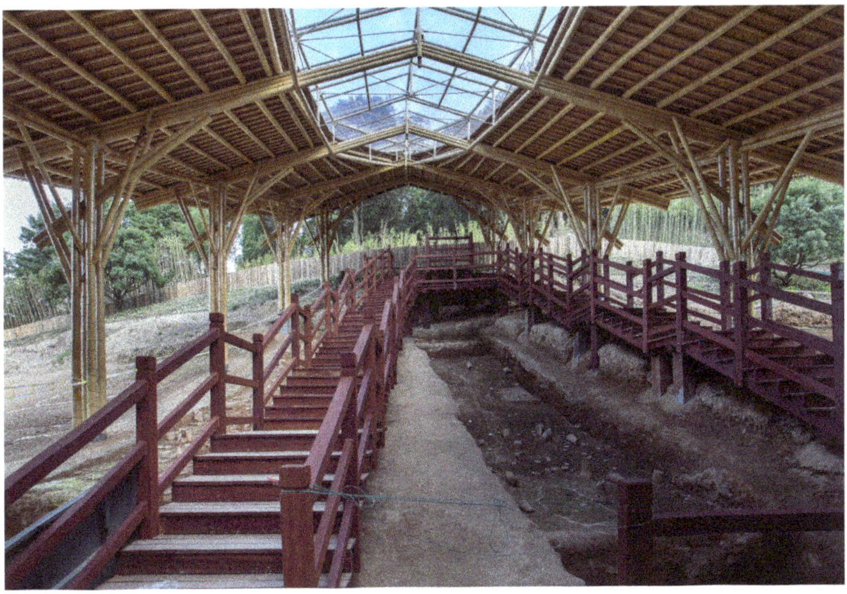

Fig. 6.82 Under the bamboo exhibition corridor

Tourist Service Center on Nantian Island in Xiangshan, Ningbo

Location: Ningbo, Zhejiang, China

Completed: 2020

Architects: Gao Wei, Zhao Xiaoyu, Li Wangjing

Located at the entrance of Xiaobaizhang Village on Nantian Island, the tourist service center has two elliptical cones. The large one serves as a multifunctional hall and the small one is a public restroom. The hall covers an area of nearly 300 m^2. It adopts a column-free structure, creating a large space. Enclosed by glass curtain walls, satisfying various needs, such as conferences, markets, weddings and catering.

Application of bamboo products: The architecture adopts a full-culm structure. The exposed bamboo fully reflects the beauty in the rhythm of structure. The visual effect is low-key and rustic, as bamboo tips are used in the roof cladding, which contrasts with the rubble masonry wall. The vertical columns use more than 2000 bamboo culms having diameters of 7–8 cm, and the top ring beam uses more than 3000, 5–6 cm diameter culms. There are 26 spans. The longest single span is 21 m, and the shortest is 15 m.

Photo source: Figs. 6.83, 6.84 and 6.85 are provided by ADR AuDi Design Studio.

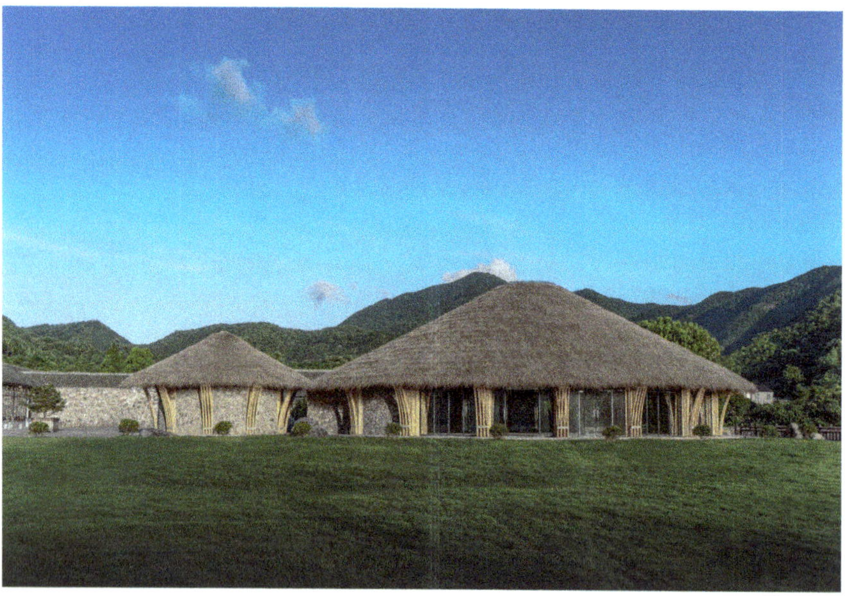

Fig. 6.83 Front view of Tourist Service Center

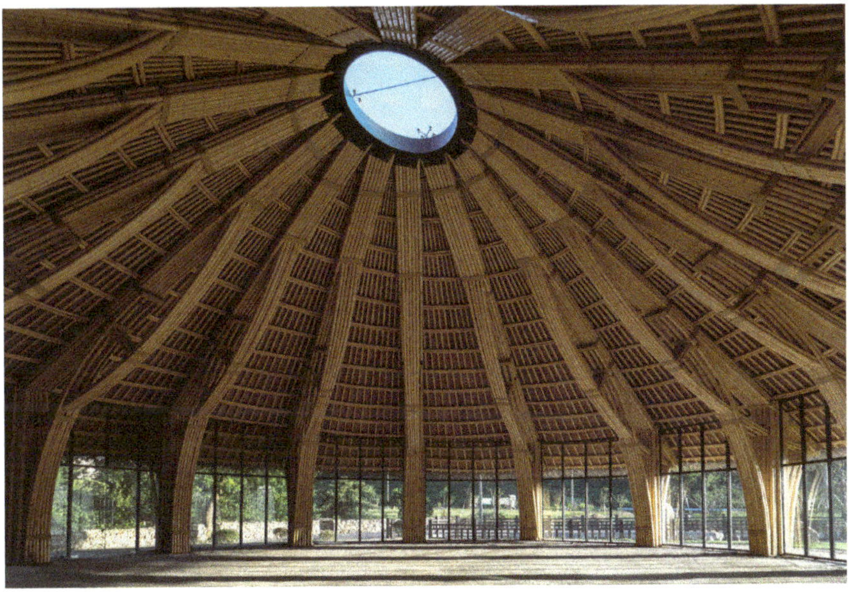

Fig. 6.84 Bamboo dome of Tourist Service Center

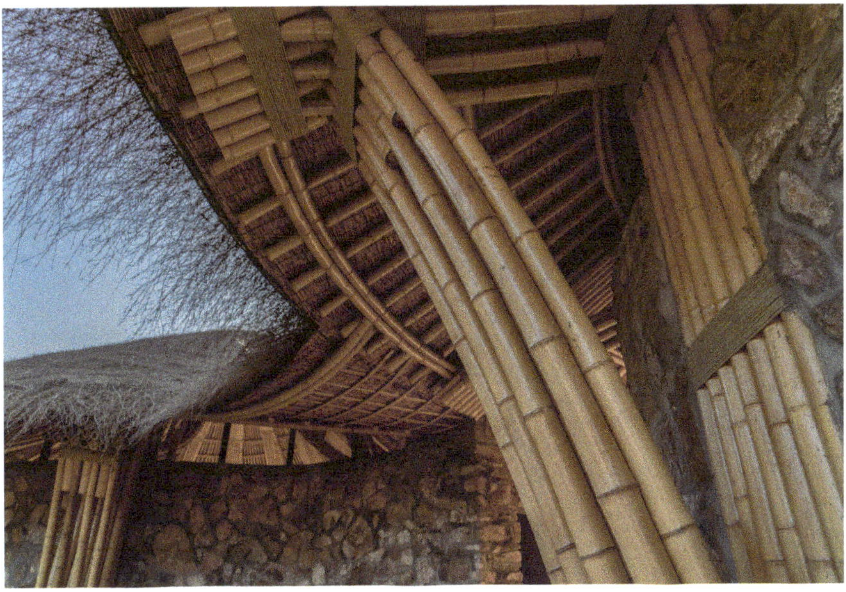

Fig. 6.85 Detail of the structure

Baili Teahouse

Location: Jiang'an, Sichuan, China

Completed: 2020

Architects: Li Wangjing

The project is located in the Baili Tianxia demonstration area for rural revitalization, covering an area of more than 400 m^2. The teahouse is a curved corridor made of full-culm bamboo. With the innovative use of bamboo, the types of space are no longer single, but gradually develop in diversity. The designer skillfully combines traditional architectural elements (such as bamboo, tiles, reed mats, etc.) with modern architectural elements (such as floor-to-ceiling glass curtain walls) to create a unique interior and semi-exterior spaces, providing local villagers with a relaxing and comfortable cultural living space.

Application of bamboo products: The main structure is all built with full-culm bamboo from Anji, Zhejiang Province. The main column structure uses large diameter curved bamboo to present the shape of a "wing". Other columns are woven in a grid with 5 cm diameter culms, presenting the shape of a "feather". Bamboo chairs, bamboo stools and creative bamboo art lamps are used in the interior decoration with exquisite details.

Photo source: Figs. 6.86, 6.87 and 6.88 are provided by Anji Zhujing Bamboo Industry Technology Co., Ltd.

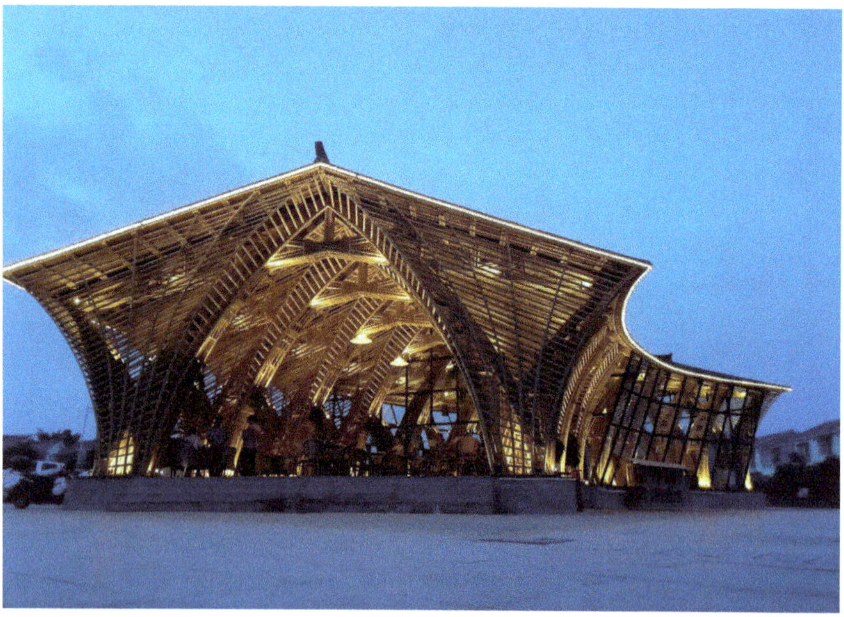

Fig. 6.86 Front view of Baili Teahouse

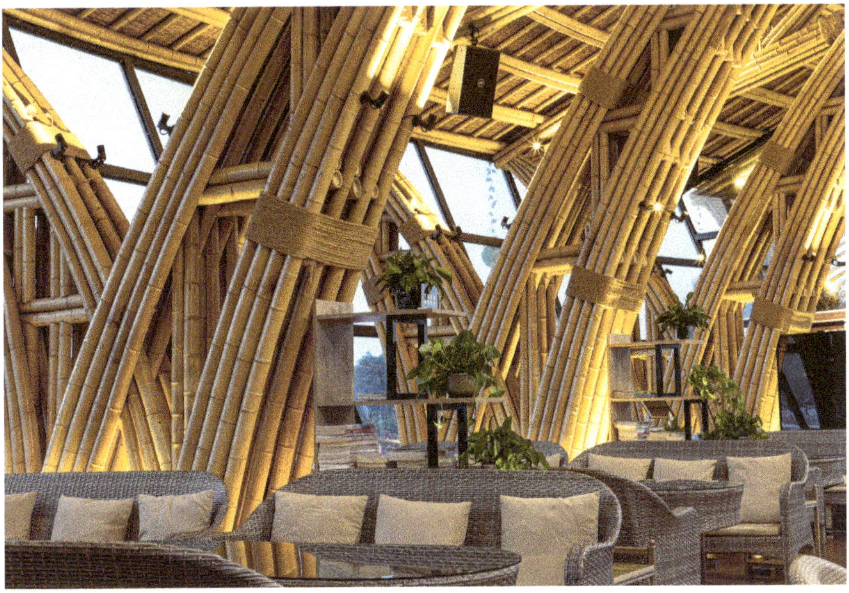

Fig. 6.87 Interior of Baili Teahouse

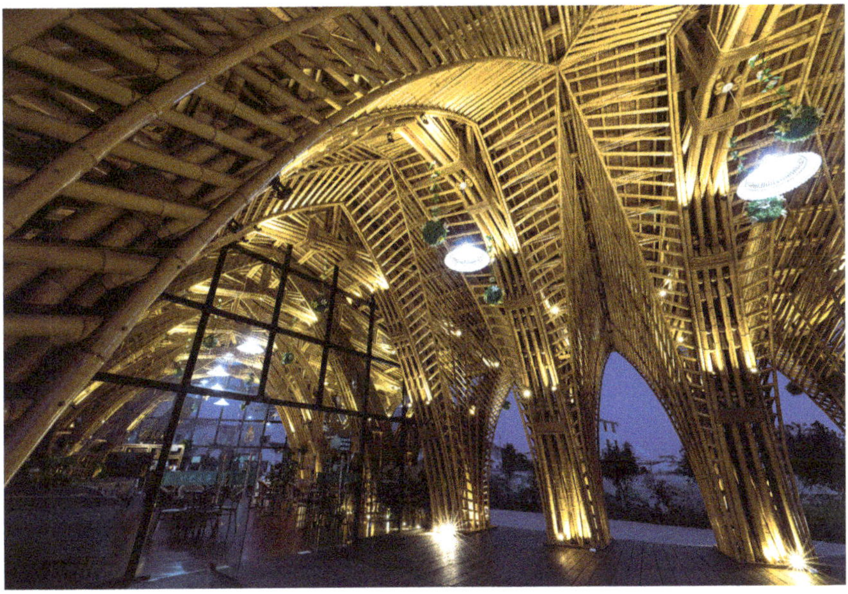

Fig. 6.88 Outdoor platform

Reception Center of Jiufeng Village

Location: Jiufeng Village, Fuzhou, Fujian, China

Completed: 2019

Architects: Cui Kai, Guo Haian et al.

Located in the North Peak of Fuzhou, Jiufeng village is surrounded by mountains and streams. The picturesque scenery attracts many tourists. A nearly deserted old house was selected for reconstruction in order to improve the accommodation capacity of the village and provide an effective public space for the villagers. In the main part of the old house, old walls were retained, the curved wooden beams were straightened and reinforced, and the damaged brick columns were replaced. Other courtyard walls, gate towers and warehouses were rebuilt using original practices. Based on a full-culm bamboo structure, a large-scale gathering room, convenient for visitors' receiving, conferences, trainings and tea parties, was created. The adjacent garden adopts the technique of "borrowing scenery" to open the originally closed courtyard walls and replace them with reeds and grass. The opposite side of the bamboo living room is ingeniously facing the ancient pavilion by a stream in the distance. It is a part of the view from the living room, complemented by the trees and faraway mountains.

Application of bamboo products: In order to avoid blocking the old house, a full-culm bamboo truss structure is used to create a large space; this is convenient for construction and helped to control costs. To ensure the stability of the structure, each vertical support column is composed of six 8 cm diameter culms, and bamboo diagonal braces are added between each truss. In order to improve the uplift-resistance of the roof, a grounded bamboo wall system is adopted on both sides. The windward side of the roof is low, and the leeward side is pulled up to the maximum height, so as to adapt to the windy weather that may be encountered. In addition, bamboo, which is most common in the mountainous areas of Fujian Province, is used to create a natural landscape in the courtyard. The scenery appears to be charming with new green bamboos and old yellow ones.

Photo source: Figs. 6.89, 6.90, 6.91 and 6.92 are provided by Rural Culture D–R–C of China Architecture Design & Research Group.

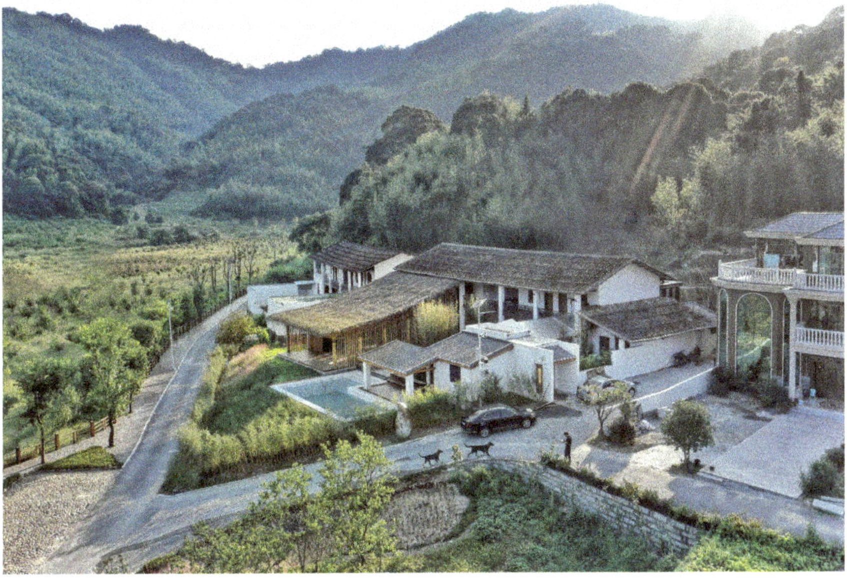

Fig. 6.89 Sunset over Jiufeng Village

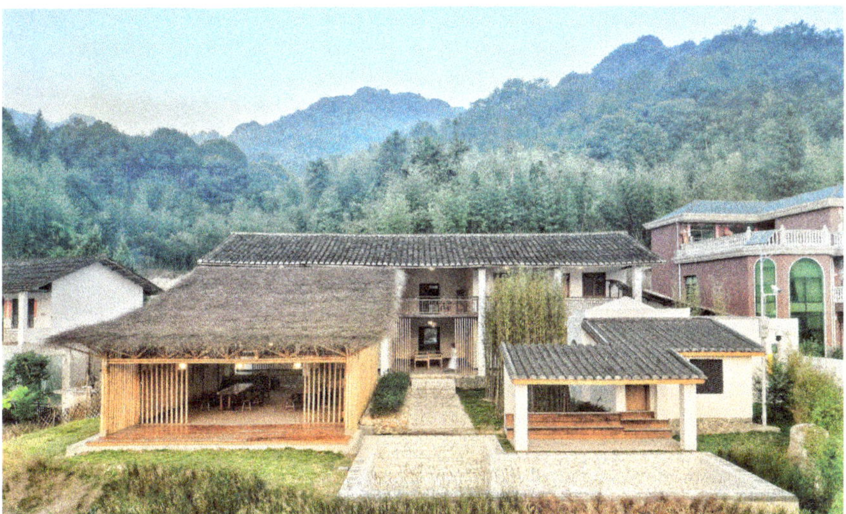

Fig. 6.90 Reception centre at the foot of the mountain

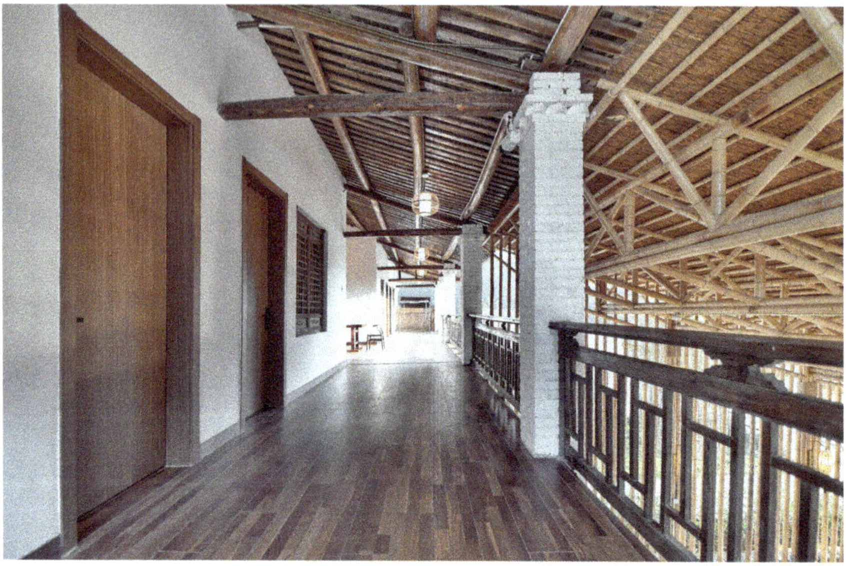

Fig. 6.91 The second floor of the old house

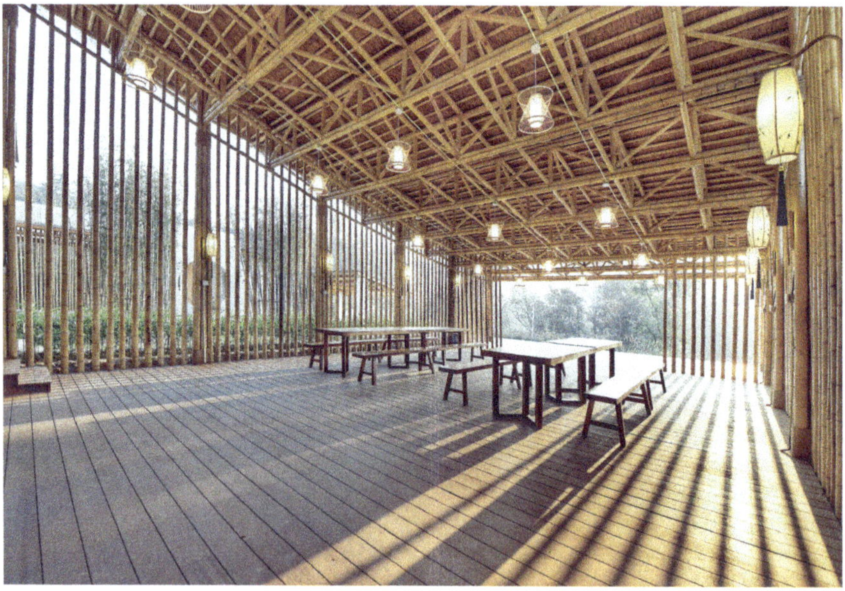

Fig. 6.92 Bamboo reception area

Liu Yun: Full-culm Bamboo Architecture in Huangjueya Old Street, Nan'an District of Chongqing

Location: Chongqing, China

Completed: 2020

Architects: Huang Yun, Wang Pingshu, Ju Qian

"Liu Yun" is one of the renovation projects of Huangjueya Old Street, next to the former residence of the famous writer Echo. The 35,000 m^2 project takes "the millennium Huangjueya, a link between the past and the present" as its main cultural thread, restoring 175 old residences and Bayu-style architecture including Echo's former residence and the Guizhou Chamber of Commerce. The project makes the lonely old street full of vitality through the exploration of innovative designs, addition of fashion elements, and combination of history and modernity. Inspired by the light and floating fluid space formed by the mist in the mountains, "Liu Yun" takes use of unique bamboo elements to create a new kind of artistic visual beauty.

Application of bamboo products: Due to the 20 m elevation differences of the site, three scattered platforms were created to provide various activities. Groups of curved bamboo poles make up the main support of the architecture, creating the beauty of natural growth with a rich vertical level and multi-dimensional viewing angle. As the core part of the architecture, the highest platform is a "tree house" built on a large tree. The curved shape of the bamboo roof conveys the elegance and ease of clouds. With its material characteristics, bamboo architecture plays an indispensable role in creating spatial experience, expressing a strong artistic tension.

Photo source: Figs. 6.93 and 6.94 are provided by Chongqing Huangyun Landscape Design Studio.

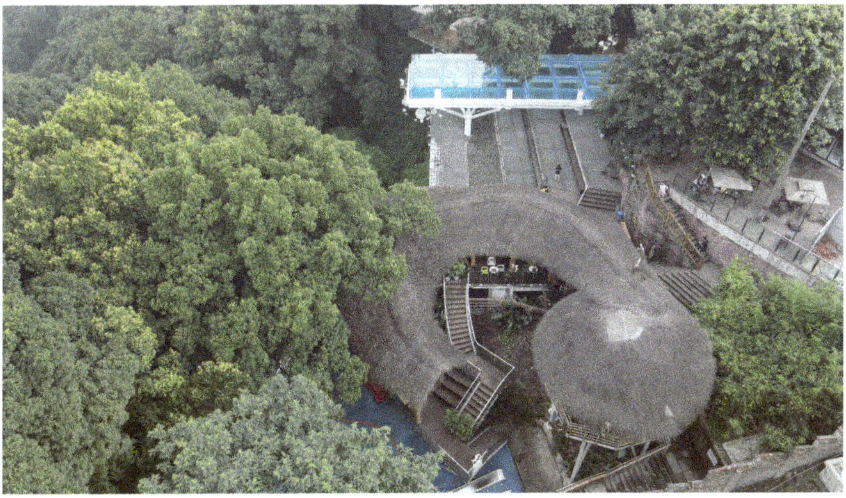

Fig. 6.93 Aerial view of "Liu Yun"

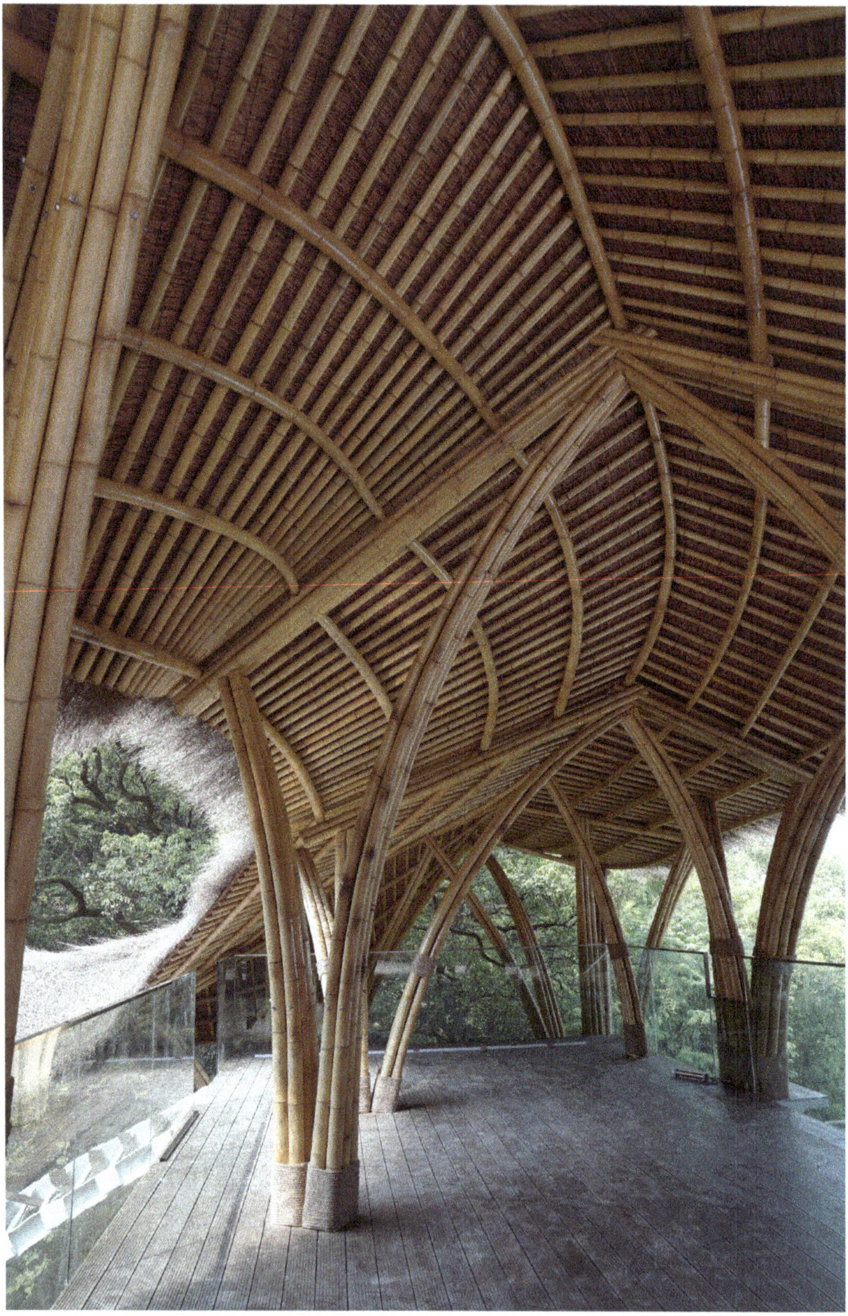

Fig. 6.94 Under the corridor of "Liu Yun"

Wing of the Wind

Location: Yibin, Sichuan, China

Completed: 2020

Architects: Huang Yun, Wang Pingshu, Ju Qian

The project is located in Changjiang Bamboo Island, Jiang'an County, Yibin City, Sichuan Province. It is an ecological environmental protection project in the Yangtze River Basin and also an ecological expo park with the theme of bamboo ecology and bamboo culture. Its main landscape architectures are two full-culm structures named "Wing of the Wind" and "Gap of the Cloud" (to be introduced in this chapter). Inspired by the river breezes at the head of the island, "Wing of the Wind" tries to turn "wind" into space through the agile characteristics of bamboo. This strong spatial form is derived from the beauty of bamboo itself, which makes it possible to see the sound of the river breeze.

Application of bamboo products: The curved bamboo forms a large-span truss structure, which has a sense of strength. The multiple rows of trusses keep "Wing of the Wind" smooth on the outside, while expressing the beauty of the complex structure on the inside. Arranged symmetrically along the axis, the eight groups of bamboo arches create a sense of sequence from large to small with an emphasis on the ritual space, thus gradually focus attention to the distance pointed by the head of the island.

Photo source: Figs. 6.95, 6.96 and 6.97 are provided by Anji Zhujing Bamboo Industry Technology Co., Ltd.

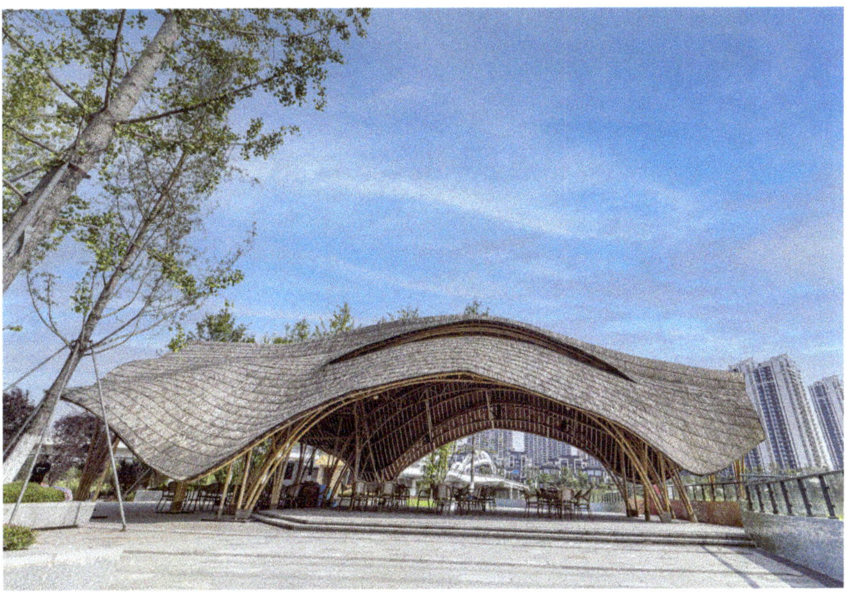

Fig. 6.95 Exterior side view of the Wing of wind

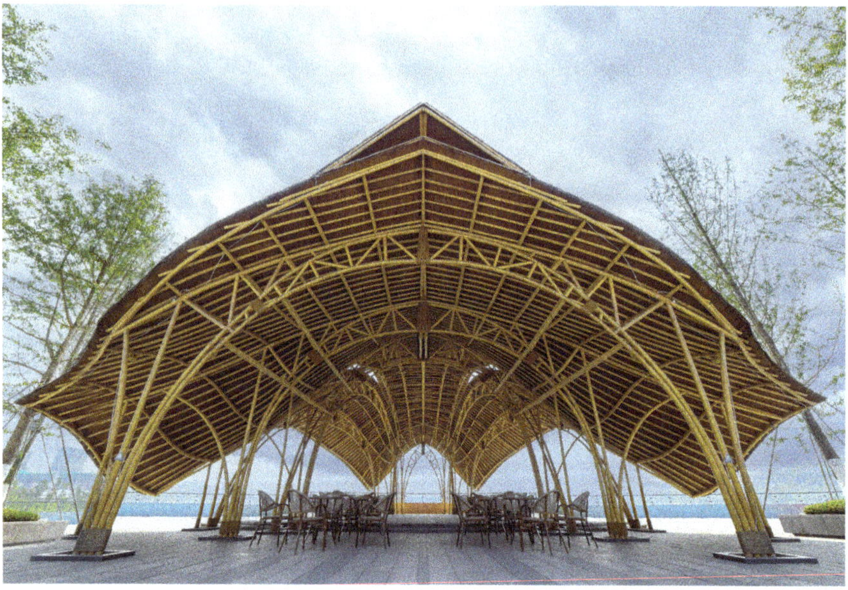

Fig. 6.96 Exterior axial view of the Wing of Wind

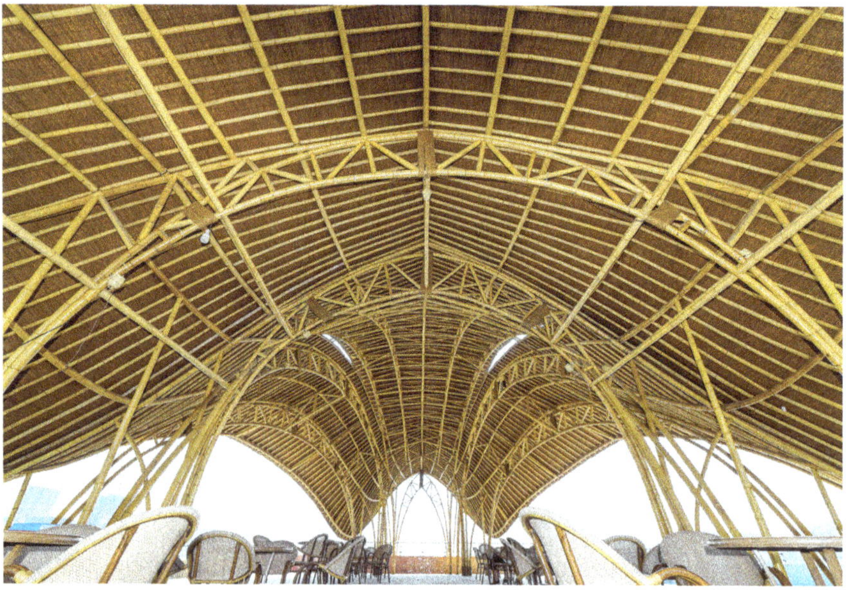

Fig. 6.97 Under the Wing of Wind

Gap of the Cloud

Location: Yibin, Sichuan, China

Completed: 2020

Architects: Huang Yun, Wang Pingshu, Ju Qian

Inspired by the visual perception of light penetrating through clouds, "Gap of the Cloud" uses multi-curved undulating roof to create the shape of cloud. The bamboo is arranged in a curl to create caves. The beauty of the multi-layer structure brings a step-by-step view when walking through the building. The holes in the roof symbolize the light passing through the dark clouds while increasing the indoor lighting, forming light spots on the ground.

Application of bamboo products: Leveraging the flexible characteristics of bamboo, waist-drum-shaped pillars with large ends and small middles are made out of multiple bamboo culms to form the main structure of "Cloud". An arched space is created in multiple directions among the interior columns to simulate a feeling of passing through a cave, and auxiliary structures are added for reinforcement. Inspired by the umbrella structure, the curved structure is pulled to increase the span of the corners, reduce the support, and prevent the change of the roof shape due to the elasticity of the bamboo itself in the face of the larger spans.

Photo source: Figs. 6.98 and 6.99 are provided by Chongqing Huangyun Landscape Design Studio and Anji Zhujing Bamboo Industry Technology Co., Ltd. respectively.

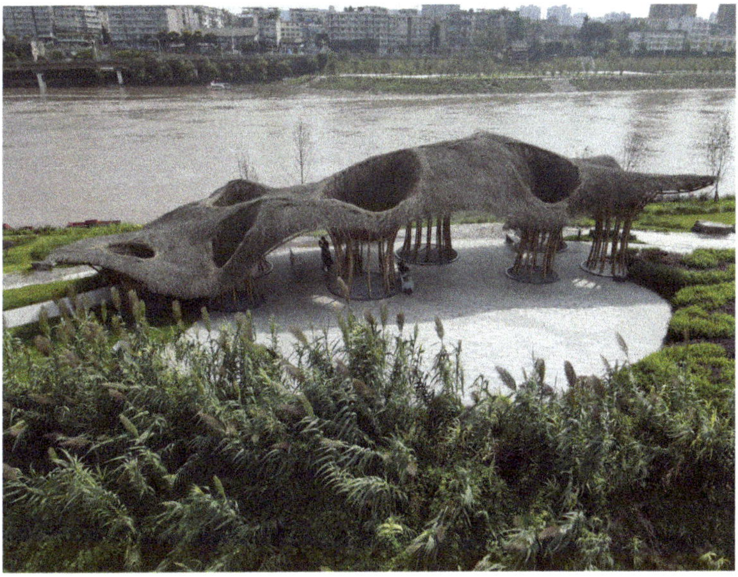

Fig. 6.98 Aerial view of the gap of the cloud

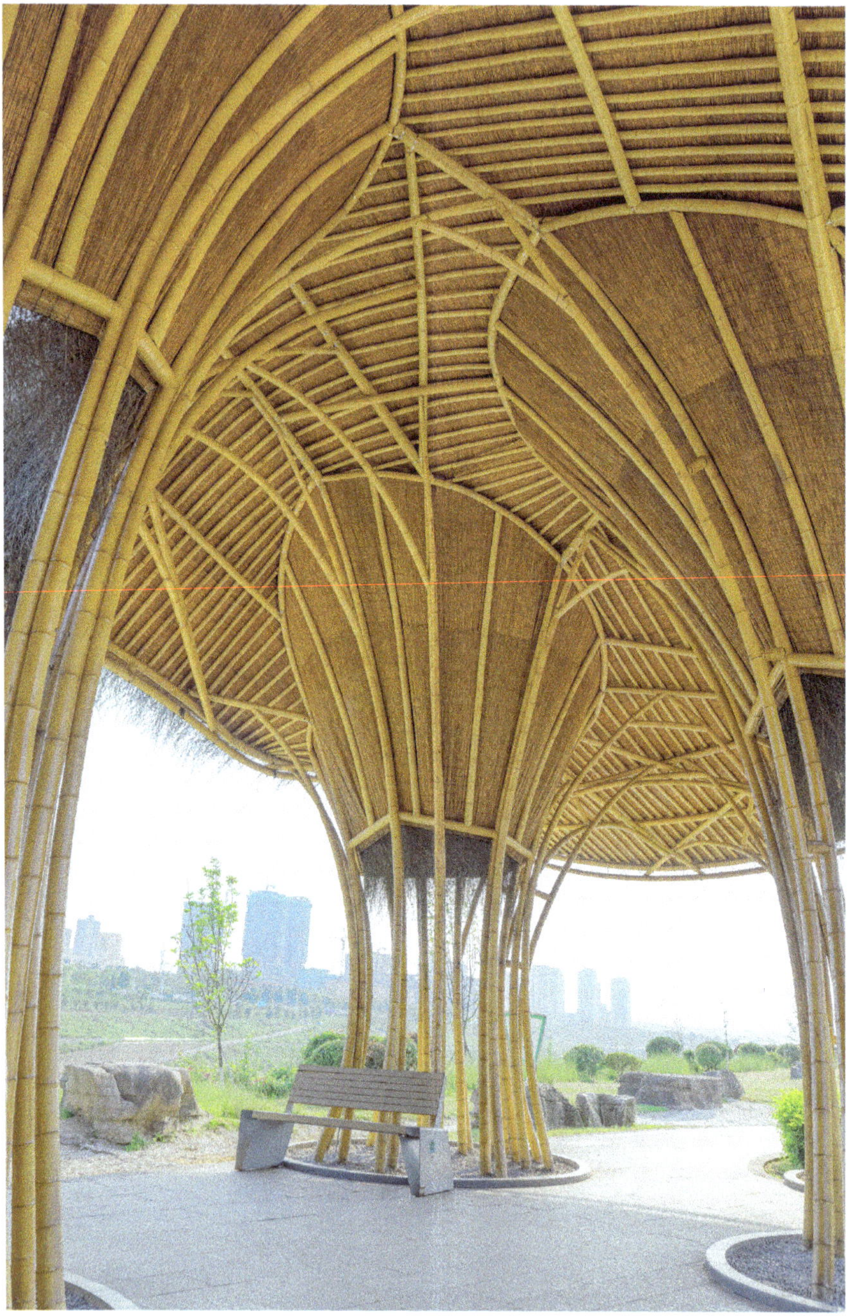

Fig. 6.99 Under the structure

Baizhishan Tourist Reception Center

Location: Tongzi, Guizhou, China

Completed: 2019

Architects: Yang Yuzhen, Qin Lin

Structural engineers: Nie Shidong, Cui Jia, Chen Mingdi

Baizhi Mountain is in Tongzi County, Guizhou Province. Entering the scenic spot from Shixi Town at the foot of the mountain, the Baizhishan Tourist Reception Center is located on a slope at an altitude of approximately 1250 m. The site is located on the east side of the climbing road with its hairpin turns and wide views. You can overlook the rolling mountains and savor the sky of clouds, sunset and stars. Scenery is the most precious resource. "Enjoying the scenery" is the fundamental starting point of this design. Instead of enclosing people in a room, it provides as much space as possible to view the scenery, so that people can communicate with nature and fellow travelers in comfort. Baizhishan Tourist Reception Center itself will also become a scene in the mountains.

Application of bamboo products: The visitor center is composed of a convex base building, a circular ramp, a stand and a bamboo structure building. The bamboo building is composed of six full-culm bamboo arches (each arch has 3 spans, which are 6, 12 and 6 m, respectively) in the middle and two spans of 12 m at the front and the back of the bamboo building. The arches are spaced 4 m apart. The main part of the base building is step-shaped, covering an area of about 500 m^2, and is covered with bamboo buildings that change with the steps.

Photo source: Figs. 6.100, 6.101, 6.102, 6.103 and 6.104 are provided by the architects.

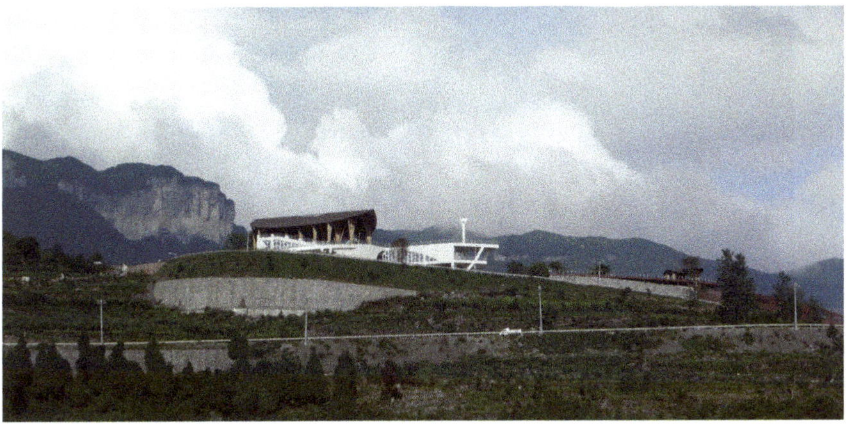

Fig. 6.100 With distant mountains (*photo by Yang Yuzhen*)

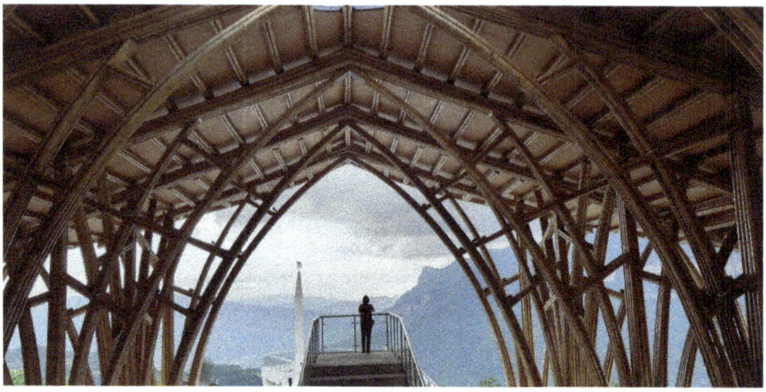

Fig. 6.101 Viewing platform and bamboo building (*photo by Yang Yuzhen*)

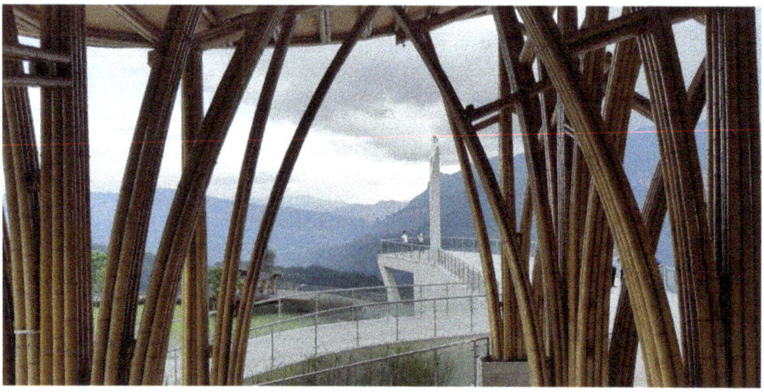

Fig. 6.102 Looking at the scenery through bamboo building (*photo by Yang Yuzhen*)

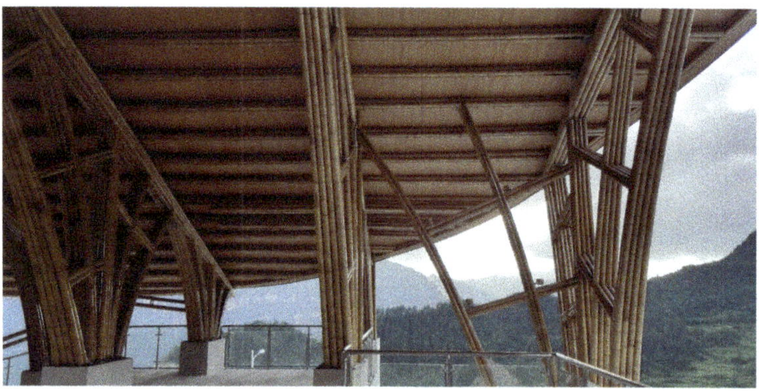

Fig. 6.103 Detail of the structure (*photo by Yang Yuzhen*)

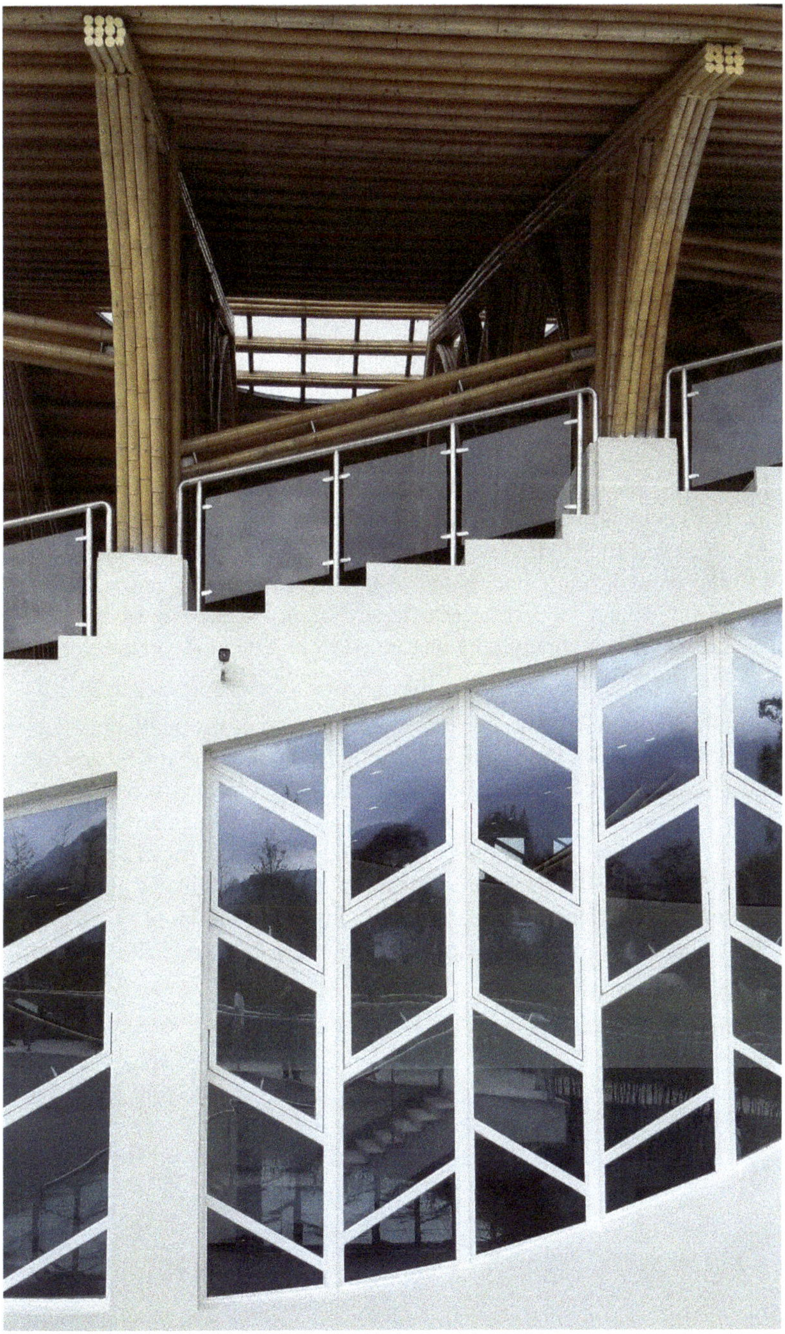

Fig. 6.104 Concrete base building supporting bamboo building (*photo by Yang Yuzhen*)

Yibin International Bamboo Products Trading Centre

Location: Yibin City, Sichuan Province, China

Completed: 2021

Architects: Guan Lang, Yang Xiaoping et al.

Structural engineers: Peng Baoning, Yin Weiqiang et al.

Yibin International Bamboo Products Trading Centre has a building area of 13,000 m^2 over two and three floors. The maximum span of the structural floor and roof is 20 m, and the maximum length of the cantilever is 16 m. The maximum span of the outdoor bamboo arches is 59 m, the arch height is 18 m and the forward inclination is 23 m. The project is the first and the largest single building in China to use a steel and full culm bamboo hybrid structure for load-bearing. The design is V-shaped in plan and extracts the concept of bamboo's node and cavity to form five ellipsoidal "bamboo cavity" pavilions. Extending from the main hall to both sides, the metallic building is simple and solemn, with light and flexible lines. When you look down from the air or when the lights are on, the building is like a foreign object in the sky, with a strong sense of future science and technology. The project perfectly integrates oriental aesthetics and modern architectural design. On the day of completion, it became a landmark building on the main road from Yibin City to the Bamboo Sea in South of Sichuan.

Application of bamboo products: The main structure of the building is steel, with large-span decorative bamboo arches. The roof trusses are hybrid steel and full culm bamboo structures. Full bamboo culms are used as the lower chord components of the roof trusses. In order to address fire resistance and to meet requirements of China's National Code for Fire Protection Design of Buildings, the project employs a fire spray cooling system. The project demonstrates fire protection design for full culm bamboo structure in China and sets a good example for similar projects. Images of fire testing conducted for this project are seen in Figs. 3.18 and 3.19.

Photo source: Figs. 6.105, 6.106 and 6.107 are provided by Xi'an Construction Engineering Green Construction Group Co., Ltd and Xi'an Architectural Design and Research Institute Co., Ltd.

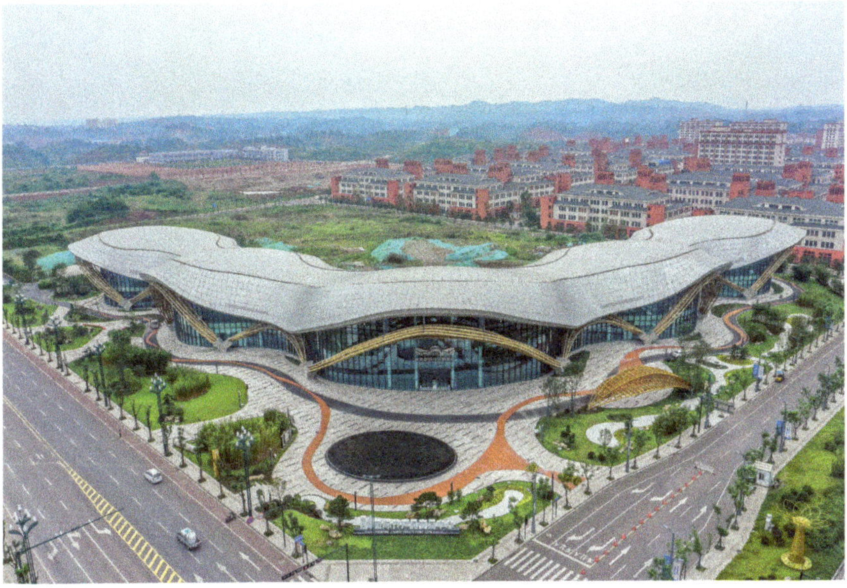

Fig. 6.105 Aerial view of International Bamboo Products Trading Centre

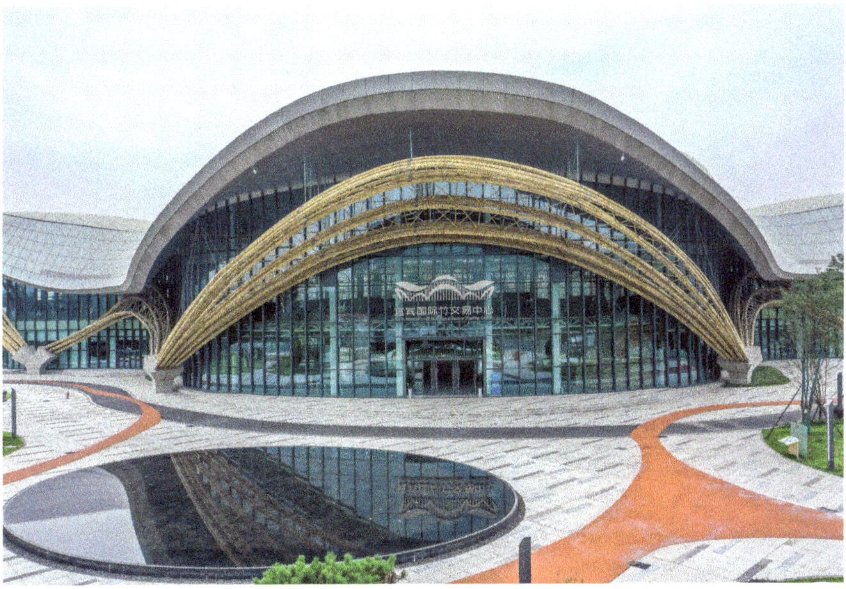

Fig. 6.106 Front façade of International Bamboo Products Trading Centre

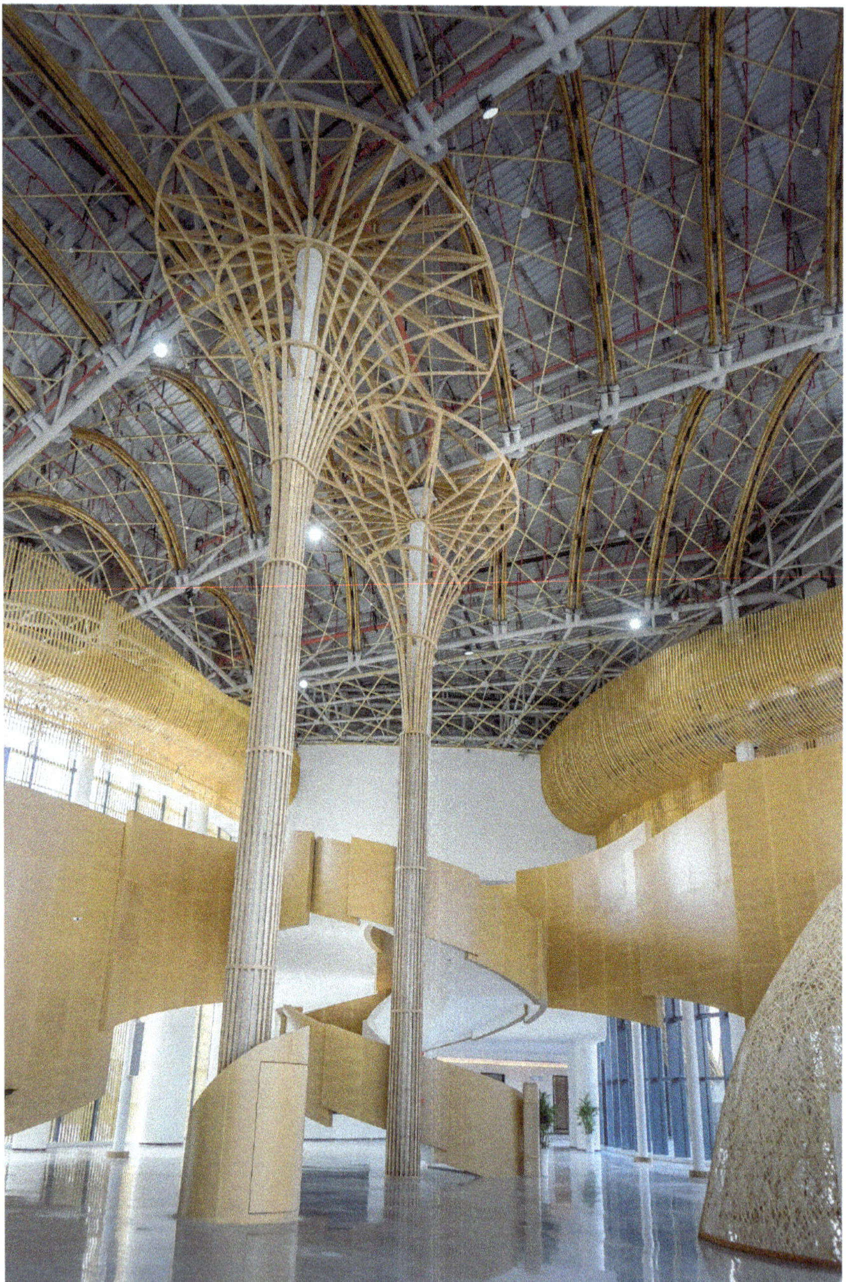

Fig. 6.107 Interior space of Yibin International Bamboo Products Trading Centre

Waterside Bamboo Pavilion of INBAR Garden at 2021 Yangzhou International Horticultural Expo (2021)

Location: Yangzhou, Jiangsu, China

Completed: 2021

Architects: Fan Xuquan, Ma Xiaoping, Zhao Jinxing et al.

The Waterside Bamboo Pavilion is built near the water, facing the rolling hills, waterfall, spring, and a six-sided bamboo structure. It is an artistic interpretation of the poem: "Unknown in the deep forest, I am accompanied by the moon". This traditional garden pavilion is built with engineered bamboo, and explores the application of modern bamboo materials in the structure, connection, production and decoration of traditional buildings.

Application of bamboo products: The pavilion takes "using engineered bamboo to replace wood to construct Chinese traditional pavilion" as a guiding principle. Glued laminated bamboo was used as both structural and decorative material in the pavilion. All elements were prefabricated in the factory and installed in 15 days on site.

Photo source: Figs. 6.108, 6.109, 6.110, 6.111 and 6.112 are provided by INBAR.

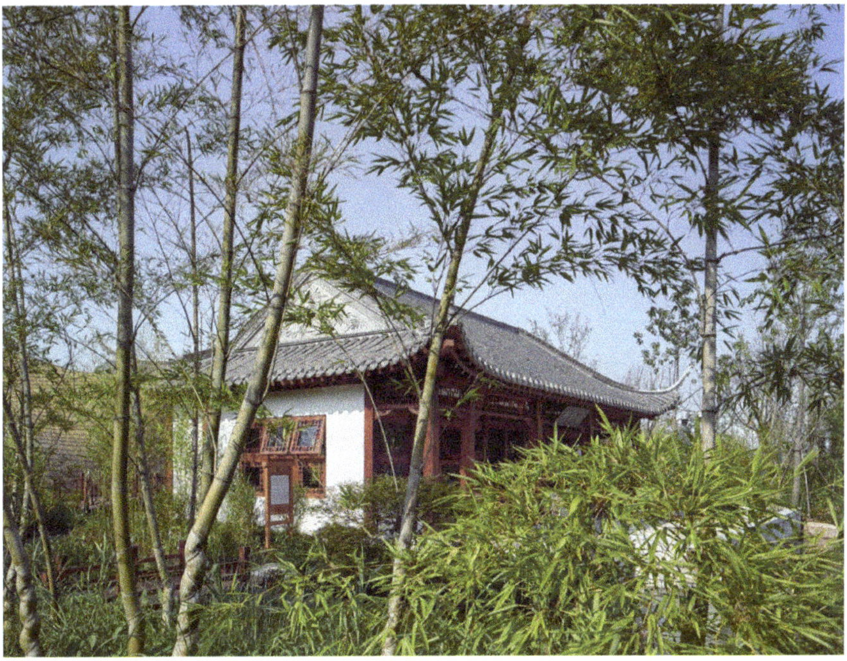

Fig. 6.108 Waterside Bamboo Pavilion surrounded by bamboo (*photo by Chu Yingnan*)

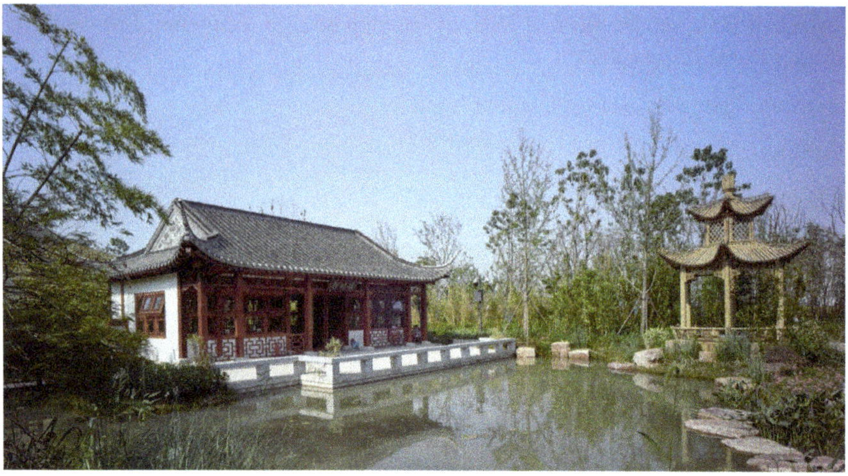

Fig. 6.109 Waterside Bamboo Pavilion with six-sided bamboo structure (*photo by Chu Yingnan*)

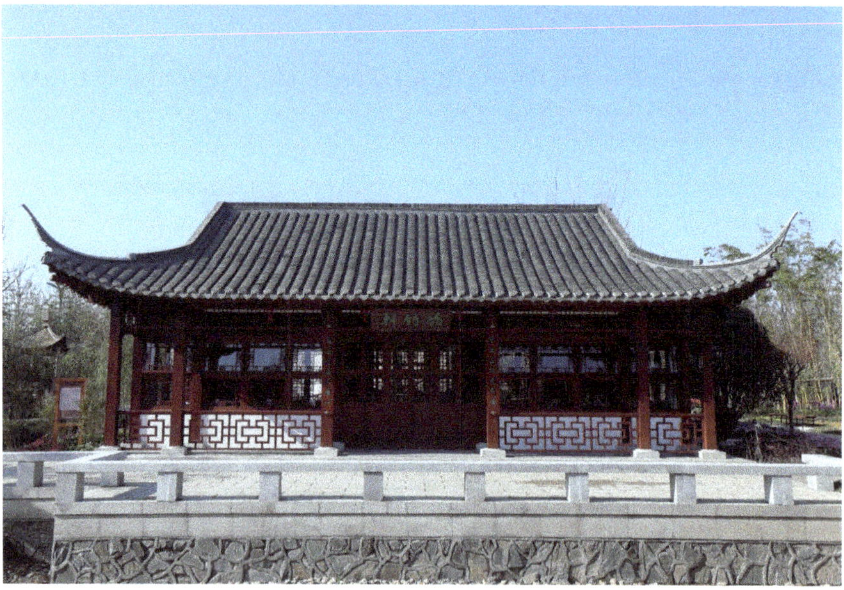

Fig. 6.110 Front façade of the Waterside Bamboo Pavilion (*photo by Wang Xudong*)

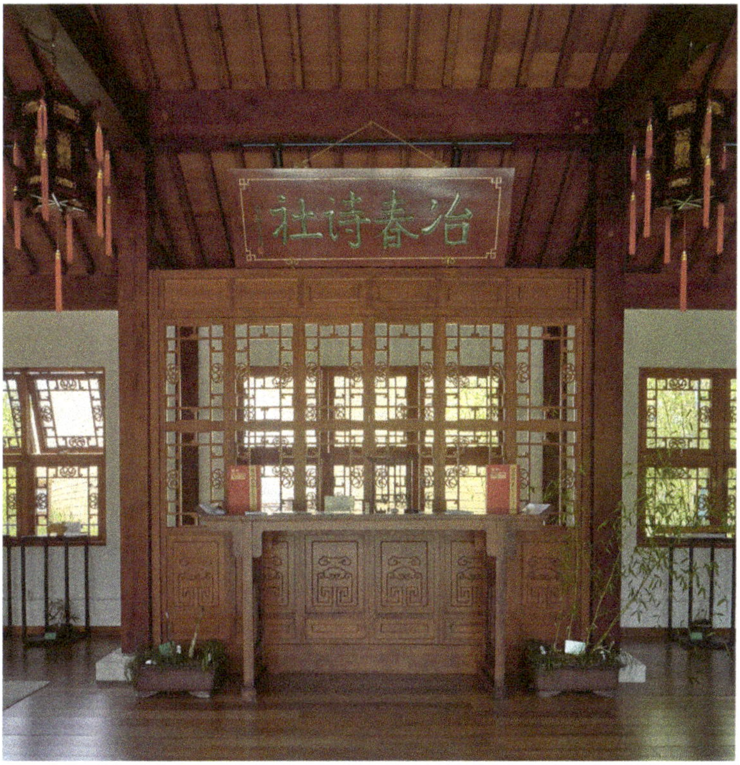

Fig. 6.111 Entrance space of Water Bamboo Pavilion (*photo by Chu Yingnan*)

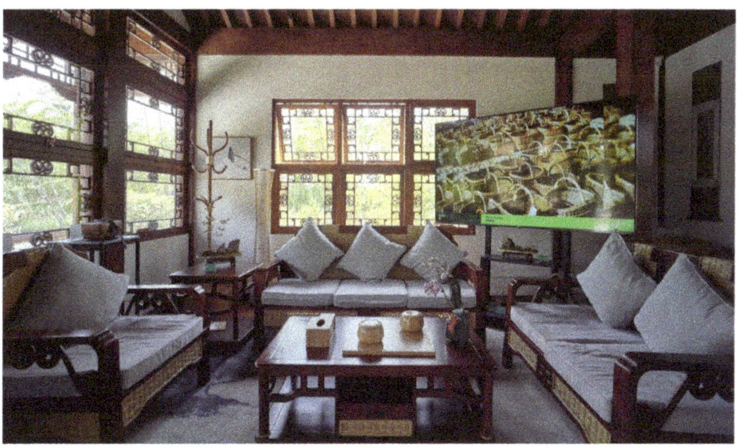

Fig. 6.112 Interior of Waterside Bamboo Pavilion (*photo by Chu Yingnan*)

Qishe Courtyard Renovation

Location: Beijing, China

Completed: 2020

Architects: Han Wenqiang

Located in the core of the old Beijing city, Qishe Courtyard is a small-sized quadrangle courtyard with three entrances. The original architecture suffered the passage of time: apart from the basic timber frame, and the partial arched door openings in the style of the Republic of China, most of its roofs, walls, doors and windows had been broken or removed. The renovation was intended to reorganize and reinforce the courtyard, preserve its history, recreate traditional buildings and integrate modern infrastructure and utilities. The architect retains the original structure while replacing the partially damaged components with pine. Bamboo scrimber is used as the new "wood" in the new veranda, doors, windows and some furniture. The veranda is supported by a frame structure with multi-ribbed beams and slabs to make it as transparent and light as possible in order to best integrate into the old environment.

Application of bamboo products: Using bamboo scrimber as both structural material and decorative material for new veranda, doors, windows and some furniture.

Photo source: Figs. 6.113, 6.114, 6.115 and 6.116 are provided by ARCHSTUDIO.

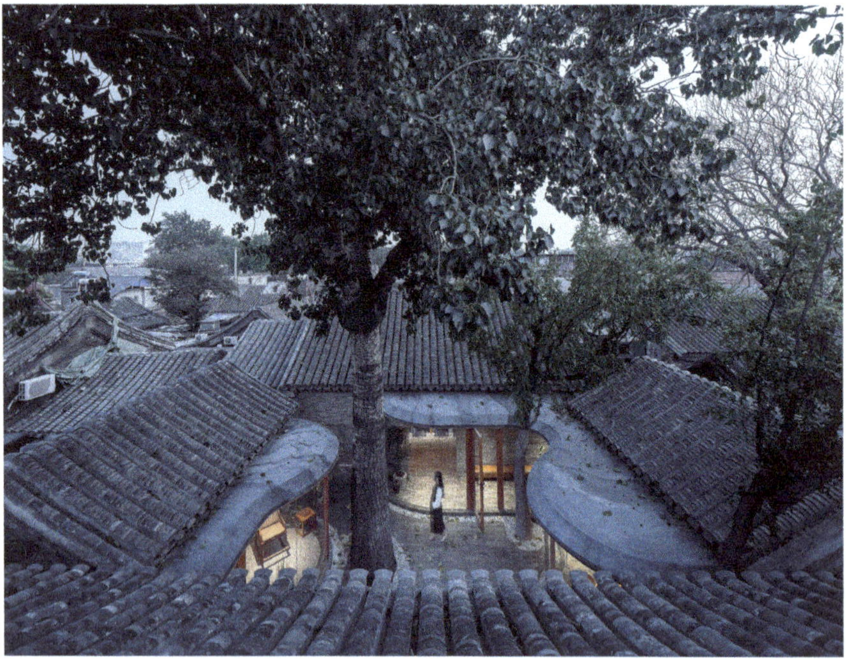

Fig. 6.113 Aerial view of Qishe Courtyard (*photo by Wu Qingshan*)

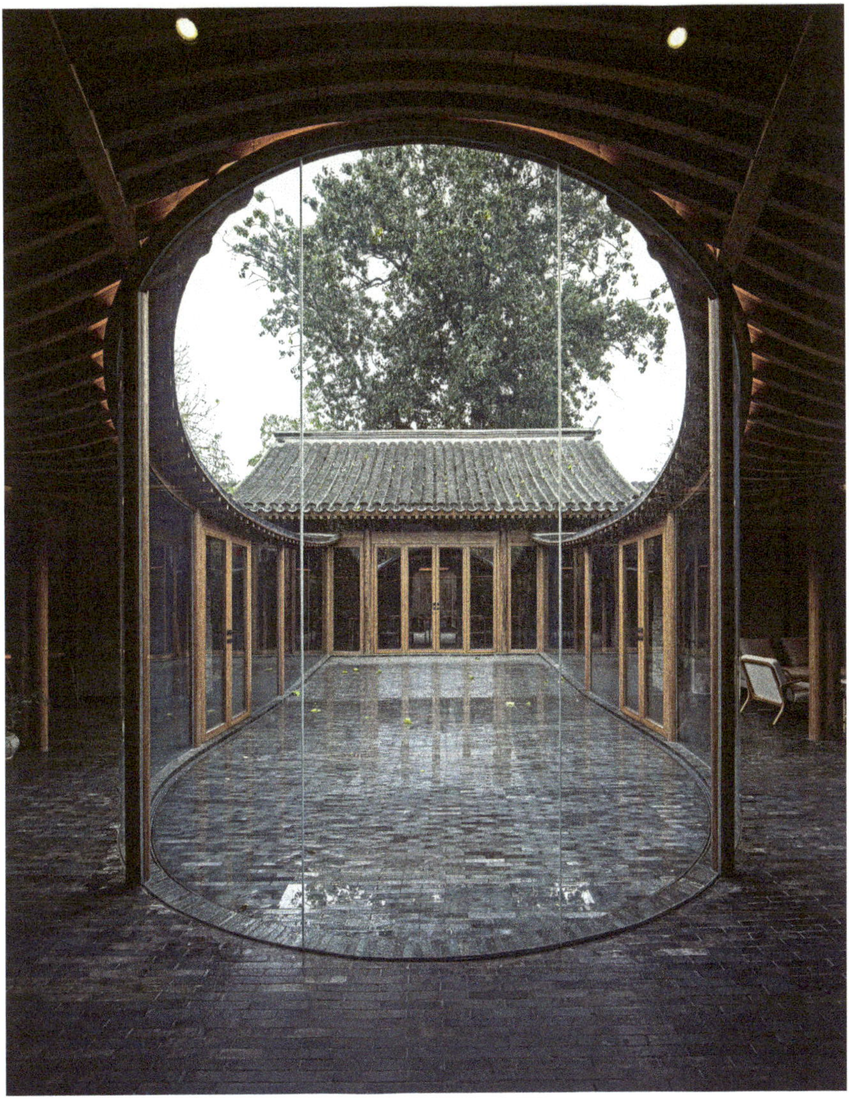

Fig. 6.114 View from entrance to courtyard (*photo by Wang Ning*)

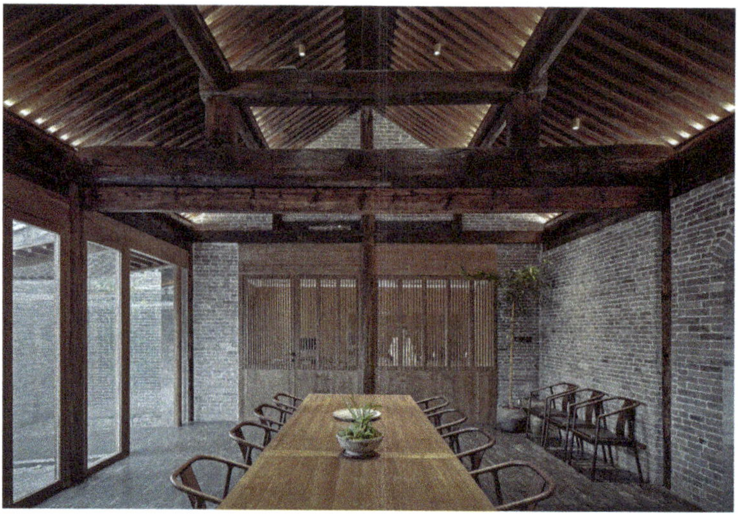

Fig. 6.115 Dining room (*photo by Wang Ning*)

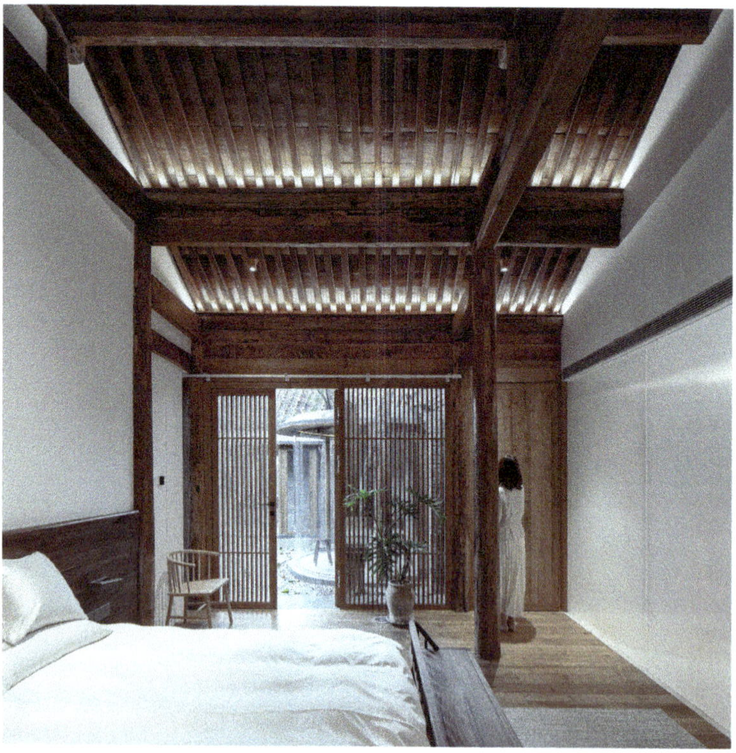

Fig. 6.116 Bedroom (*photo by Wang Ning*)

Office Building of Ganzhou Sentai Bamboo & Wood Co., Ltd.

Location: Ganzhou, Jiangxi, China

Completed: 2020

Architects: Ganzhou Sentai Bamboo & Wood Co., Ltd.

The office building of Ganzhou Sentai Bamboo & Wood Co., Ltd (Sentai) is located at Shangyou industry Zone, Ganzhou, China, covering an area of 570 m^2 (27 m × 21 m). As a manufacturer of structural glued laminated bamboo material, Sentai collaborated with the International Research Center for Bio-Composite Building Materials and Structures, Nanjing Forestry University to build the office building for itself. It takes "maximizing the use of natural materials" as a guiding principle, a 3-storey glued laminated bamboo structure was built within 3 months by using prefabricated technologies. The biggest beam has a span of 12.5 m. It is well proved that glued laminated bamboo can be widely used in architectural construction and decoration industry.

Application of bamboo products: Glued laminated bamboo was used as both structural materials and decorative materials, including varies sizes of bamboo beams, columns, rafters, wallboards and flooring.

Photo source: Figs. 6.117, 6.118 and 6.119 are provided by Ganzhou Sentai Bamboo & Wood Co., Ltd.

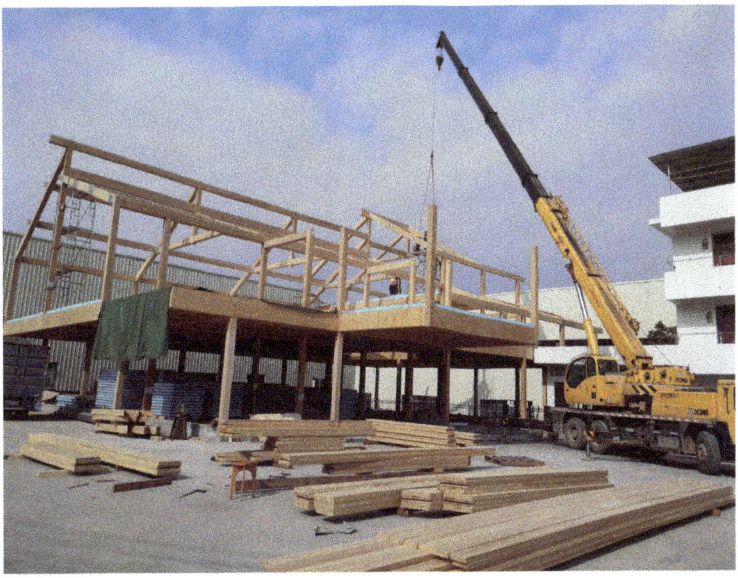

Fig. 6.117 Construction with glued laminated bamboo

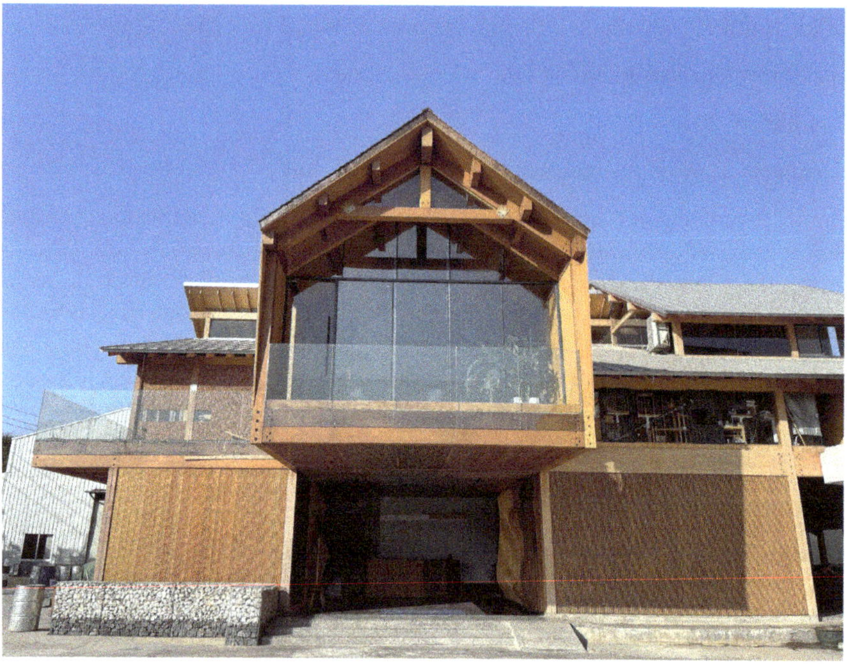

Fig. 6.118 Front façade of the office building

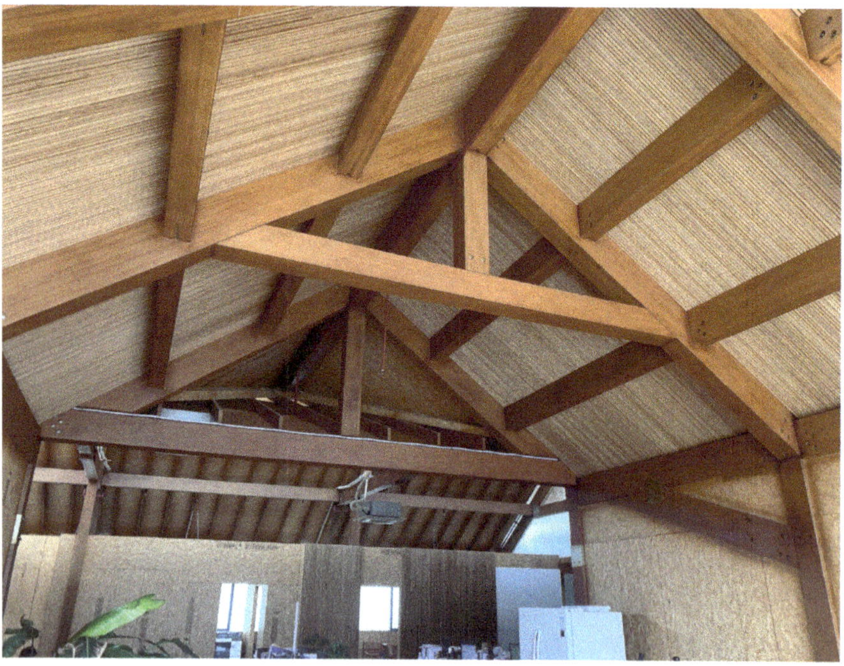

Fig. 6.119 Glued laminated bamboo trusses

Tourist Service Room for Tangshan Mine Camp

Location: Nanjing, Jiangsu Province, China

Completed: 2019

Architects: Li Zhu, Xu Xiao

The project is located in the Tangshan Mine Park, Nanjing, at the southern foot of the Tangshan Mountain. Transformed from a number of abandoned mine pits, Tangshan Mine Park is used for science, education, entertainment and parent–child activities. The site is in a small mine pit in the park, providing supporting services for tourists. Taking the natural mine as the boundary, architects make full use of the inherent spatial characteristics of the mine and reduce human activities to the smallest unit, the functional rooms, to minimize the impact of people on nature. The "roof" of the architecture uses seven umbrella-shaped bamboo structures of different sizes, ranging from 7 to 19 m in diameter. Scattered in a pattern, like a cluster of wild mushrooms, the roof protects tourists from wind and rain, while fully integrating with the environment. One facade faces tourist activities and uses glue laminated bamboo to create a relaxed atmosphere, while the facade facing the rock wall is wrapped with stainless steel plates like staggered diamond mirror to blur the boundary and weaken the sense of distance between the architecture and the rock wall. Also, the multi-angle mirror reflection conforms to the broken rock surface texture, corresponding to the calcite crystals scattered on the site.

Application of bamboo products: Bamboo umbrellas are composed of three parts: crown, stem and base. Connected by custom-made steel nodes, the crown and the stem are made of glue laminated bamboo. Stem is connected to anchored steel base.

Photo source: Figs. 6.120, 6.121 and 6.122 are provided by Architecture Technology and Art Studio, Architects & Engineers Co., Ltd of Southeast University.

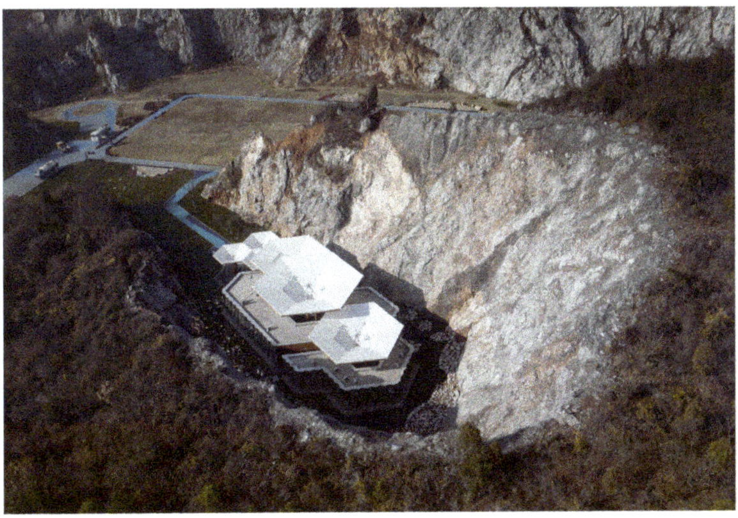

Fig. 6.120 Aerial view of tourist service room (*photo by Timeraw Studio*)

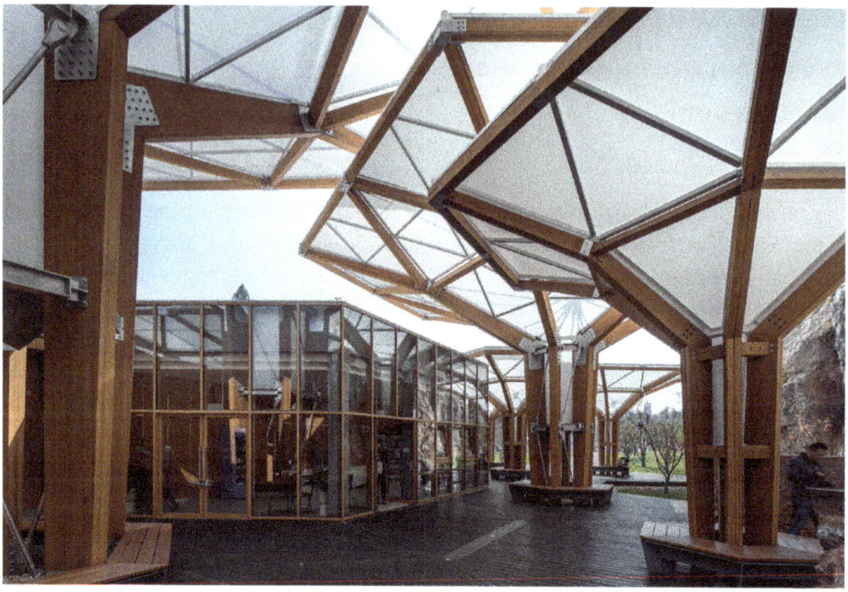

Fig. 6.121 Umbrella-shaped bamboo structures (*photo by Timeraw Studio*)

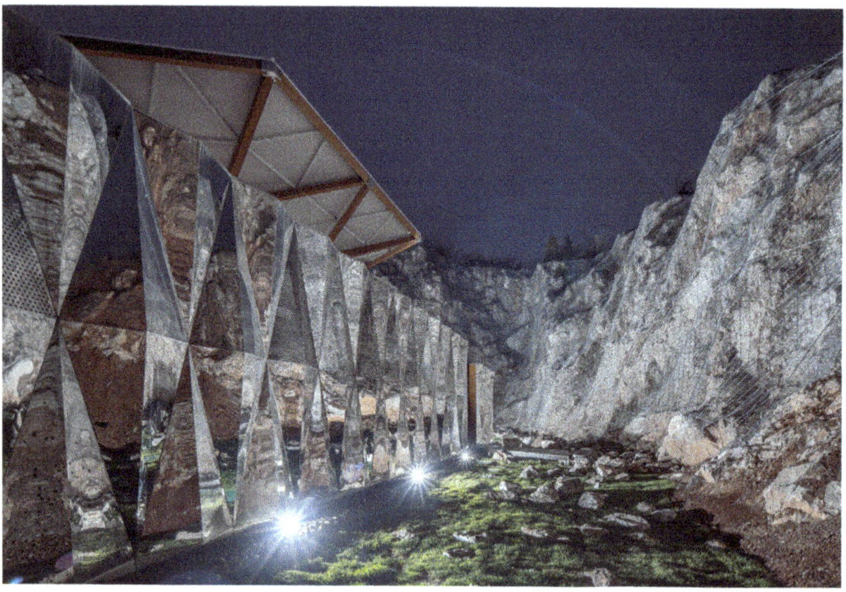

Fig. 6.122 Facade with stainless steel plates facing the rock (*photo by Timeraw Studio*)

Zhaojun Museum

Location: Huhhot, Inner Mongolia, China

Completed: 2018

Architects: Cao Xiaoxin

The original scenic spot of Zhaojun Museum took the Han Dynasty Zhaojun Green Tomb, built more than 2000 years ago, as its visual center. On both sides of the Central Avenue, there are original and kinship gardens, Xiongnu Museum, Zhaojun's hometown garden, and other buildings, such as a large tent. The newly built Zhaojun Museum is located at the southern starting point of the Avenue Axis, forming a dialogue between distance, time and space with the Zhaojun Green Tomb. According to archaeological studies, Zhaojun Green Tomb was constructed by tamping with artificial soil, and it was inferred that there were wooden structures on it. Civil engineering is not only an abstract symbol of Chinese traditional culture, but also a representation of concrete building materials. The design of the new pavilion expresses the respect of future generations for Zhaojun Green Tomb by reproducing "the image of civil engineering" and "the feeling of civil engineering".

Application of bamboo products: The design aims to translate the ancient civil construction methods in China, using concrete as the new rammed earth and bamboo scrimber as new wood. Bamboo scrimber was chose as structural material of entrance canopy. The design is simple and light. The overall stability of structure is improved by the umbrella structure as well as the positive and negative conical truss structure.

Photo source: Figs. 6.123, 6.124 and 6.125 are provided by China Architectural Design & Research Group.

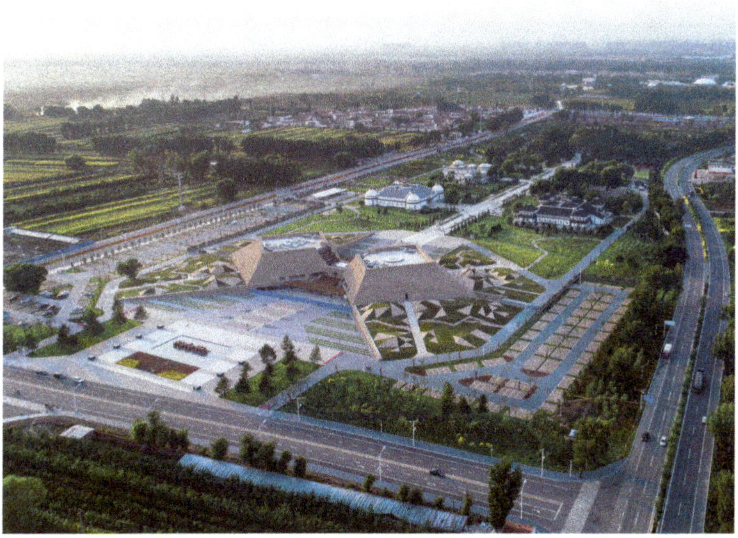

Fig. 6.123 Aerial view of Zhaojun Museum (*photo by Liang Bin*)

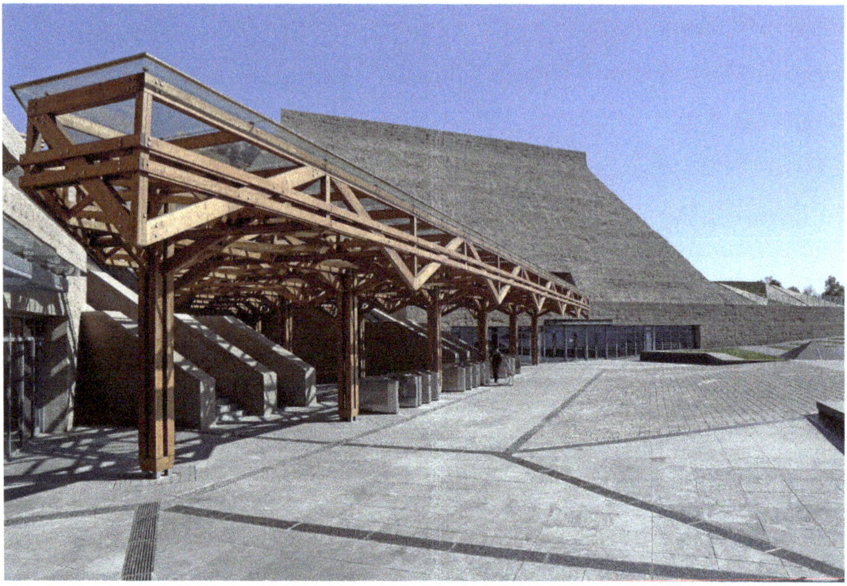

Fig. 6.124 Northwest view of entrance canopy

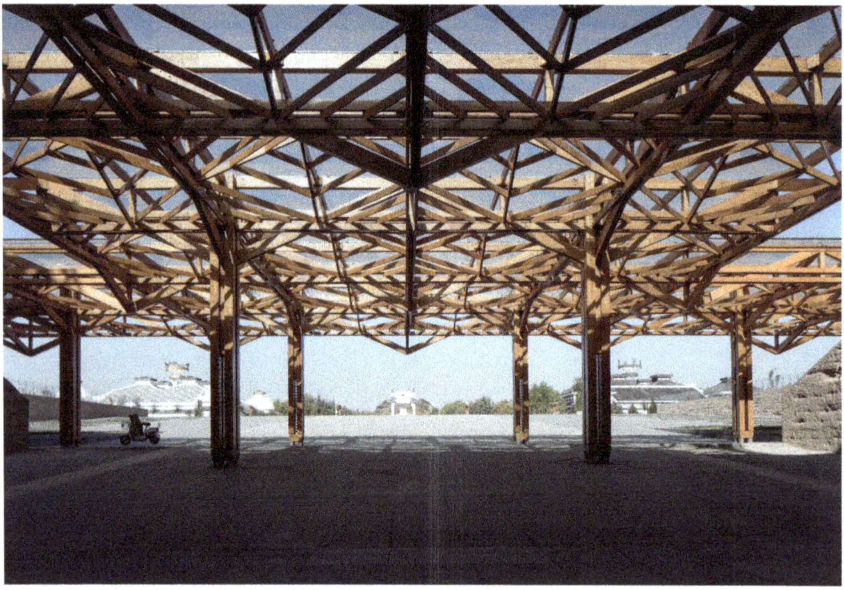

Fig. 6.125 Entrance canopy

Youth Hostel at Niubeishan Mountain

Location: Ganzi, Sichuan, China

Completed: 2014

Architects: Li Daode

Known as the most suitable place for enjoying a sea of clouds in China, there are many travellers come to Niubeishan Mountain for hiking. Because of the region's remoteness, inadequate infrastructure and underdevelopment, such travel had potential risks. Pumaidi Village is a small village near to the top of Niubeishan Mountain in which many houses have been in disrepair for a long time. The project was a public welfare renovation project and provided a public welfare practice base for young volunteers. Apart from helping travelers in need, it also provides services and assistance for the elderly and children of the village. Before renovation, the house was a traditional house in poor repair. The design team improved the basic function of the building and made it more open and public. The original wooden structure was strengthened and an open first-floor public space created. The steel lattice glass door is used to store firewood and connect to the outdoors. Kitchens, showers and bathrooms are provided. The digital design method and generation logic was used to build an organic roof which corresponds to the mountains and the sea of clouds.

Application of bamboo products: Bamboo scrimber (*D. affinis*) produced in Sichuan was used as the structural material of the organically shaped roof and viewing platform.

Photo source: Figs. 6.126, 6.127, 6.128 and 6.129 are provided by dEEP Architects.

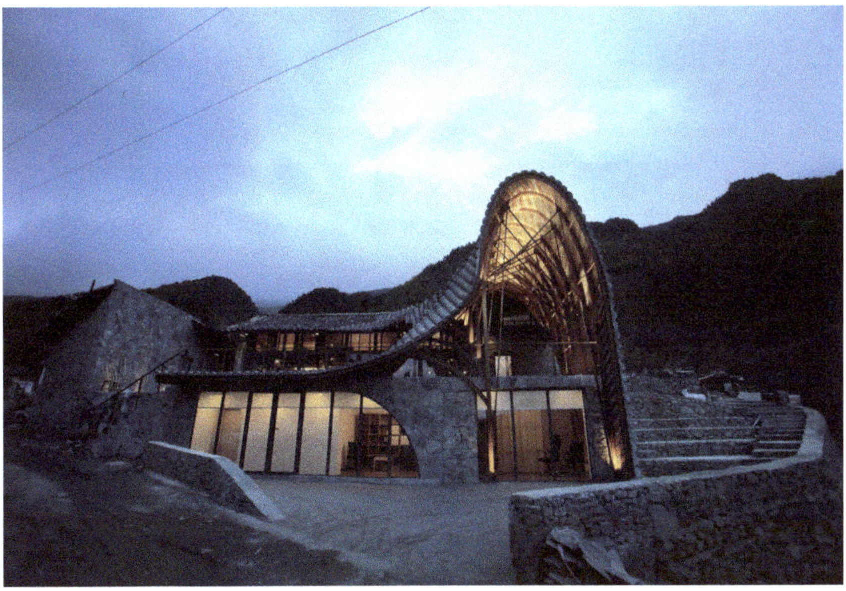

Fig. 6.126 Exterior view of Youth Hostel at Niubeishan Mountain

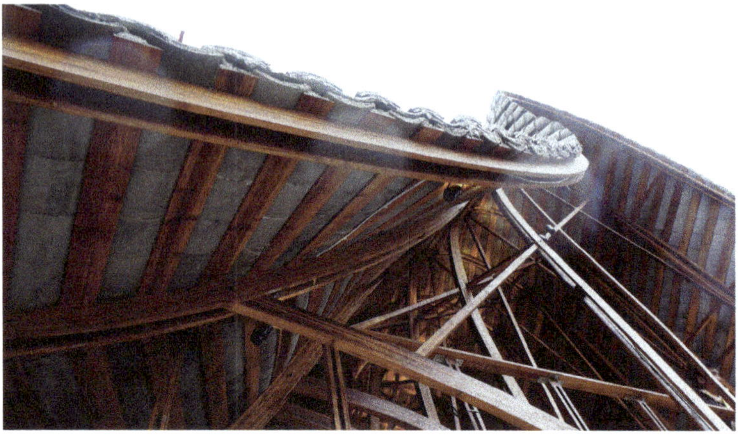

Fig. 6.127 Detail of the bamboo structure

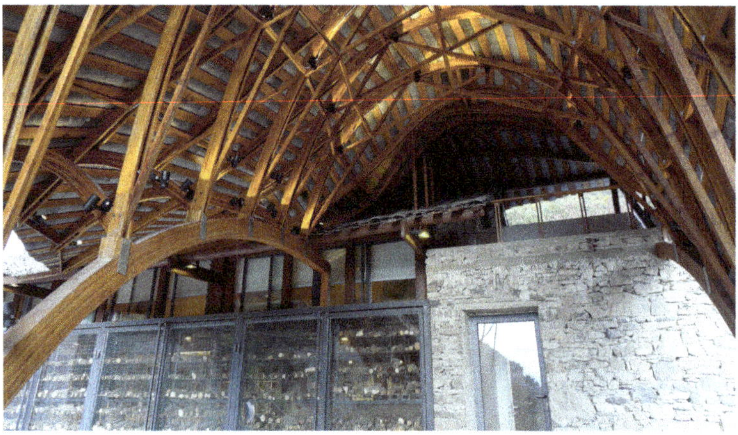

Fig. 6.128 Under the canopy

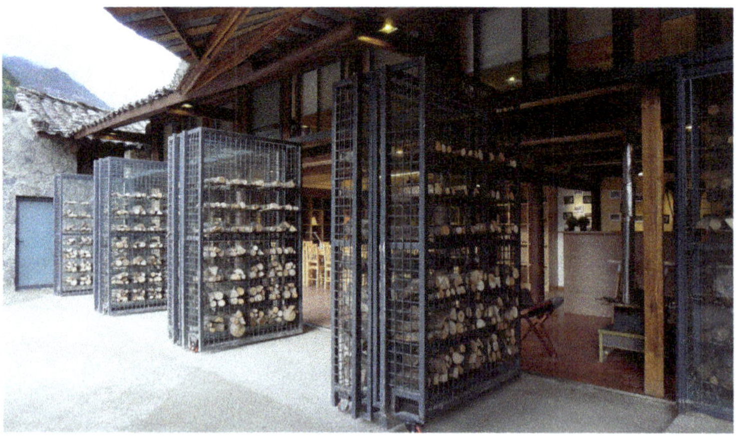

Fig. 6.129 Openable facade with firewood

Site of Chu River and Han Border World King of Chinese Chess Tournament Contest Pavilion of World King of Chinese Chess

Location: Xiangyang, He'nan, China

Completed: 2018

Architects: Zhang Wei, Xie Lei, Zhang Zhuohan et al.

Located in the relics of 2000-year-old Chu and Han dynasty battlefields, the project faces the Xingyang Gap to the north. The site has unique features, an historical background, and special significance as "the sacred place of chess circle". The aim of the project was to create a "divide for the boundary, Chu and Han rivalry" king confrontation scene of Chinese chess, and create the permanent site of Chu River and Han Border World King of Chinese Chess Tournament and contest pavilion of World King of Chinese Chess.

Application of bamboo products: Starting from loess as the base, glued laminated bamboo with similar colour and texture is selected for the contest pavilion, to meet the requirements of building appearance, structural safety, and durability in the harsh outdoor environment. All the columns are inclined at 60° facing the wide gap, and the horizontal crossbeams are fixed by bolts. More than 70 columns and beams are used in the whole building. The column and beam elements are 15 cm × 30 cm in section and as long as 15 m; a single column may weigh up to 400 kg. The dry construction method of factory prefabrication and field installation is adopted, with the field assembly taking only 10 days.

Photo source: Figs. 6.130, 6.131 and 6.132 are provided by Zhejiang Green Architecture Design Corporation, Ltd. (gad).

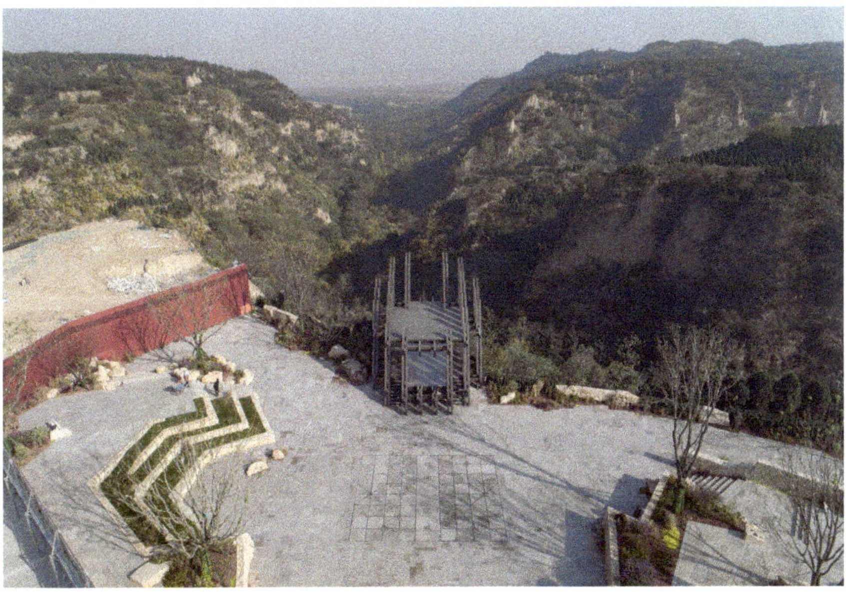

Fig. 6.130 Aerial view from the central axis

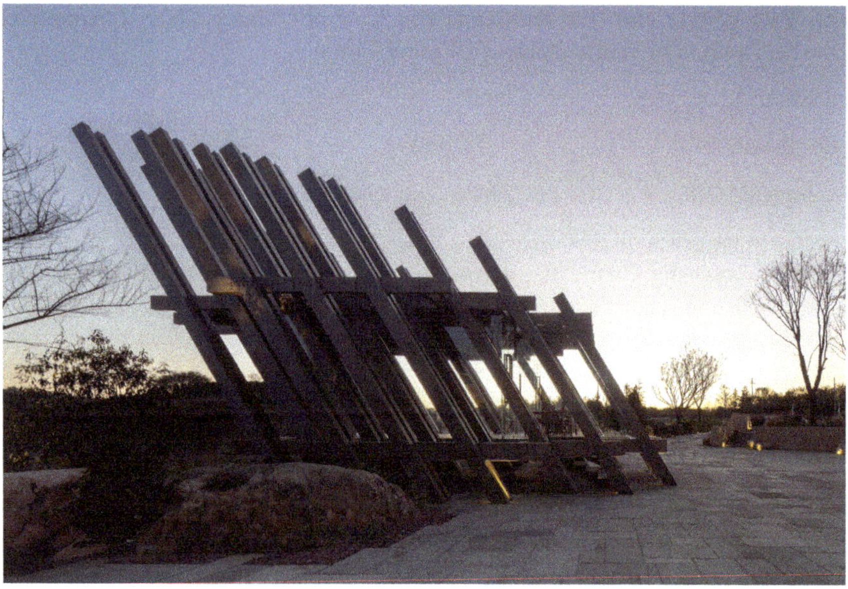

Fig. 6.131 Elevation

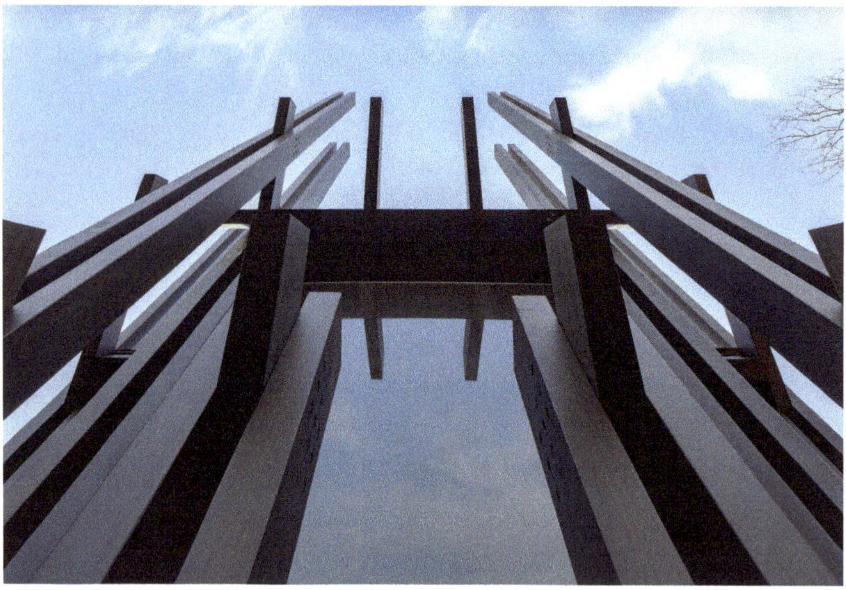

Fig. 6.132 Detail of structure

Zen Centre at Baoguo Temple in Lezhi

Location: Lezhi, Sichuan, China

Completed: 2016

Architects: Liang Jingyu, Ye Siyu

The Zen Center of Baoguo Temple in Lezhi is located in a county town surrounded by mountains in Sichuan Province. The history of the site can be traced back to the Sui Dynasty (581–618 A.D.). In order to meet the increasing demands of visitors, a new meditation centre was built. The meditation center is a two-storey building with its facade hidden among dense woods. Bamboo and glass echo the surrounding natural environment. The whole building fits well with the site.

Application of bamboo products: Because the site is located in the mountains and forests, no heavy construction machinery was used; only human transport and small construction equipment was used. Architects desired light, easily spliced poles to form a space grid structure to construct the large-span roof of the meditation center. Bamboo scrimber, having good strength and durability, was chosen. Each element was only 2 cm × 4 cm in section and connected with metal joints. The individual elements were made smaller than was possible with timber.

Photo source: Figs. 6.133, 6.134 and 6.135 are provided by Approach Architecture Studio.

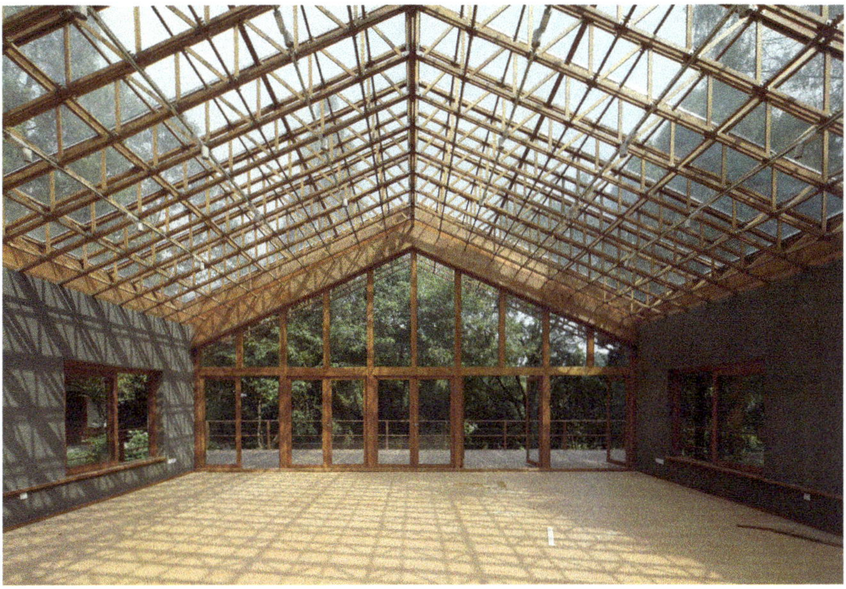

Fig. 6.133 Interior of the Zen Center

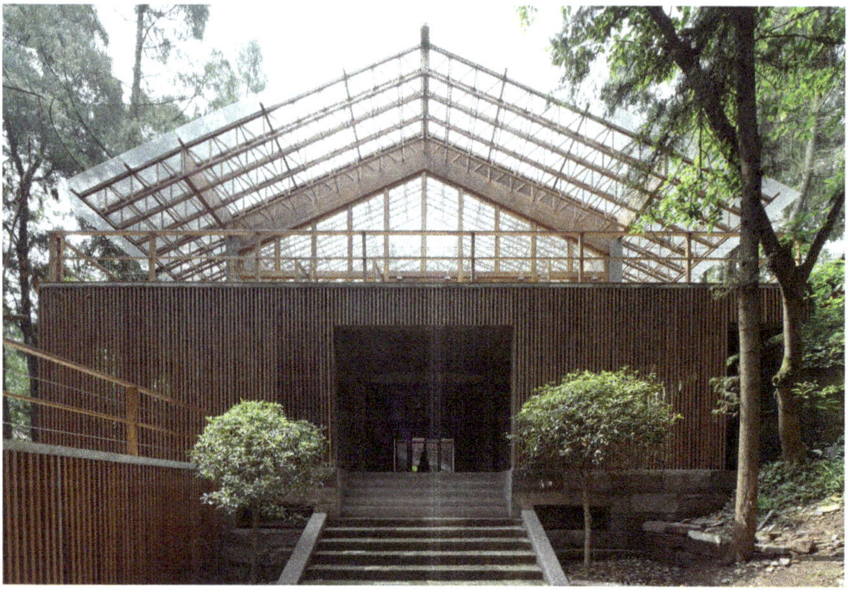

Fig. 6.134 Main entrance of the Zen Centre

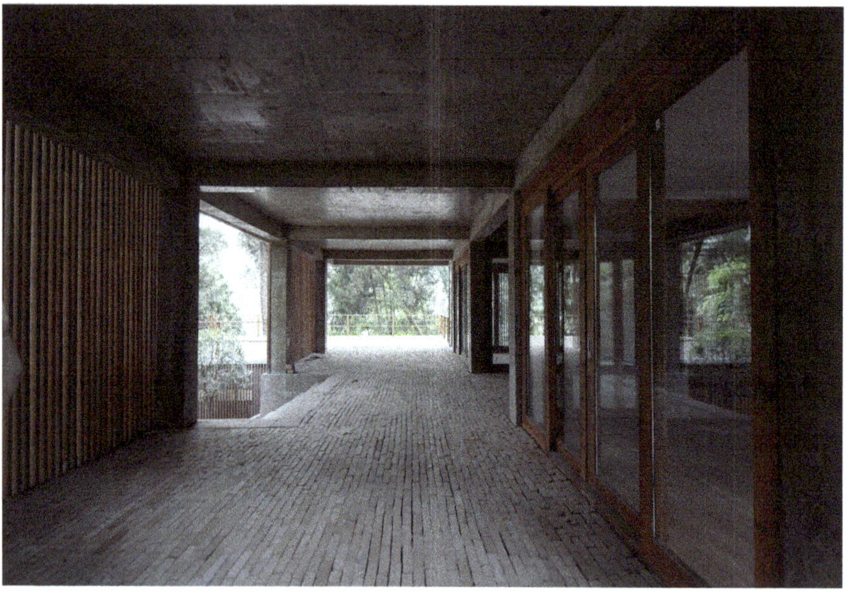

Fig. 6.135 Gray space of the first floor

Sanhe Canteen

Location: Sanhe, Hebei, China

Completed: 2016

Architect: Wang Gang

Because of its use for wedding banquets and large-scale activities, Sanhe Canteen, a 2000 m^2 building, required as much open space as possible. Rather than a conventional frame or truss structure, a tree crown structure system was created that achieved a 17.8 m span. This structural system was naturally adaptable to bamboo scrimber, resulting in strong, durable, lightweight and flexible rural architecture as the project required. Outside the building, an eave up to 6 m deep provides a great expansion of the interior space. Combining with a large square on the south side of the main building, a rural market square and the community centre are formed. Another creative approach is that the designer arranges a large stove and brick chimney under the eaves of the main facade of the south building, which not only improves the overall stability of the structure, but also makes the smoke and aroma an integral part of the overall form of the building.

Application of bamboo products: Bamboo scrimber as the main structural material.

Photo source: Figs. 6.136 and 6.137 are provided by Master Studio of Beijing University of Engineering and Architecture.

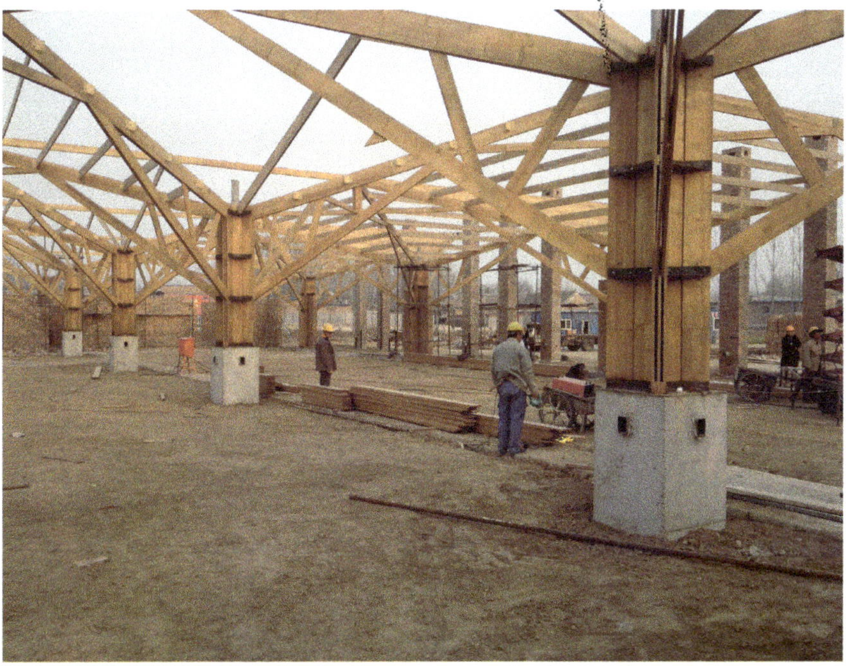

Fig. 6.136 Tree crown structure system

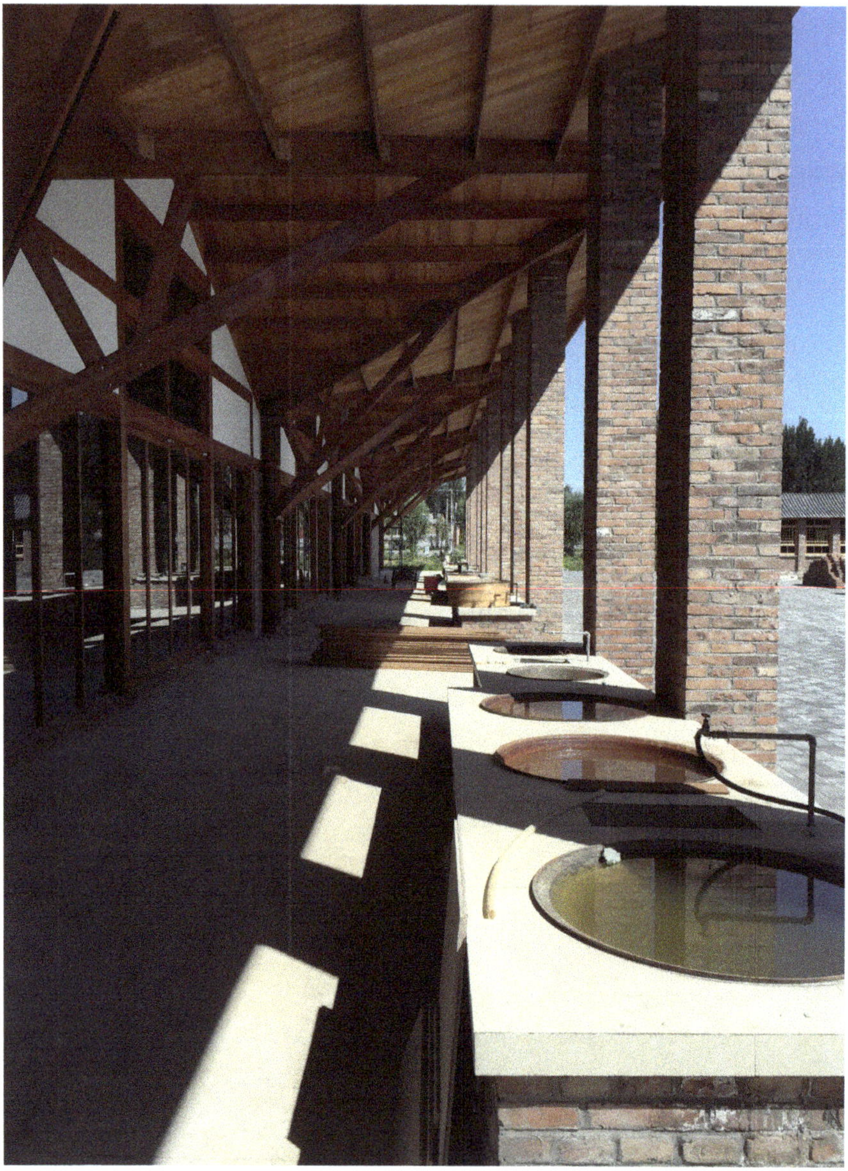

Fig. 6.137 Grand colonnade

Villas in Saudi Arabia

Location: Jeddah, Saudi Arabia

Completed: 2016

Architect: Ganzhou Sentai Bamboo & Wood Co., Ltd.

The building is the first prefabricated bamboo house in the Middle East, having a construction area of 223 m^2. All components were designed and manufactured by the factory, and construction was guided by the factory. The house can be delivered and assembled in 30 days. 109 m^3 of bamboo structural materials are used in the building, including 763 m^2 of bamboo wallboard and 112 m^2 of bamboo windows.

Application of bamboo products: Prefabricated bamboo structure and customised bamboo products (doors, windows, wallboard, indoor and outdoor flooring, etc.).

Photo source: Figs. 6.138 and 6.139 are provided by Ganzhou Sentai Bamboo & Wood Co., Ltd.

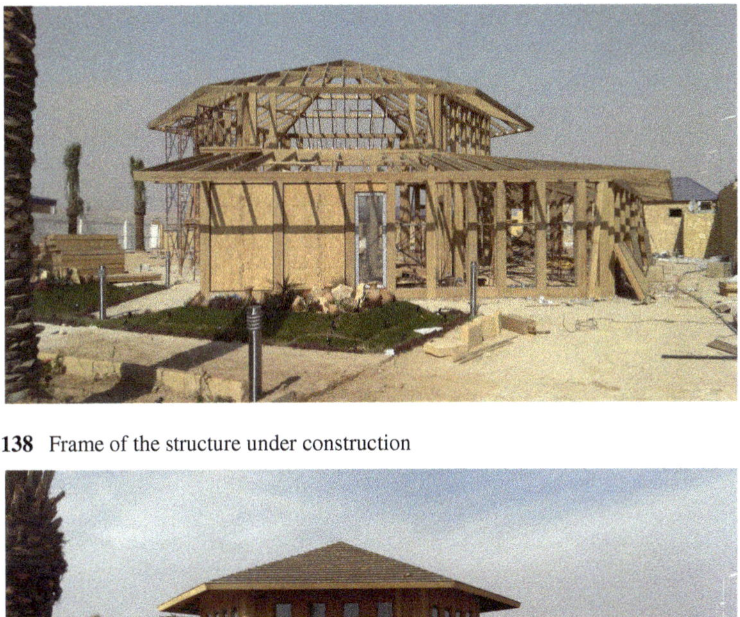

Fig. 6.138 Frame of the structure under construction

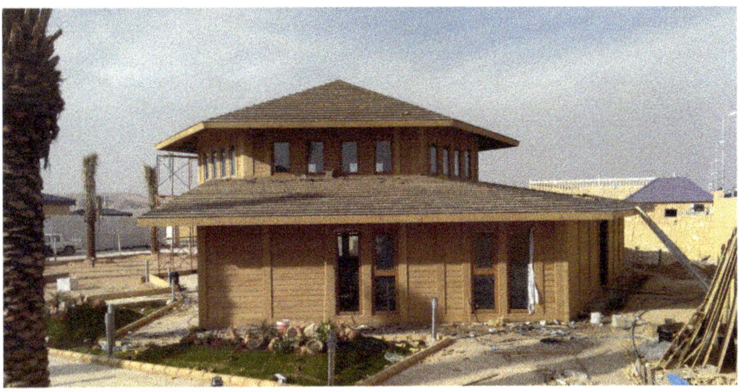

Fig. 6.139 Appearance of the villas near completion

Caishan Urban Vegetable Bar for Parents and Children

Location: Changsha, Hunan, China

Completed: 2015

Architect: Zhu Hongbo

The project attempts to build a living style for the future, an innovative urban green ecological community, a pleasant urban experience farm bar and a comfortable urban neighbourhood centre. Using the bending characteristics of bamboo scrimber products, combined with the surrounding overall environment, the design forms the building into a mountain shape. The internal space includes assembly, education and catering areas. Vegetables and plants are planted inside and outside of the building. The green vines spreading on the top integrate the building with the mountain and forest. People living in the city can experience the pleasure of farming civilization here.

Application of bamboo products: The main body of the building is constructed with bamboo scrimber. Bamboo is both structural and decorative materials, avoiding secondary decoration. Bamboo scrimber is used for customised design and production of doors and windows, shelves and furniture.

Photo source: Figs. 6.140, 6.141 and 6.142 are provided by Planting Design.

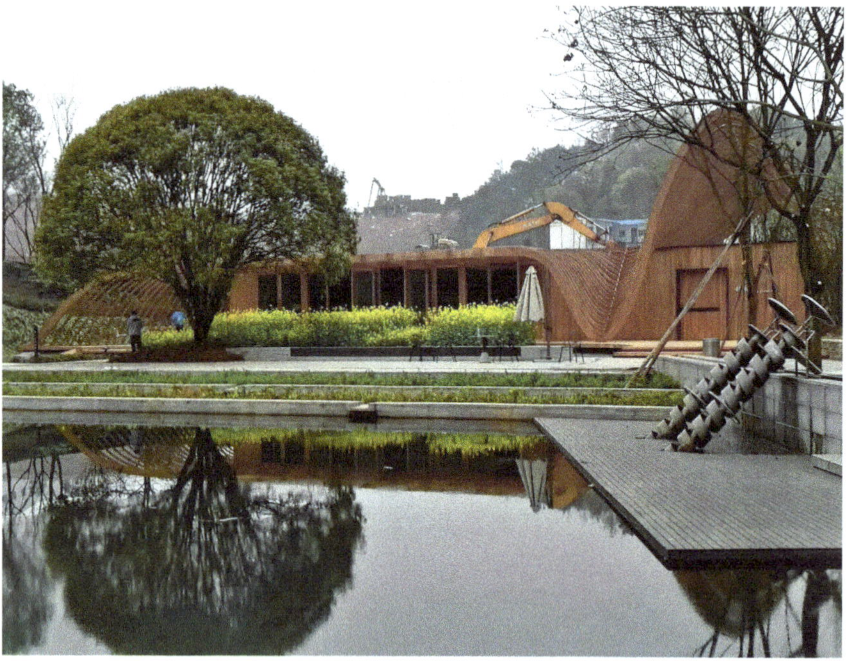

Fig. 6.140 Surroundings of Caishan Urban Vegetable Bar

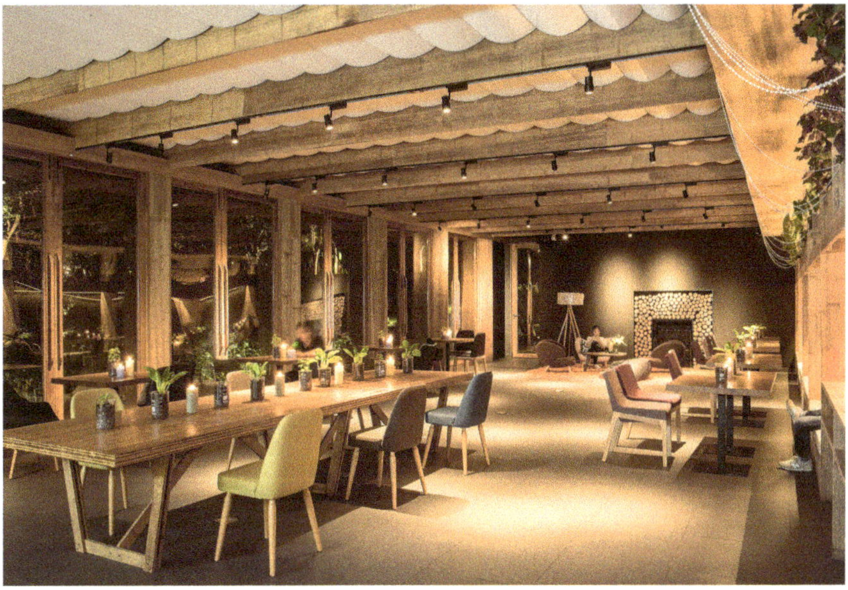

Fig. 6.141 Interior space of Caishan Urban Vegetable Bar

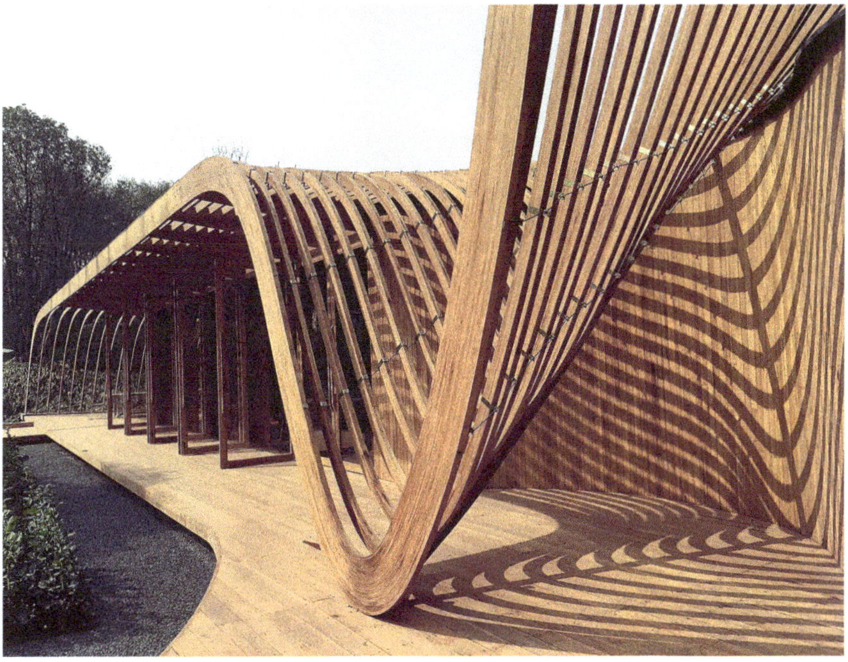

Fig. 6.142 Bending bamboo formed a mountain shape

Beijing Youth Camp International

Location: Beijing, China

Completed: 2013

Architect: Wang Gang

The site of Beijing Youth Camp International was originally a barren hill, located at the south foot of Beijing Miyun Reservoir. After being selected as the outdoor sports training base of the Youth League Central Committee, architects began to think about how to express the vigorous spirit of young people and the tenacity of outdoor sports through a house. The architect chose a fault in the mountain to expose the foundation of the building to the outside of the section. A pitched roof house became the first course for all the participants in the camp: training the basic movements of climbing the inclined roof. As a result, the house became part of the mountain. The 100 m^2 house with huge eaves, and the connected indoor and outdoor space also has many functions: reception centre, classroom, dining room, conference room, bar, campsite and equipment room.

Application of bamboo products: Bamboo scrimber as the main structural material.

Photo source: Figs. 6.143 and 6.144 are provided by Master Studio of Beijing University of Engineering and Architecture.

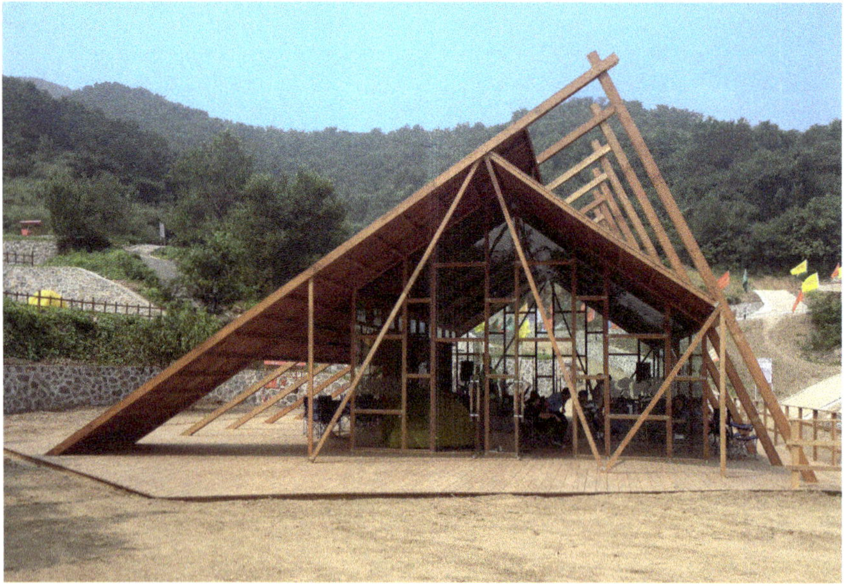

Fig. 6.143 West elevation of Beijing Youth Camp International

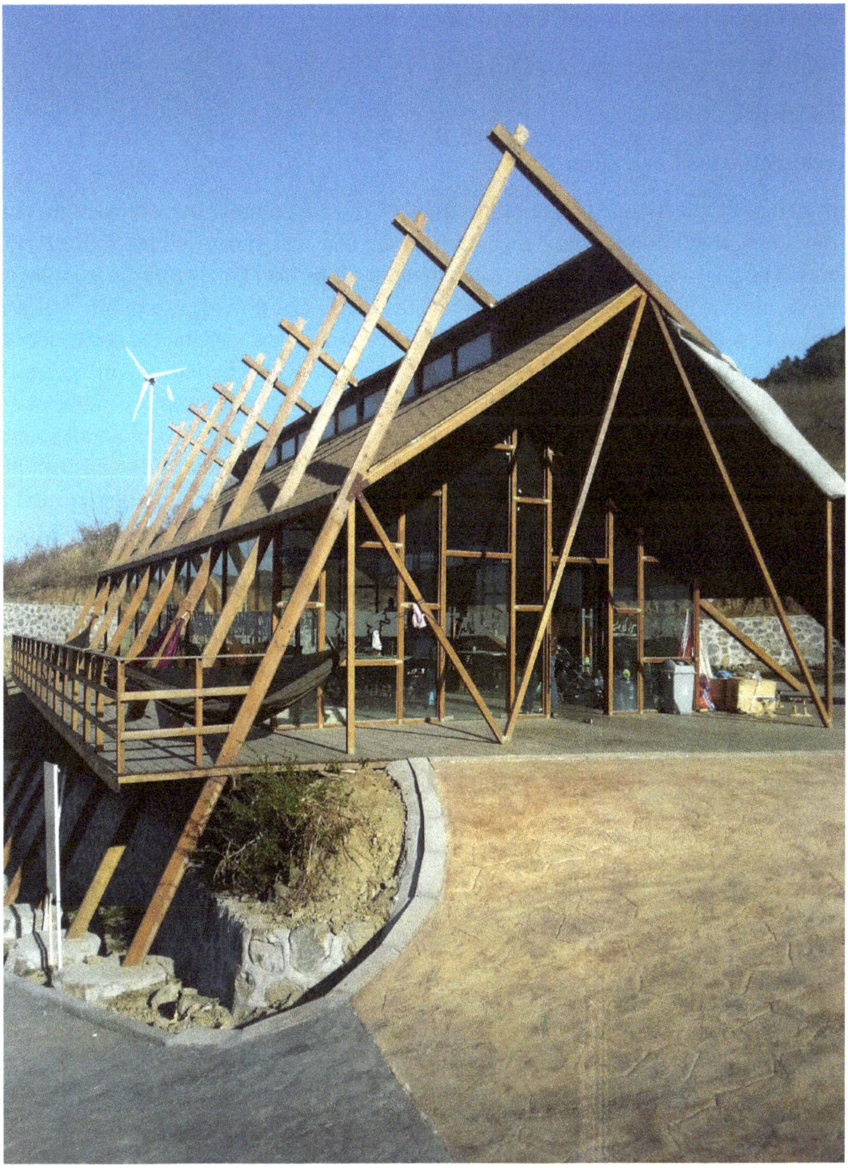

Fig. 6.144 Main entrance

Swire Properties Bamboo Community Centre

Location: Pengzhou, Sichuan, China

Completed: 2017

Architects: Hao Lin, Peaker Chu

Located in Shuanghong Village of Pengzhou City, the community centre originated from reconstruction after the Ya'an earthquake and was supported by Swire Properties. The project, collaborating with Chengdu Show You Non-profit Development Centre, aims at serving children, women and elderly and providing a public welfare platform for the community. The centre also promotes local sustainable construction and community development in the rural area. The project is built over a hilly terrain and is composed of three buildings. The three buildings form community public open spaces with a strong sense of place. Supporting the neighbourhood in an all-round way, the centre is equipped as community library, training classroom and activity centre to serve local people and for them to learn a wide range of skills through a variety of training programs. The main structure of the three buildings imitates a tree, which works reflects local climate and lifestyle. The structure forms three semi-open rooftop spaces. The overall expression has been integrated into the rural setting and surrounding environment with a unique character.

Application of bamboo products: Based on abundant bamboo resources in Sichuan Province, the community centre adopts bamboo scrimber sourced from local *D. affinis* to achieve environmental objectives and respect local conditions.

Photo source: Figs. 6.145, 6.146 and 6.147 are provided by Swire Properties.

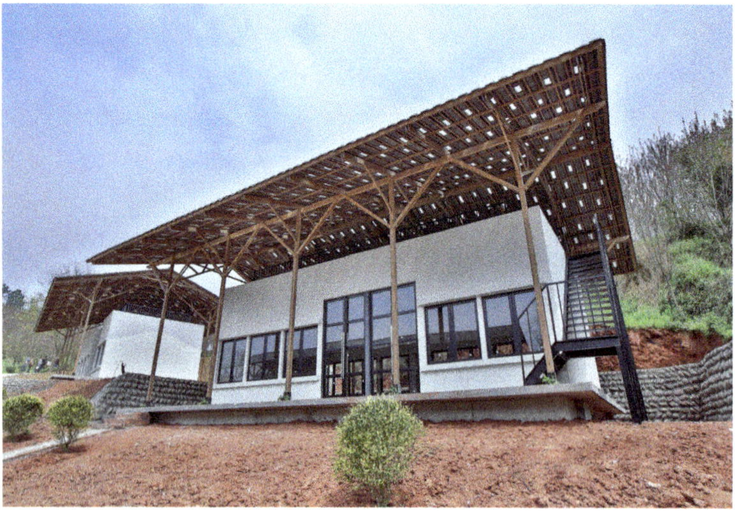

Fig. 6.145 Community Library, Training Classroom and Activity Centre

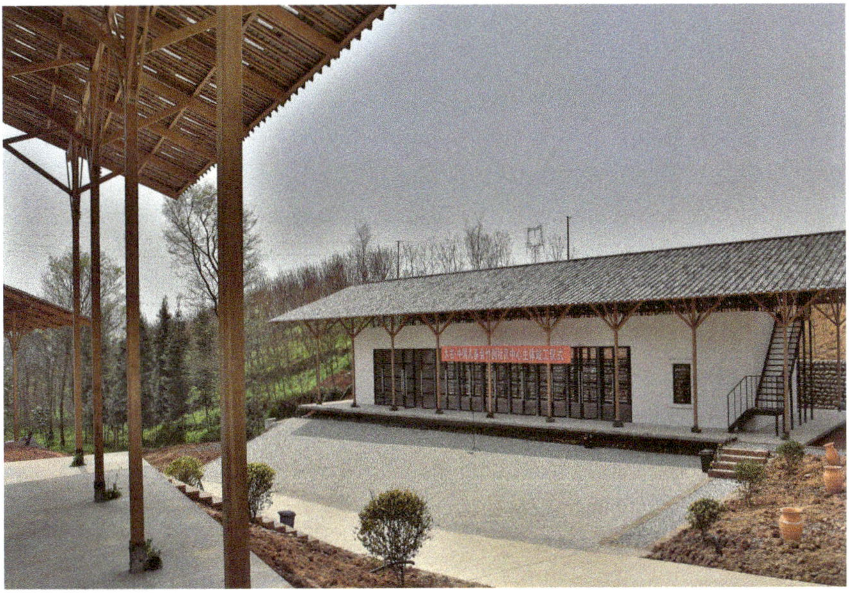

Fig. 6.146 Square of Community Centre

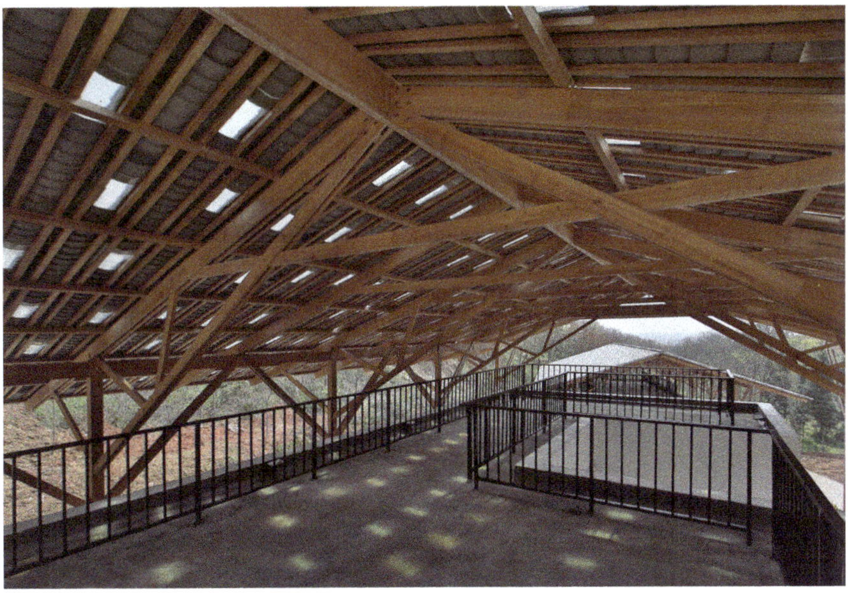

Fig. 6.147 Bamboo structural roof and semi-open rooftop space

No. 37 Luanqing Hutong—"Quad of Gourd"

Location: Beijing, China

Completed: 2017

Architects: Meng Yan, Li Jing, Li Yongcai et al.

The project is one of seven conservation and restoration courtyards located between East Damochang Street and Luanqing Hutong. The original No. 37 Courtyard of Luanqing Hutong is a three-section compound courtyard, which was in disrepair and seriously damaged for a long time. The roof collapsed in the middle of the South House, and the two Chinese toon trees reaching to the sky in the interior became the entry point of the protection and restoration design. The two bays in the middle of South House were split into a "middle courtyard" to hold trees, while the roofs on both sides were restored according to the ancient method. The two sides of the chamber were appropriately enlarged to the courtyard, forming a reed-shaped courtyard in the middle.

Application of bamboo products: The original two sides of the grey brick gable were filled with a bamboo scrimber courtyard wall in the middle, and the small three-section compound courtyard formed a "gourd courtyard" pattern. In the middle, there is a transparent vertical bamboo grille sliding door, which passes through the narrow curved doorway and enters an open circular courtyard. The grey tile roof and the warm bamboo scrimber curving wall around it delineate a quiet world. The roofs of the two bays in the middle of the south house are overlapped with the original structural size of the bamboo scrimber beams, and the rafters are combined with the cross-shaped bamboo scrimber components. On the premise of retaining the texture of the city, the quadrangle courtyard with bamboo scrimber has been implanted with new functions, which glows its second new life.

Photo source: Figs. 6.148, 6.149 and 6.150 are provided by Urbanus Architecture & Design Inc.

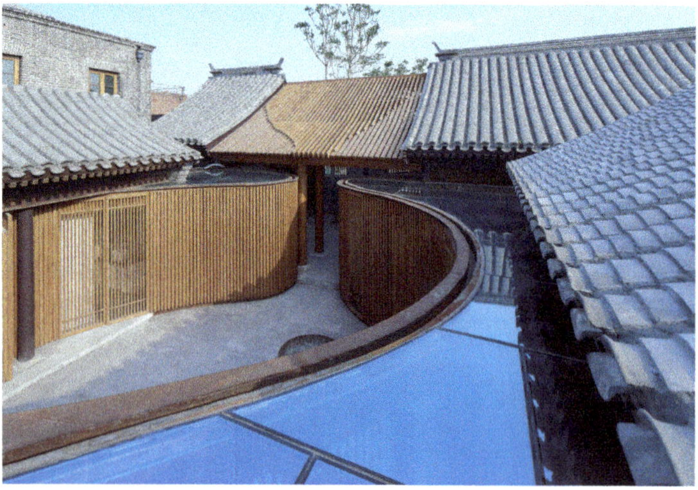

Fig. 6.148 Aerial view (*photo by Yang Chaoying*)

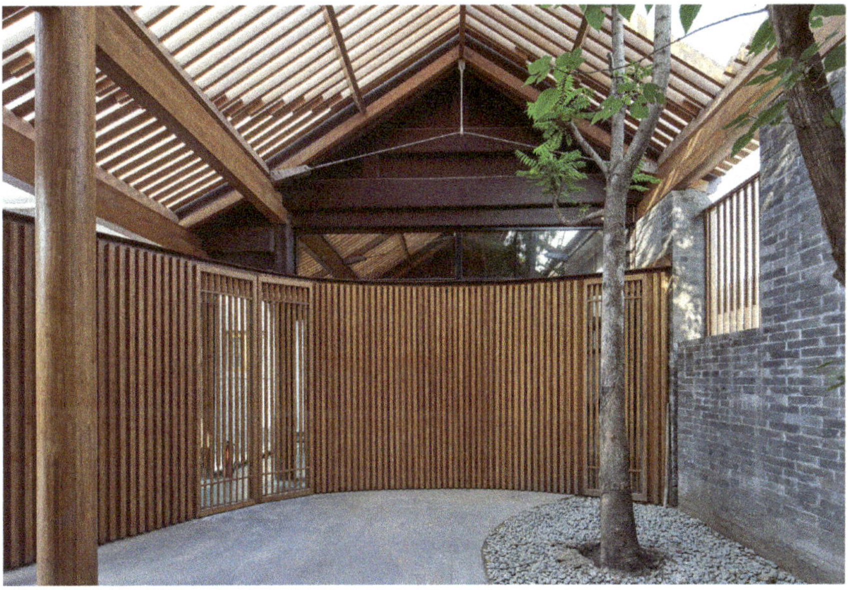

Fig. 6.149 South courtyard (*photo by Yang Chaoying*)

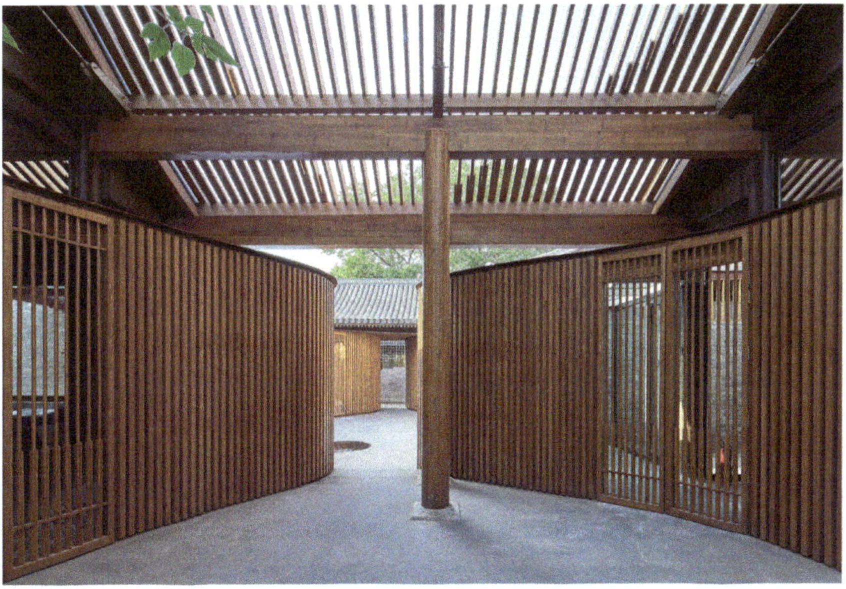

Fig. 6.150 Facing north courtyard from the south (*photo by Yang Chaoying*)

Shuangxi Academy

Location: Pingjiang, Hunan, China

Completed: 2018

Architect: Li Liang

Shuangxi Academy is a project for holidays which integrates leisure, nourishment of life and reading. It fully reflects the spirit of southern literati and the feelings of the mountains and plains. Shuangxi Academy group is composed of seven small buildings, which are distributed in several branch-like valleys that more than 10 m wide. In addition to the reception lobby (which also serves as a restaurant) and the tea room, the other buildings are hidden in four open valleys. In terms of architectural style, they used modern building materials and structural technologies to reflect the regional traditional architectural intention. From the structural layout of the inner patio to the bamboo frame, from the continuous grey tile slope top to the stone, bamboo, tile and other materials texture, the image of these architectural designs not only comes from the local traditional residential buildings, but also extracts and abstracts the fragments of the architectural symbols of the wider southern Chinese residential buildings.

Application of bamboo products: Bamboo scrimber is used as the structural material of the reception hall and tea room; it is exposed to the outside, fresh and elegant.

Photo source: Figs. 6.151, 6.152, 6.153 and 6.154 are provided by Duoxiangjie Architectural Design.

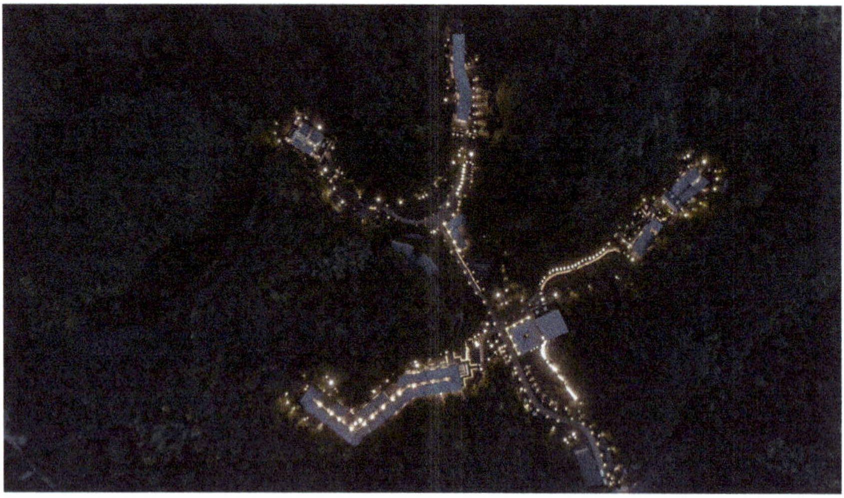

Fig. 6.151 Aerial view of Shuangxi Academy at night

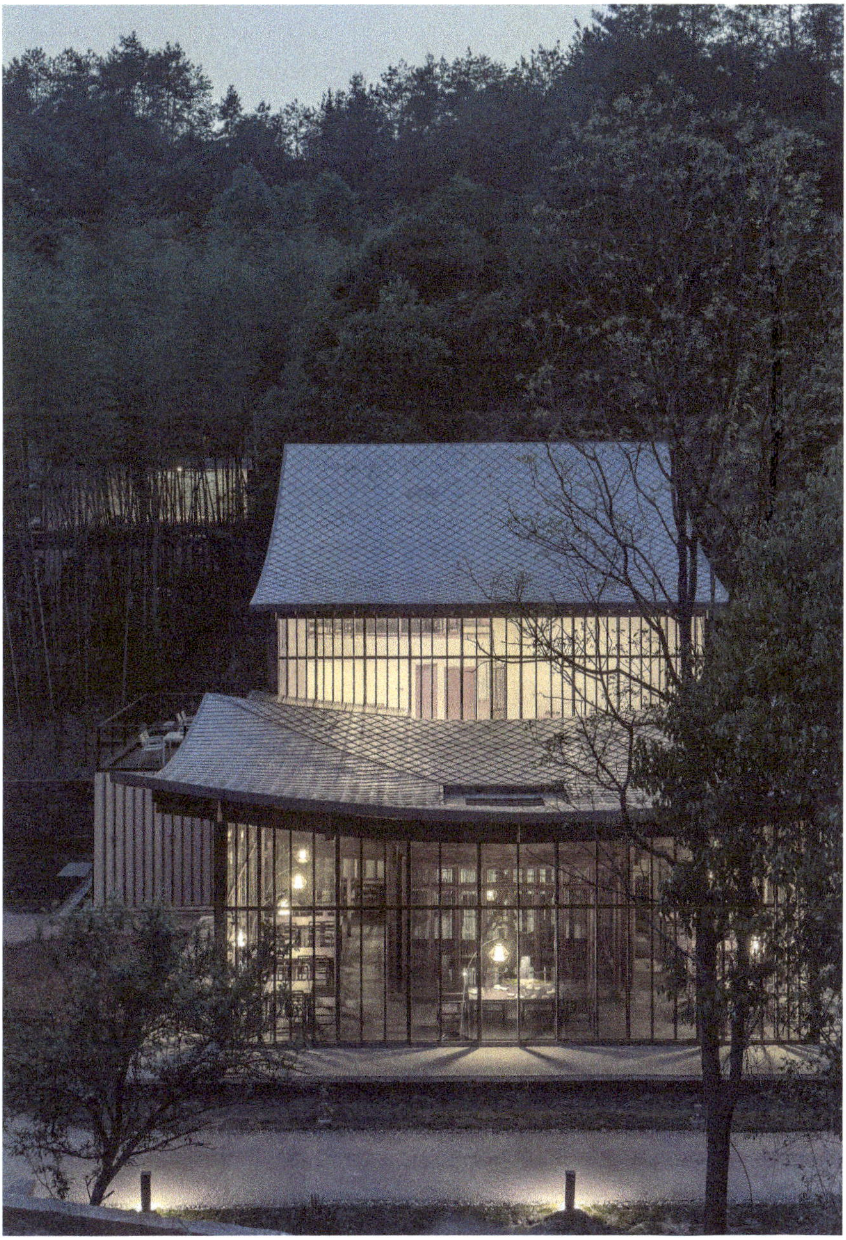

Fig. 6.152 Reception hall

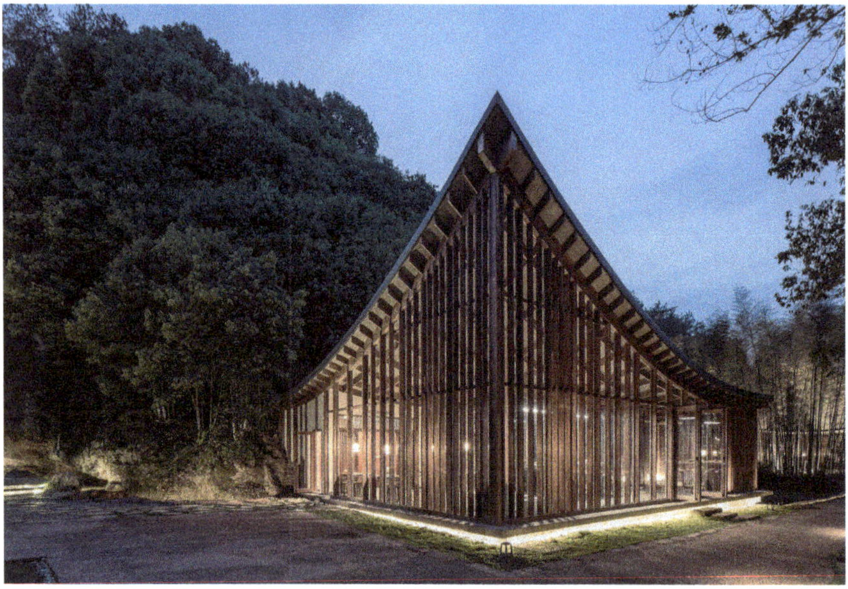

Fig. 6.153 Tea room

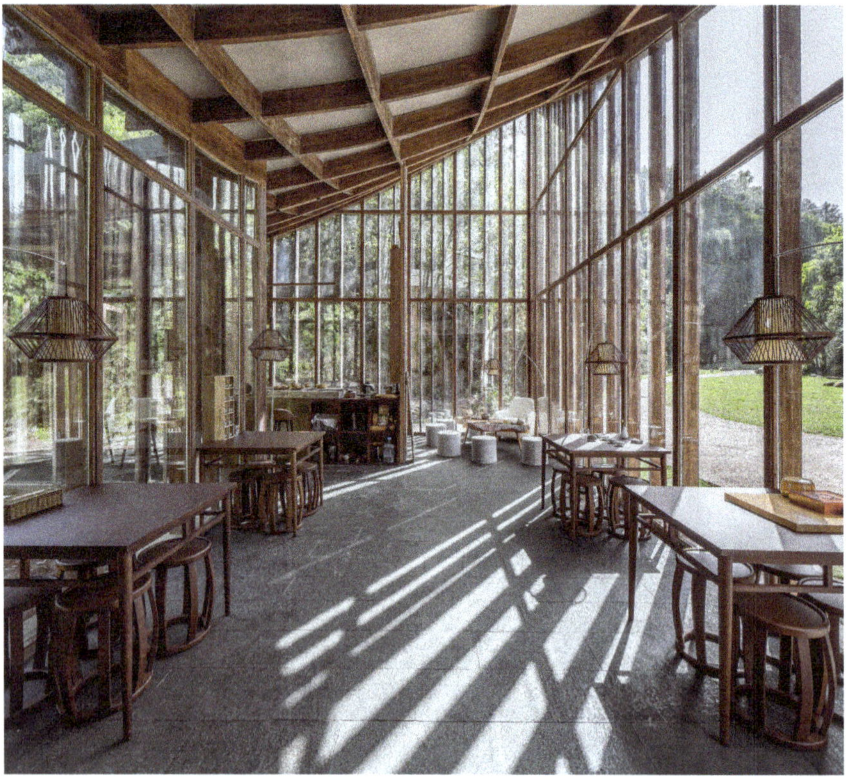

Fig. 6.154 Interior of tea room

Qing Bridge—Public Facility

Location: Nanjing, Jiangsu, China

Completed: 2017

Architects: Li Zhu, Wang Jiajun, Han Qing et al.

Located in Nanjing Zidong International Creative Park, the project is a 240 m^2 public toilet with rest facilities. As a common name for toilets in the Tang Dynasty, Qing means closing the door and cleaning up the filth, reflecting the creativity of ancient people in making characters. The structure is sited in a depression in a cedar forest. To comply with the ancient concept of shelter, the toilet is shaped into a corridor bridge. Except for the basic function, three rest zones are set around the toilet. The rest zone on the open corridor bridge faces the cedar forest. People can sit and meditate towards the space of forest. Shaped as a staircase, the view point rest zone is at the end of the bridge, offering views to the ruins of the ancient city walls, which are down the slope. The treetop rest zone lies on the flat roof of the third bathroom. People can climb up the steps and get close contact with the leaves on the roof. With the collocation of three kinds of material, glue laminated bamboo, fair-faced concrete slab, and mirror stainless steel plate, the style of the architecture appears to be minimalistic yet gentle. Large areas of honeycomb structure mirror surface stainless steel plate reflect the changes of light in the forest, which is so magnificent that the entire architecture seems to disappear, leaving the suspending body of the bridge alone. Spanning the forest and the dry stream, the architecture allows people to sit, to climb, and to enjoy the scenery in a convenient way.

Application of bamboo products: Beams, columns, roof trusses and other components are made of glue laminated bamboo, connected by steel nodes, prefabricated in the factory, and assembled on site quickly. In addition, bamboo plywood is also used for exterior walls.

Photo source: Figs. 6.155, 6.156, 6.157, 6.158 and 6.159 are provided by Architecture Technology and Art Studio, Architects & Engineers Co., Ltd of Southeast University.

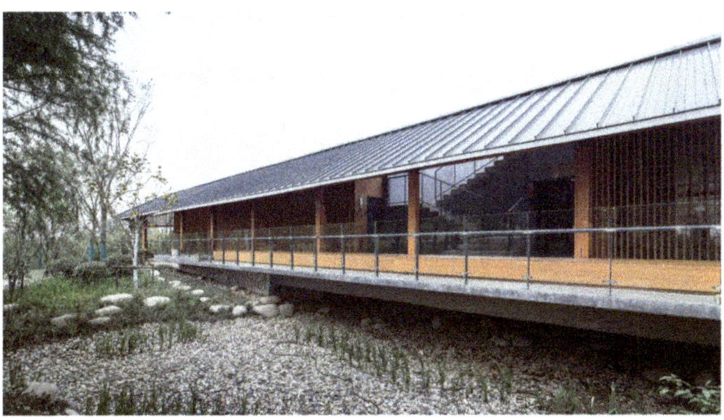

Fig. 6.155 Qing Bridge (*photo by Zhong Ning*)

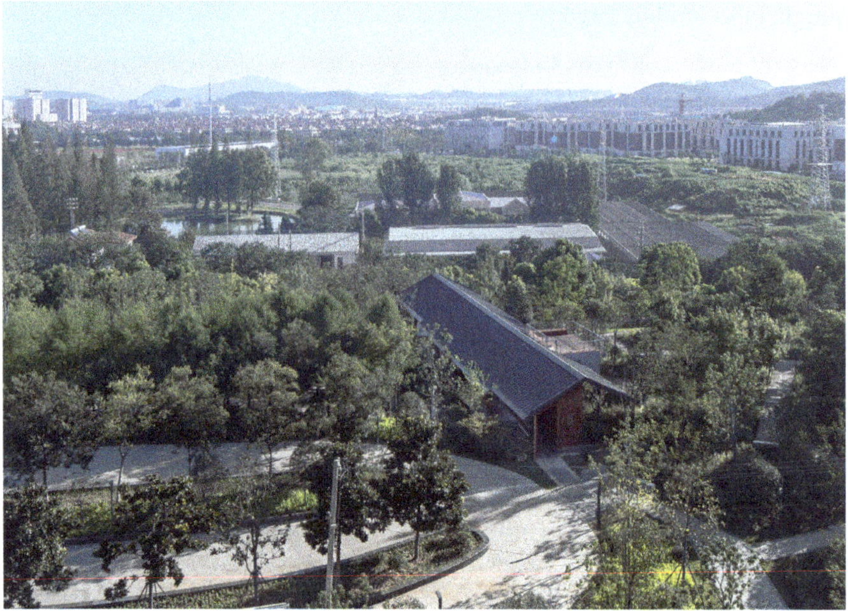

Fig. 6.156 Aerial view of Qing Bridge (*photo by Zhong Ning*)

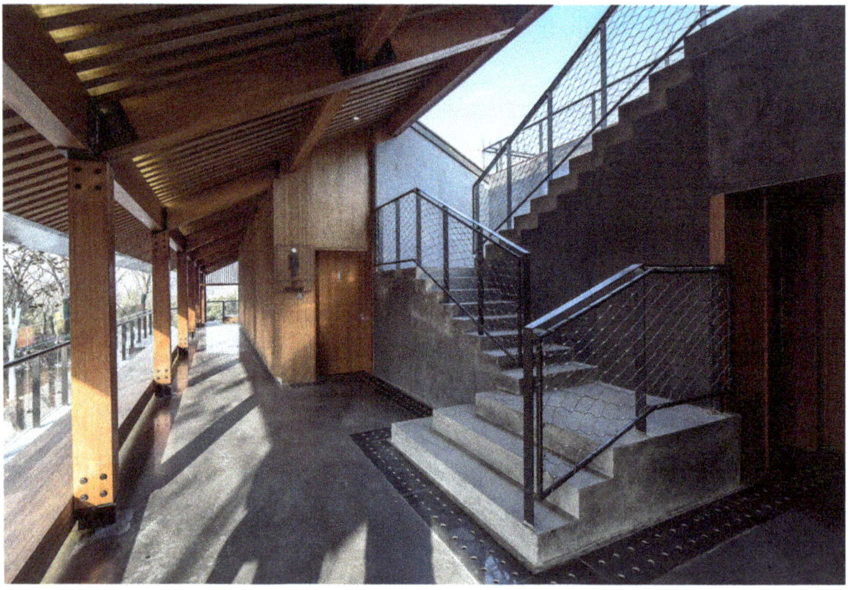

Fig. 6.157 Rest zone and traffic space of corridor bridge (*photo by Zhong Ning*)

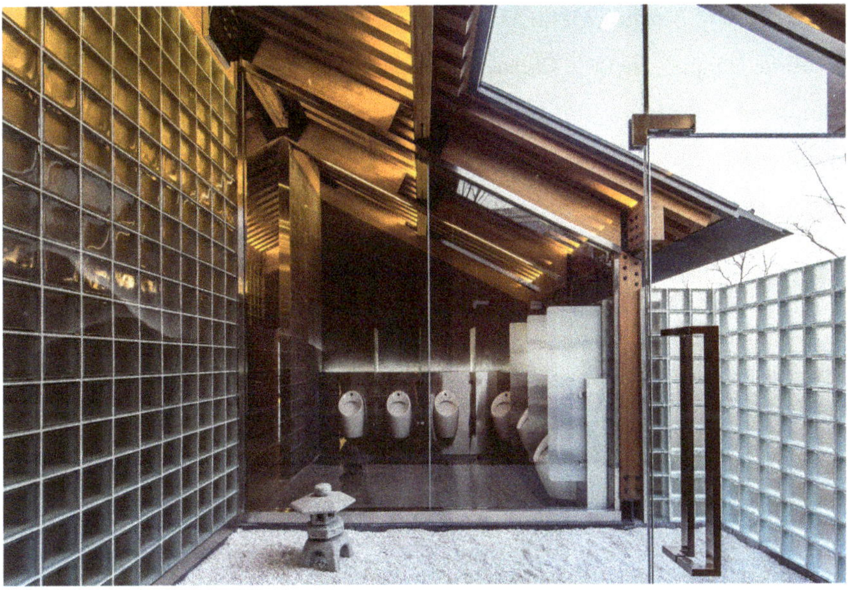

Fig. 6.158 Men's room (*photo by Zhong Ning*)

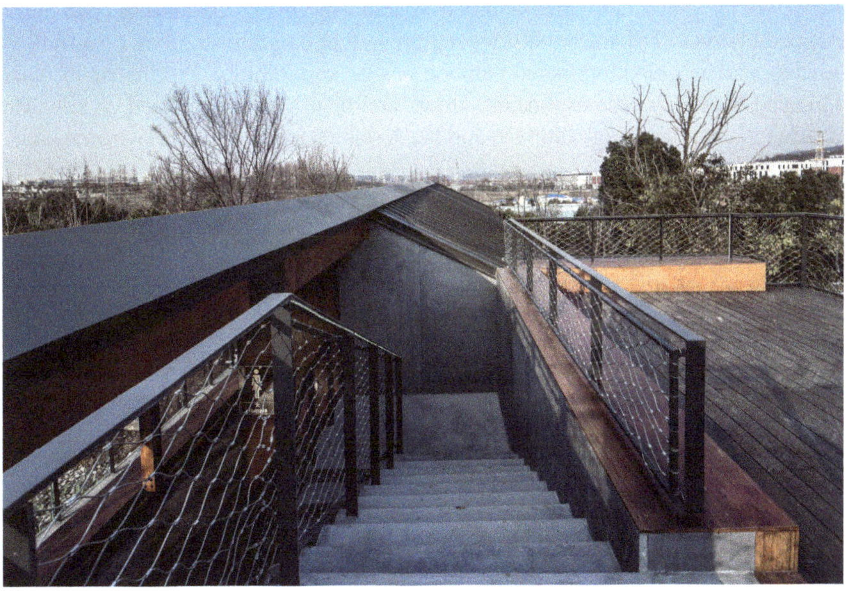

Fig. 6.159 Treetop rest zone (*photo by Zhong Ning*)

Rest Station of Nanjing Qixia Mountain Scenic Area

Location: Nanjing, Jiangsu, China

Completed: 2018

Architects: Li Zhu, Yin Yue, Yang Zixuan et al.

Nanjing Qixia Mountain Scenic Area is a famous Buddhist holy land in China. The Ancient Qixia Temple is hidden among the towering trees in the mountain. With an increasing number of tourists, improved rest stations are required. Six rest stations have been built on sites around the mountain. The sites are located at the foot of the mountains or beside streams, surrounded by tall trees with difficult construction conditions. The stations need to have modern facilities and equipment for tourists, as well as space for rest, accommodation and sightseeing. The architectural style needs to be coordinated with the ancient temples in the mountains. At the same time, the building also needs to adapt to the humid and large temperature variation in the mountains to facilitate daily maintenance and management.

Application of bamboo products: There are two applications of bamboo products. One is to hide the characteristics of concrete material through decorative glued laminated bamboo boards, rafters and grilles outside the concrete frame structure, so as to integrate it with the surrounding temples and ancient trees. The second is to build a new type of glued laminated bamboo structure by using the narrow open space wrapped by trees. The site conditions require a light building structure on a shallow foundation. The glued laminated bamboo has good durability, and the small glued laminated bamboo columns, beams, rafters and boards can be hoisted on site. The construction is convenient and fast, and has less interference to the environment.

Photo source: Figs. 6.160, 6.161, 6.162, 6.163 and 6.164 are provided by Architecture Technology and Art Studio, Architects & Engineers Co., Ltd of Southeast University.

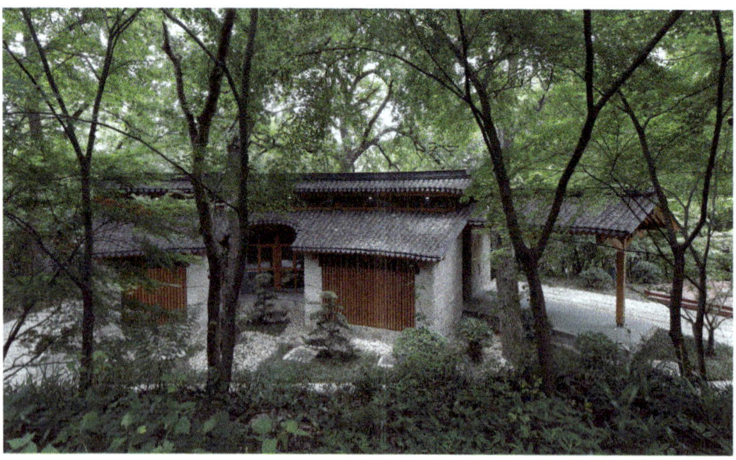

Fig. 6.160 Peach Blossom Fan Rest Station hidden among the trees (*photo by Zhong Ning*)

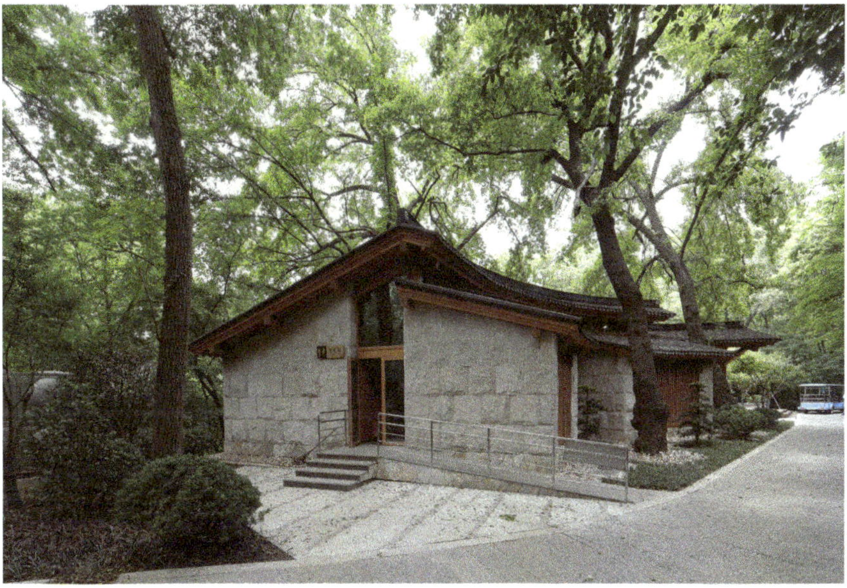

Fig. 6.161 Shallow foundation of glued laminated bamboo structure protects roots of trees (*photo by Zhong Ning*)

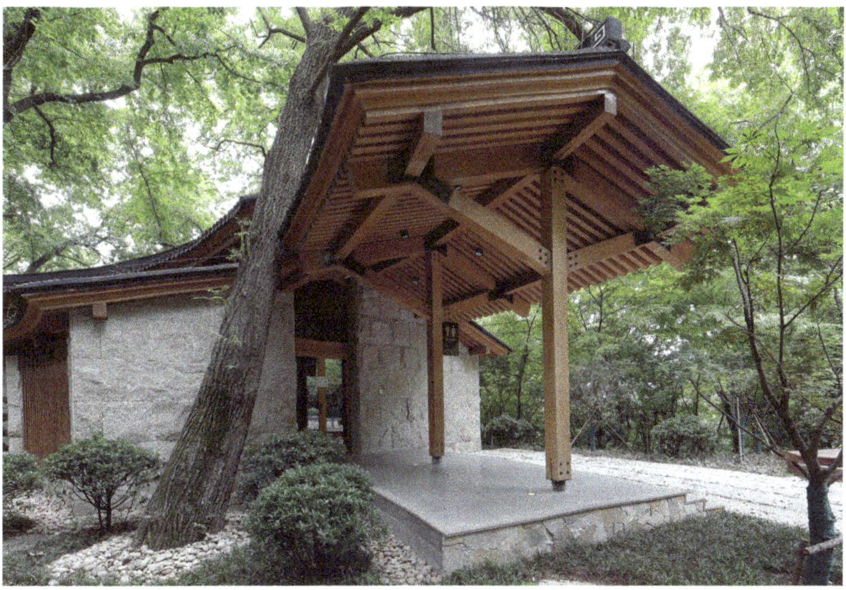

Fig. 6.162 Rest station gives way to a tree on site (*photo by Zhong Ning*)

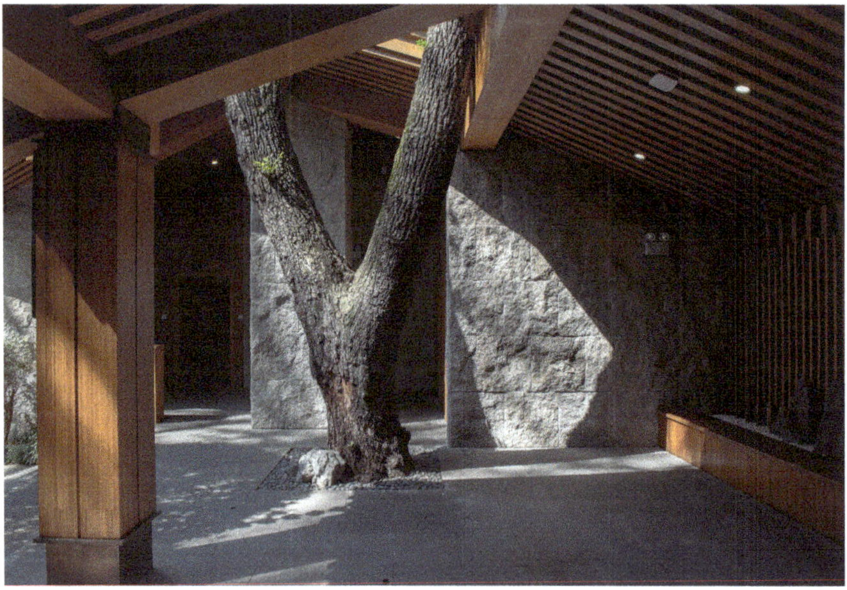

Fig. 6.163 An existing tree is the protagonist of the space (*photo by Zhong Ning*)

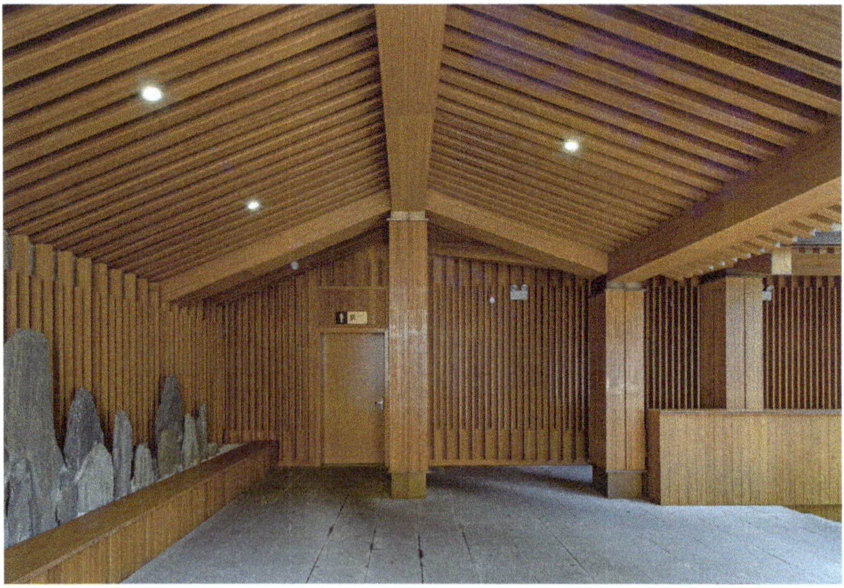

Fig. 6.164 Rest space of Gulou Rest Station (*photo by Zhong Ning*)

Courtyard N4, Lixiang Village

Location: Nanjing, Jiangsu, China

Completed: 2019

Architects: Li Zhu, Wang Jiajun, Wu Wei et al.

Lixiang Village is a suburban village about 80 km from Nanjing, China. With the development of rural revitalization, Lixiang Village attracts tourists for blueberry planting and Red Tourism resources. In order to solve the problem of centralized accommodation for tourists and the lack of public space for villagers, the village committee decided to transform the original headquarters into a "N4 courtyard" integrating several functions for tourist reception, villagers' activities and rural hotels. The architecture needs to organise villagers' daily life and tourists' sightseeing in a new rural public space, and improve the quality of rural life by retaining rural customs and humanistic characteristics.

Application of bamboo products: The architect does not conform to the local traditional architectural style, but draws inspiration from local construction and uses modern technical means for processing and construction, aiming at improving the quality of rural architecture and construction technology. The combination of glued laminated bamboo and steel members realises the installation feasibility of small-size structural members under the condition of rural construction, and uses these to form a large indoor space. The design explores a way of public space construction in China's suburban villages, which not only realises the modernisation of rural construction, but also retains the cultural characteristics of the countryside, reflecting the contemporary thinking of rural construction.

Photo source: Figs. 6.165, 6.166, 6.167, 6.168 and 6.169 are provided by Architecture Technology and Art Studio, Architects & Engineers Co., Ltd. of Southeast University.

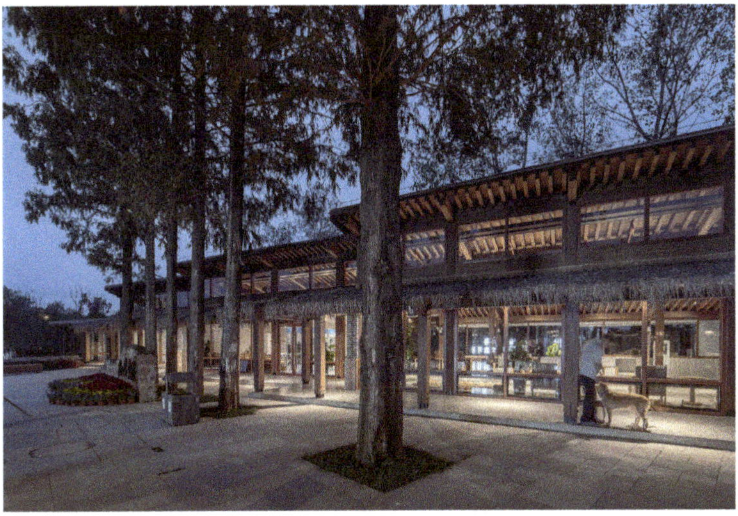

Fig. 6.165 Night view of Villagers' Activity Centre (*photo by Timeraw Studio*)

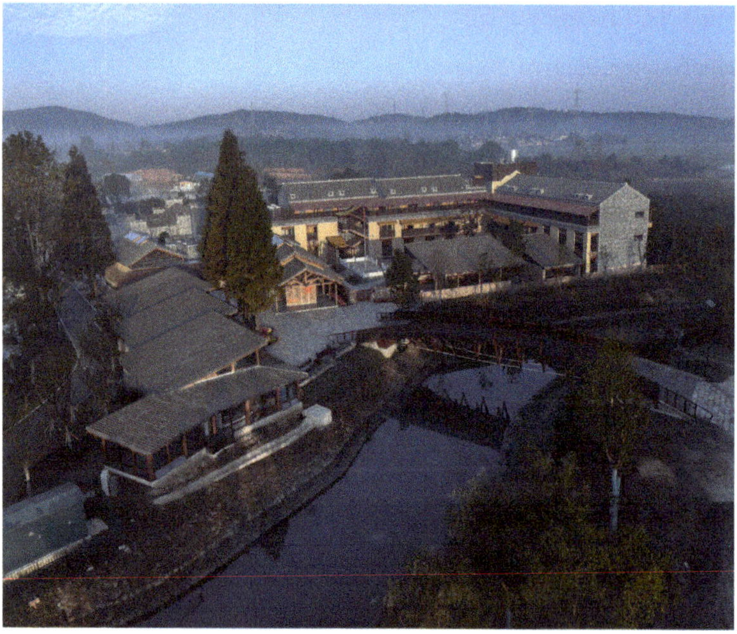

Fig. 6.166 Aerial view of Courtyard N4 (*photo by Timeraw Studio*)

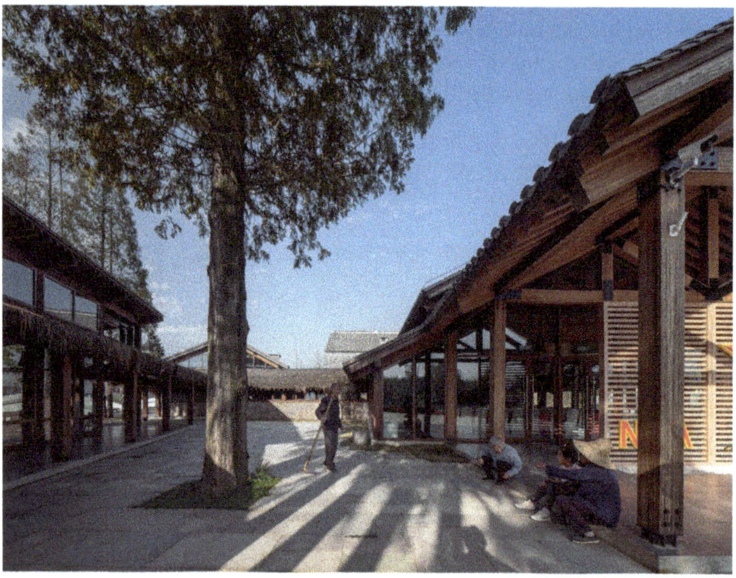

Fig. 6.167 New buildings and reserved trees (*photo by Timeraw Studio*)

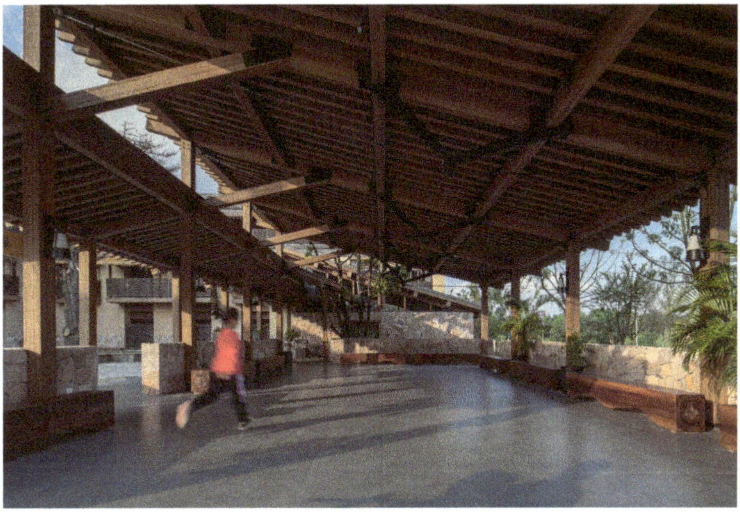

Fig. 6.168 Glued laminated bamboo shed (*photo by Timeraw Studio*)

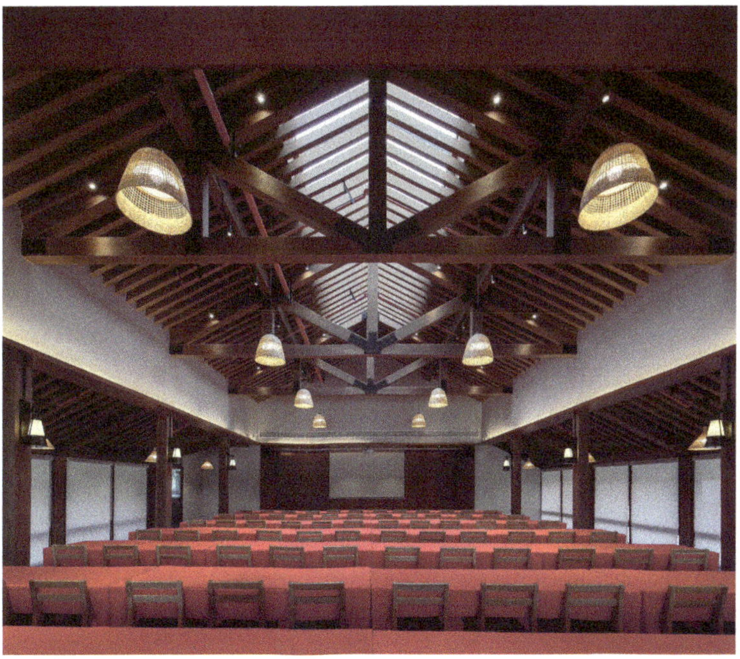

Fig. 6.169 Multifunctional hall for villagers constructed of small scale glued laminated bamboo components (*photo by Wang Haihua*)

Bamboo Chalets in the West Historic Block of Gongchen Bridge, Hangzhou Section, Grand Canal, China

Location: Hangzhou, Zhejiang, China

Completed: 2017

Designer: Zhou Chungui

Embedded in the long history of the Grand Canal, Gongchen Bridge is an invaluable heritage site and also a touchstone for local people. During the renovation, the architect added a lot of modern elements, but at the same time retained the traditional memory. Bamboo decks using outdoor bamboo scrimber material were laid along the Zhijie street, along with 18 bamboo chalets. The bamboo surface is made of old technology, trying to retain the memory of old Hangzhou. The use of traditional mortise and tenon technology makes the perfect combination of modern bamboo and traditional assembly techniques, reflecting the overlapping of the rapid development of Hangzhou and historical memory.

Application of bamboo products: Outdoor bamboo scrimber is used as the floors, exterior decking and bamboo chalets. This material has excellent weather resistance, is suitably strong, durable and corrosion, insect and mildew resistant, and as a B1 fire-proofing grade.

Photo source: Figs. 6.170, 6.171 and 6.172 are provided by Hangzhou Runzhu Technology Co., Ltd.

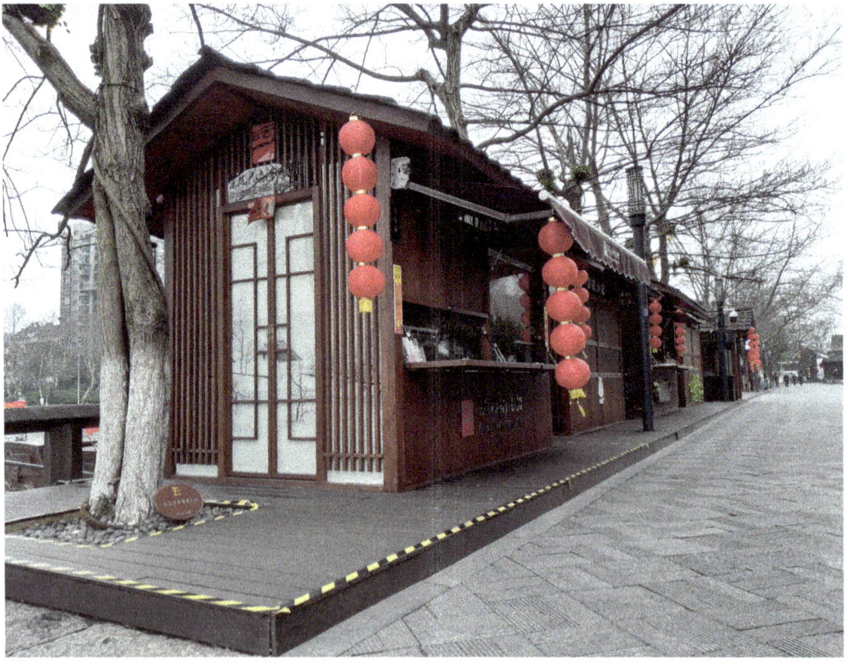

Fig. 6.170 Bamboo chalets along the grand canal

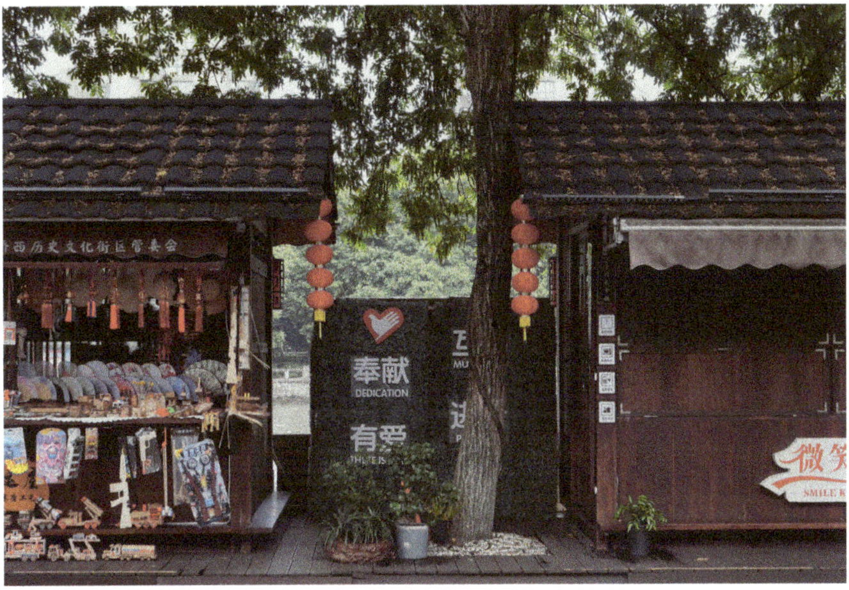

Fig. 6.171 Bamboo chalets along the grand canal

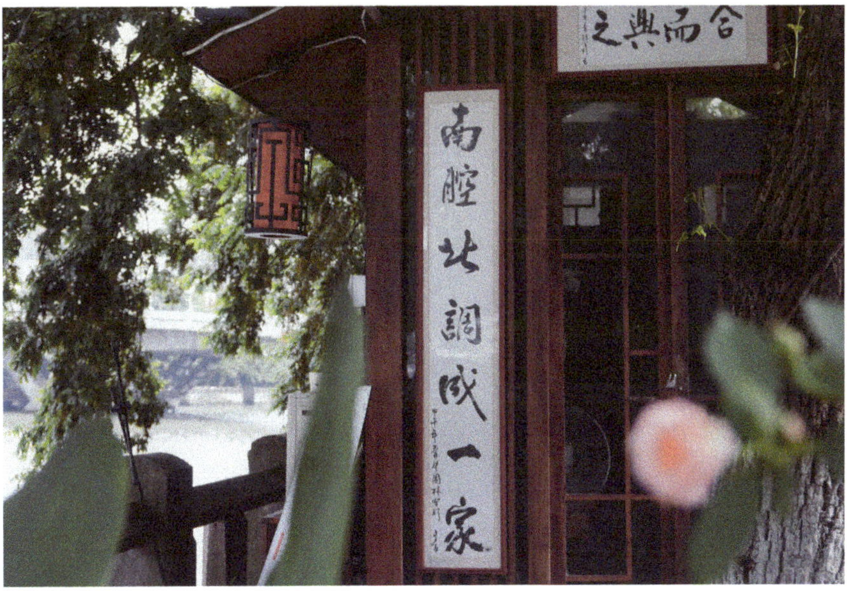

Fig. 6.172 Bamboo chalets along the grand canal

6.3 Landscapes

Full culm bamboo in Landscape:

Hilltop Gallery (2018)

Shenyang Panda Pavilion (2017)

ZCB Bamboo Pavilion (2015)

Swirling Cloud: Bulletin Pavilion (2018)

UNESCO Garden of 2019 Beijing Expo (2019).

Woven bamboo in landscape:

Bamboo Canopy and Pavilions, Impression Liu Sanjie (2020)

Engineered bamboo in Landscape:

Peaceful Time, Lingering Charm (2017)

Flexible Landscape (2015)

Barcelona Re.Set—Identity Pavilion (2014)

Outdoor Bamboo Landscape of Shanghai Qinsen Group at 2019 Beijing Expo (2019)

Baiguo Garden of Beijing Landscape Bureau at 2019 Beijing Expo (2019)

China House Vision—Life in the Future: Living Garden (2018)

Nanjing Youth Olympic Games Sports Park (2015)

Hong Kong-Zhuhai-Macao Bridge Landscape Island (2018)

Xiong'an New District Citizen Service Centre (2018).

Hilltop Gallery

Location: Chengde, Hebei, China

Completed: 2018

Architects: Li Daode, Wu Yiying, Kuai Ding

Hilltop Gallery is located on the Yanshan Mountains at the junction of Beijing and Chengde, overlooking the Great Wall at Jinshan Mountains. As an art gallery, in addition to exhibiting works of art and related cultural activities, it also undertakes the reception and display functions of the overall development project of Hilltop Valley. The architect's thoughts and extension of Chinese classical spirit are embodied in the form: inheriting and deducing the "terrain" with the form of architecture, reflecting the surrounding continuous mountain lines and natural form, and integrated with it. The first floor is the exhibition area and the second floor is the banquet area. The banquet area can be directly connected to the roof. The roof bamboo walkway links the terrace, the roof and the surrounding hillside.

Application of bamboo products: The roof bamboo walkway is connected with the mountain through the rise and fall of the roof, which blurs the boundary between natural and man-made parts, and symbolically echoes the shape and narrow spatial characteristics of the ancient Great Wall of Jinshan Mountains winding along the mountain in the distance. The application of bamboo, together with other traditional materials, integrates the overall atmosphere of the building with the Chinese classical spirit.

Photo source: Figs. 6.173, 6.174, 6.175, 6.176 and 6.177 are provided by dEEP Architects.

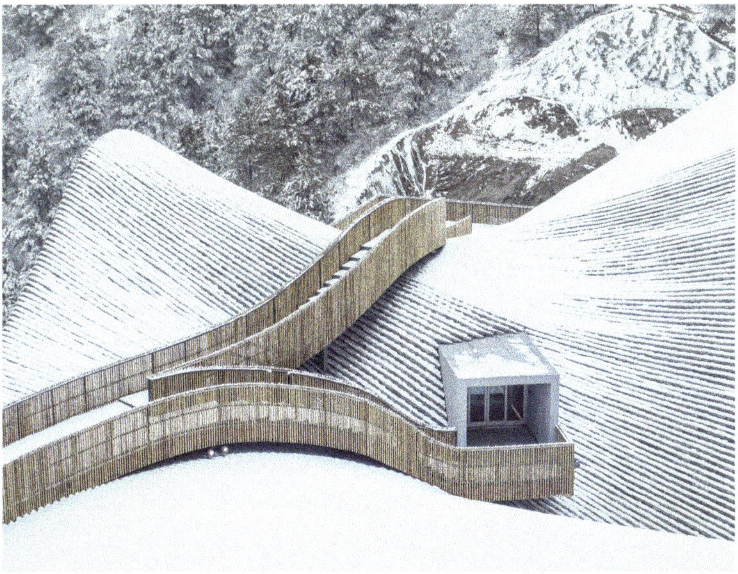

Fig. 6.173 Rooftop bamboo walkway after snow

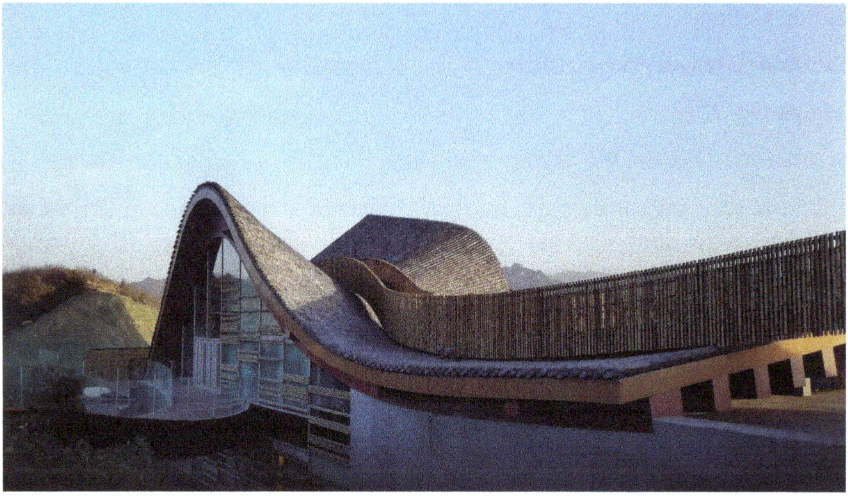

Fig. 6.174 Roof shape

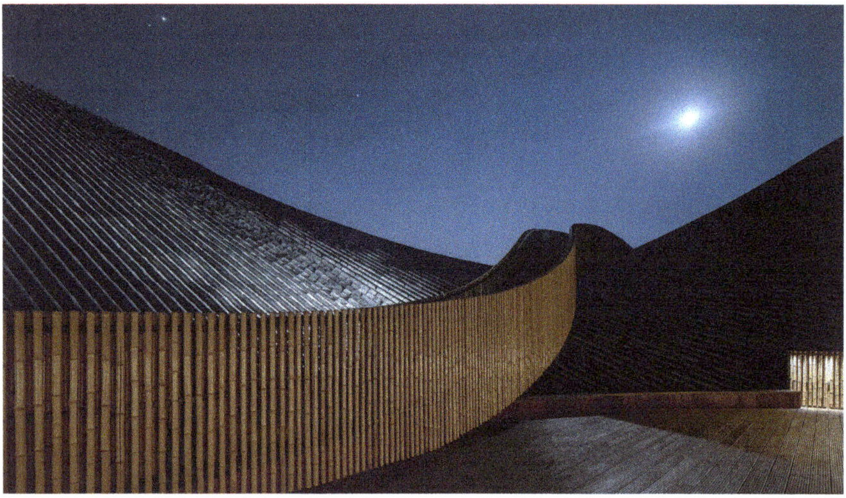

Fig. 6.175 Night scene

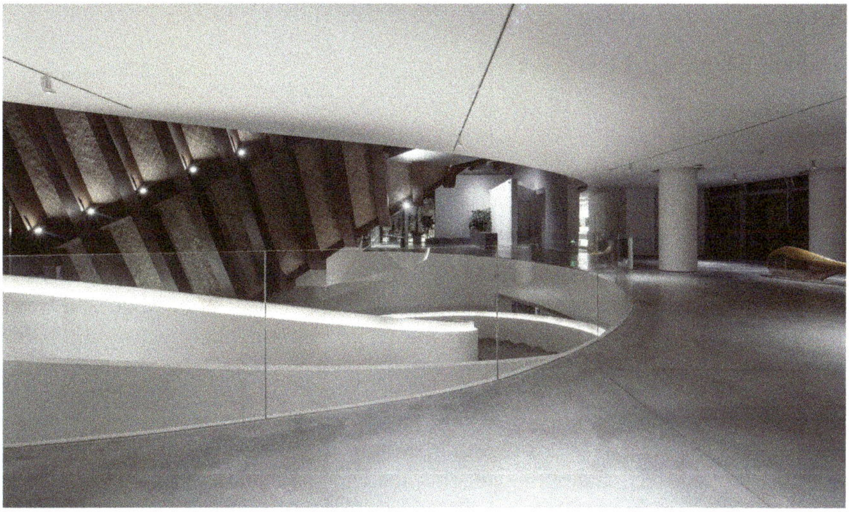

Fig. 6.176 Hilltop Gallery

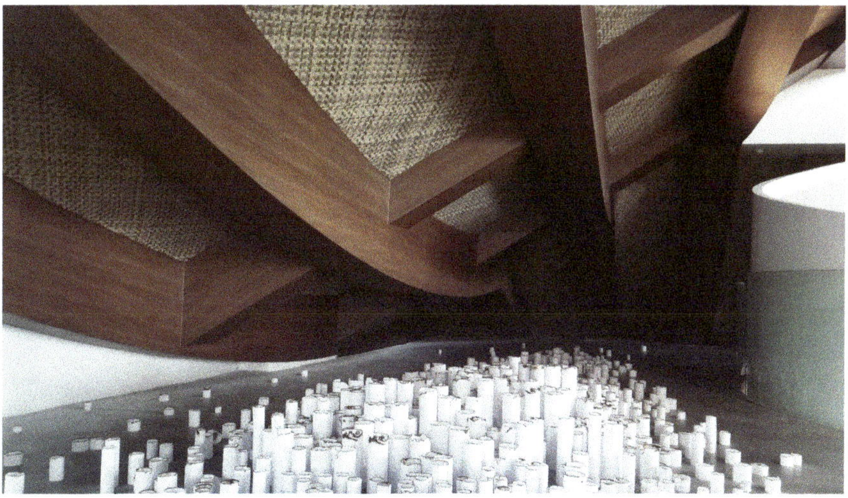

Fig. 6.177 Roof structure

Shenyang Panda Pavilion

Location: Shenyang, Liaoning, China

Completed: 2017

Architects: Zhu Ling, Liu Yida, Li Bohao

Taking "Bamboo Ecological Environment" as the design concept, architects used the method of dissolution to hide the buildings in the mountains and forests, avoiding the destruction of the environment and landscape caused by a large number of buildings settled in the forest. Architecture retreated and surrounded by bamboo and trees, visitors can enjoy the bamboo echoing the lie of the mountain, feeling the living environment of national treasures.

Application of bamboo products: The facade of the main venue and the landscape shed are a steel structure with decoration made of full-culm bamboo. The bamboo canopy is the main outdoor decoration and application structure in the park. It connects all the functional zones of the park and strengthens the integrity of the interior function of the site. Like a forest in the site, the pillars supporting the bamboo canopy, makes tourists enjoy walking and have fun among them.

Photo source: Figs. 6.178, 6.179 and 6.180 are provided by Shenyang Jianzhu University.

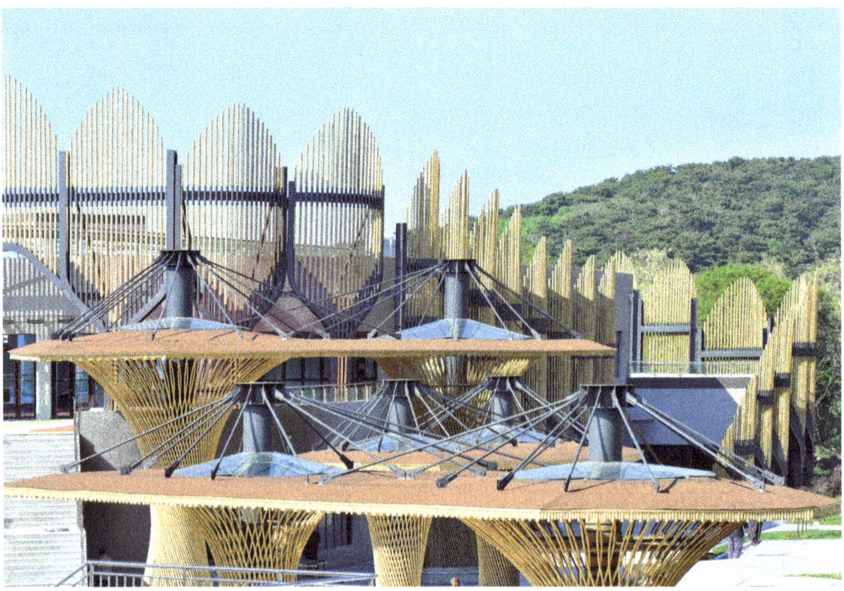

Fig. 6.178 View of architectural group from the entrance road

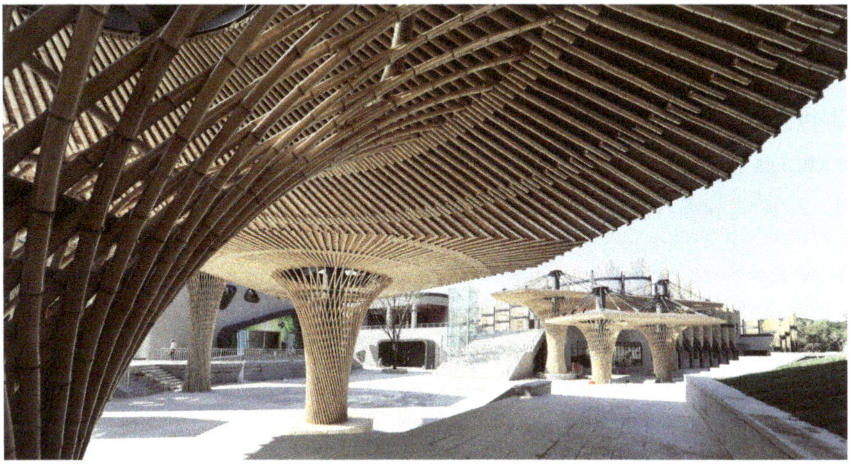

Fig. 6.179 View of panda pavilion from the square

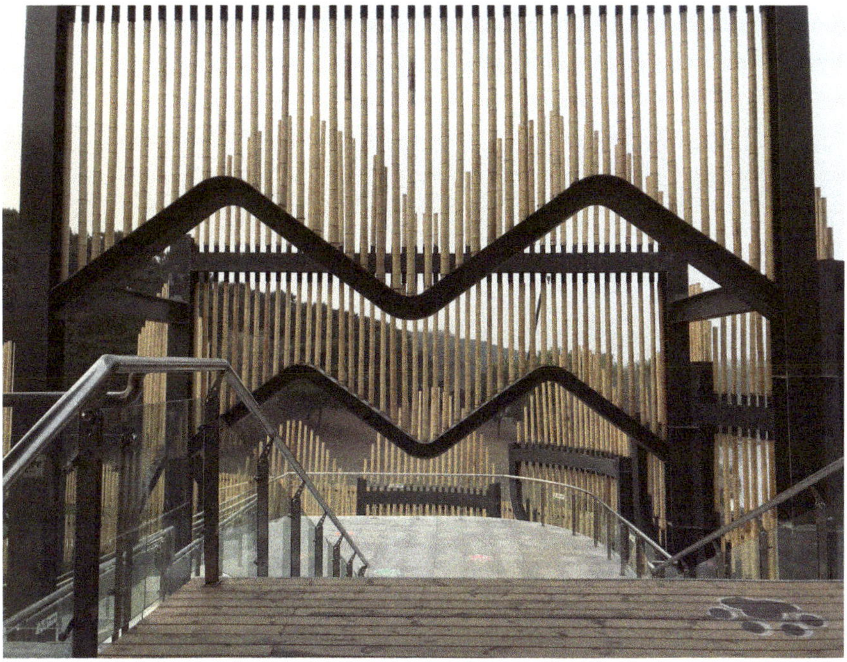

Fig. 6.180 Visit corridor

ZCB Bamboo Pavilion

Location: Hong Kong, China

Completed: 2015

Architect: Kristof Crolla

The ZCB Bamboo Pavilion is a public event space built for the Construction Industry Council's Zero Carbon Building (ZCB). It is a four-storey-high long-span bending-active bamboo gridshell structure with a footprint of approximately 350 m^2 and a seating capacity of 200 people.

Application of bamboo products: The project is built from 475 large full-culm bamboo poles that are cold-bent onsite to shape the structure and that are hand-tied together with metal wire using techniques from Cantonese bamboo scaffolding building. The project was designed using digital physics simulation engines and seeks ways for traditional Cantonese craftsmanship to adapt and evolve for novel twenty-first century applications.

Photo source: Figs. 6.181, 6.182 and 6.183 are provided by Chinese University of Hong Kong.

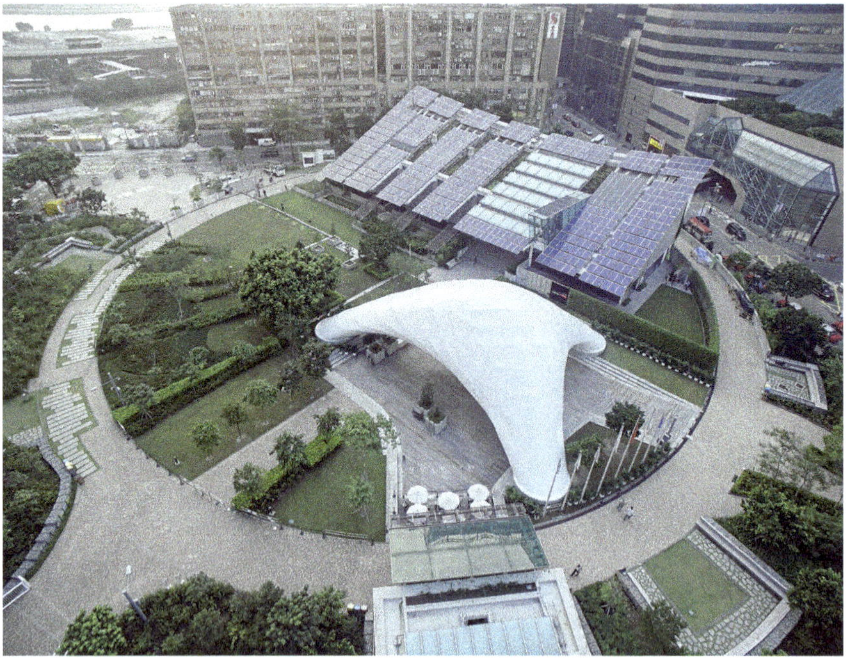

Fig. 6.181 Aerial view by day (*photo by Michael Law*)

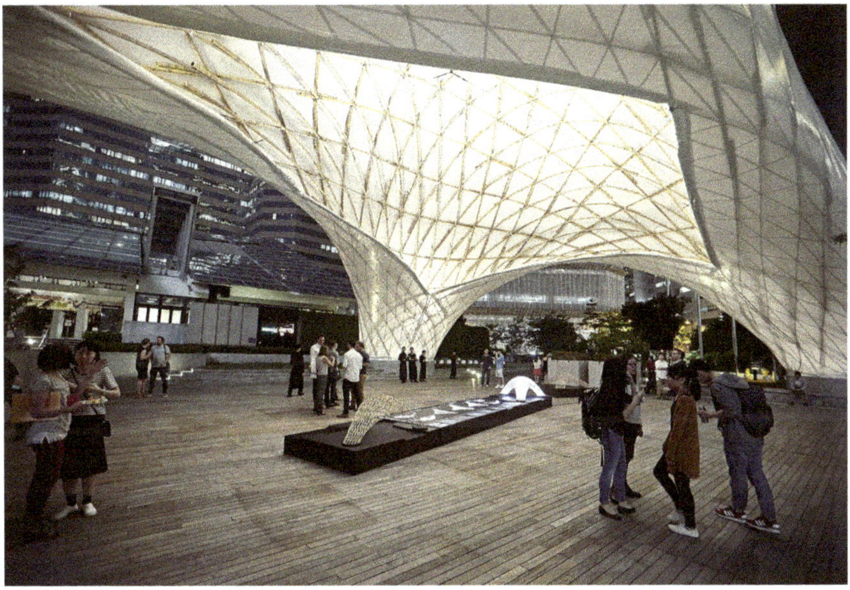

Fig. 6.182 Interior by night (*photo by Michael Law*)

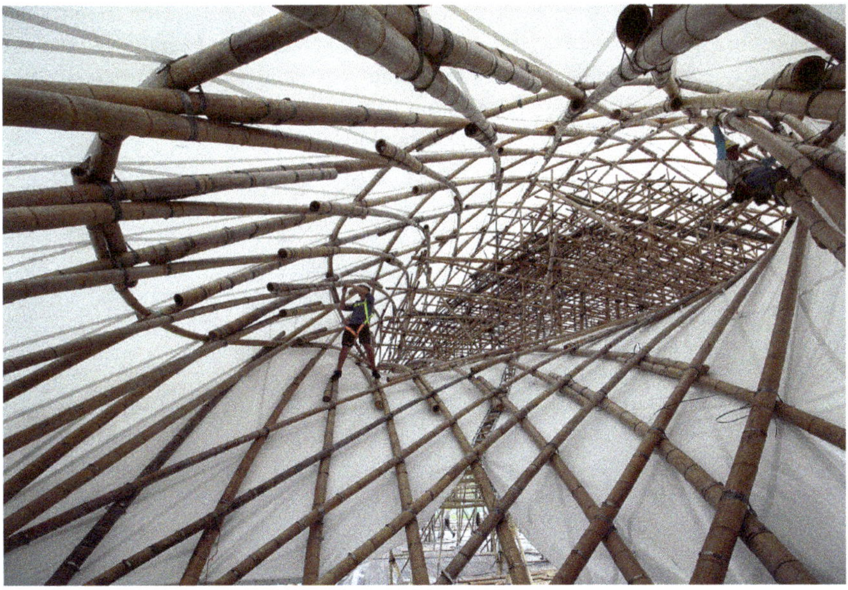

Fig. 6.183 Under construction—fabric tied to bamboo structure (*photo by Kevin Ng*)

Swirling Cloud: Bulletin Pavilion

Location: Beijing, China

Completed: 2018

Architects: Song Yehao, Chen Xiaojuan, Sun Jingfen et al.

Inspired by the Chinese poem, "The wind blows and the clouds fly" and located in a grove on the campus of Beijing Forestry University, this 120 m^2 structure is an information pavilion designed for the construction of "1st BFU International Garden-making Festival" in 2018. After the activity was completed, it has been used as an outdoor place for the daily leisure and gathering of teachers and students. Starting from the characteristics of bamboo, the design gives full play to the flexibility and bending resistance of full-culm bamboo, and adopts free and easy curved surfaces, in harmony with the campus environment and the atmosphere of garden festival. The curved openings with different heights are arranged to facilitate the free passage of people from all directions. Some of the original hedgerows are retained, extending into the pavilion. The raised beam at the top is set as a circular lighting outlet. The wind, sunshine and green plants are led into the interior, which breaks the dull sense of the traditional shade and forms the image of cloud rolling and flying by.

Application of bamboo products: The design combines full-culm bamboo construction with modern building technologies. Designers used digital design to introduce the conceptual scheme into a digital model, so as to generate the overall structure and guide the construction. The curved surface and key bamboo beam structure curve generated by mathematical logic control reflect resistance to stress. Bamboo beams were prefabricated in factories and then assembled on site.

Photo source: Figs. 6.184, 6.185, 6.186 and 6.187 are provided by SUP Atelier. Photos by Chen Su and Fang Chun.

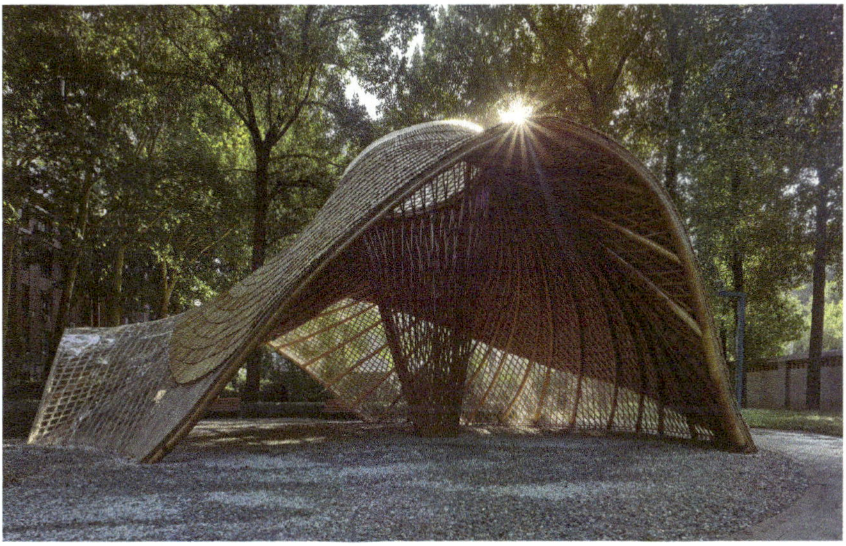

Fig. 6.184 Open bamboo pavilion

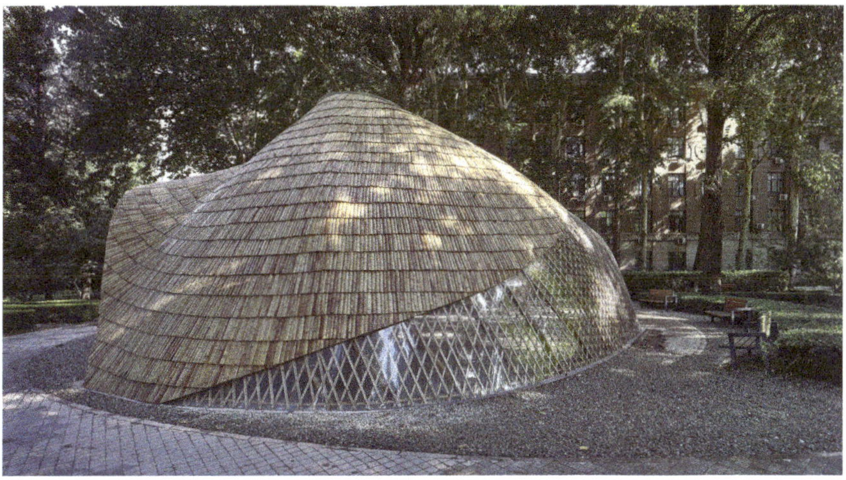

Fig. 6.185 Closed surface facing noise

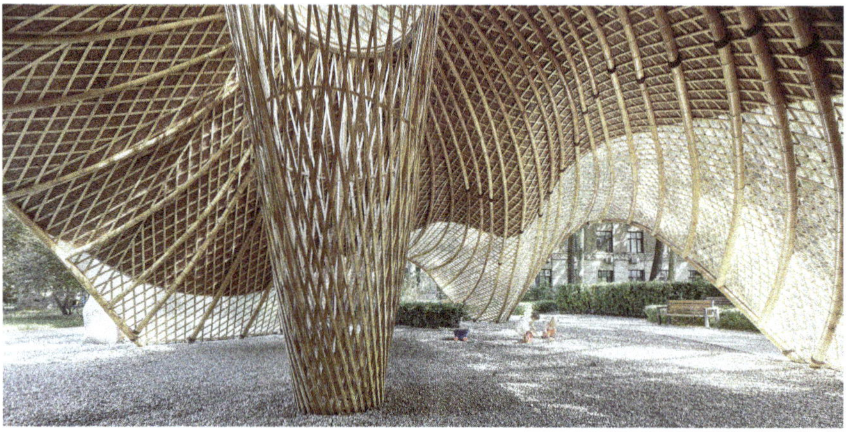

Fig. 6.186 Inside of pavilion

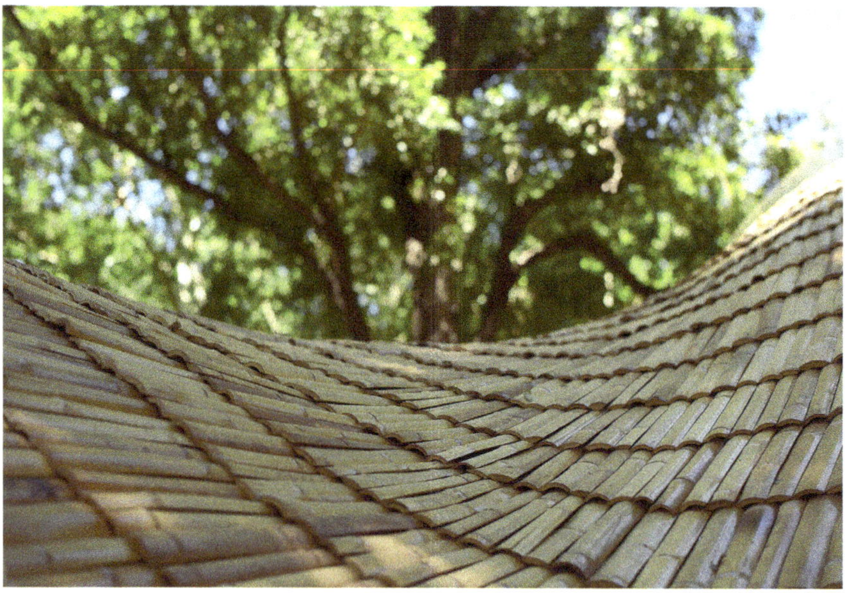

Fig. 6.187 Bamboo tiles

UNESCO Exhibition Garden at 2019 Beijing Expo

Location: Beijing, China

Completed: 2019

Architect: Nicolas Godelet

UNESCO Garden is located in the European Garden Area, covering an area of 1250 m^2. As a promoter of sustainable natural heritage conservation and development strategy, UNESCO hopes to build a gallery connecting the North and South entrances to guide visitors to understand its mission, actions and partners. The design aims to relate the following two points: respect the "earth civilization" of the environment requires everyone's participation, and each of us is the gardener of the planet; and, secondly, reduction of the global carbon footprint and new twenty-first century construction with natural materials such as grass, bamboo and wood. The Gallery has the functions of sun shading and rain proof, and provides rest seats. Along the ramp of the exhibition gallery, one can see the landscape map of the whole park. People can feel nature calmly while staying in this space.

Application of bamboo products: The gallery is constructed from *D. giganteus*. The staggered support of giant full-culm bamboo poles is parabolic, forming a complex spatial geometry. The joints are made of prefabricated steel connections and covered with white film on the top. Seen from a distance, the joints are magnificent and great; approaching closer, they are delicate and elegant.

Photo source: Figs. 6.188 and 6.189 are provided by Nicolas Godelet Architects.

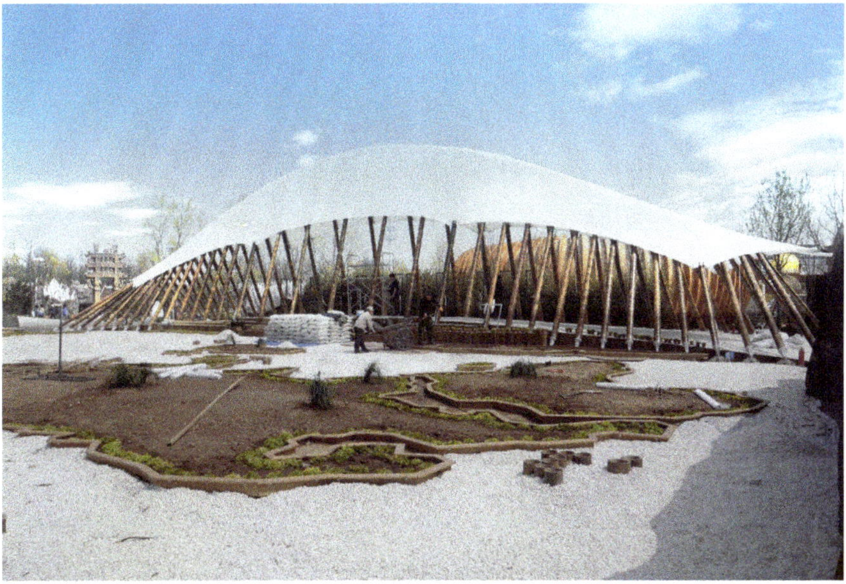

Fig. 6.188 UNESCO exhibition garden at 2019 Beijing Expo

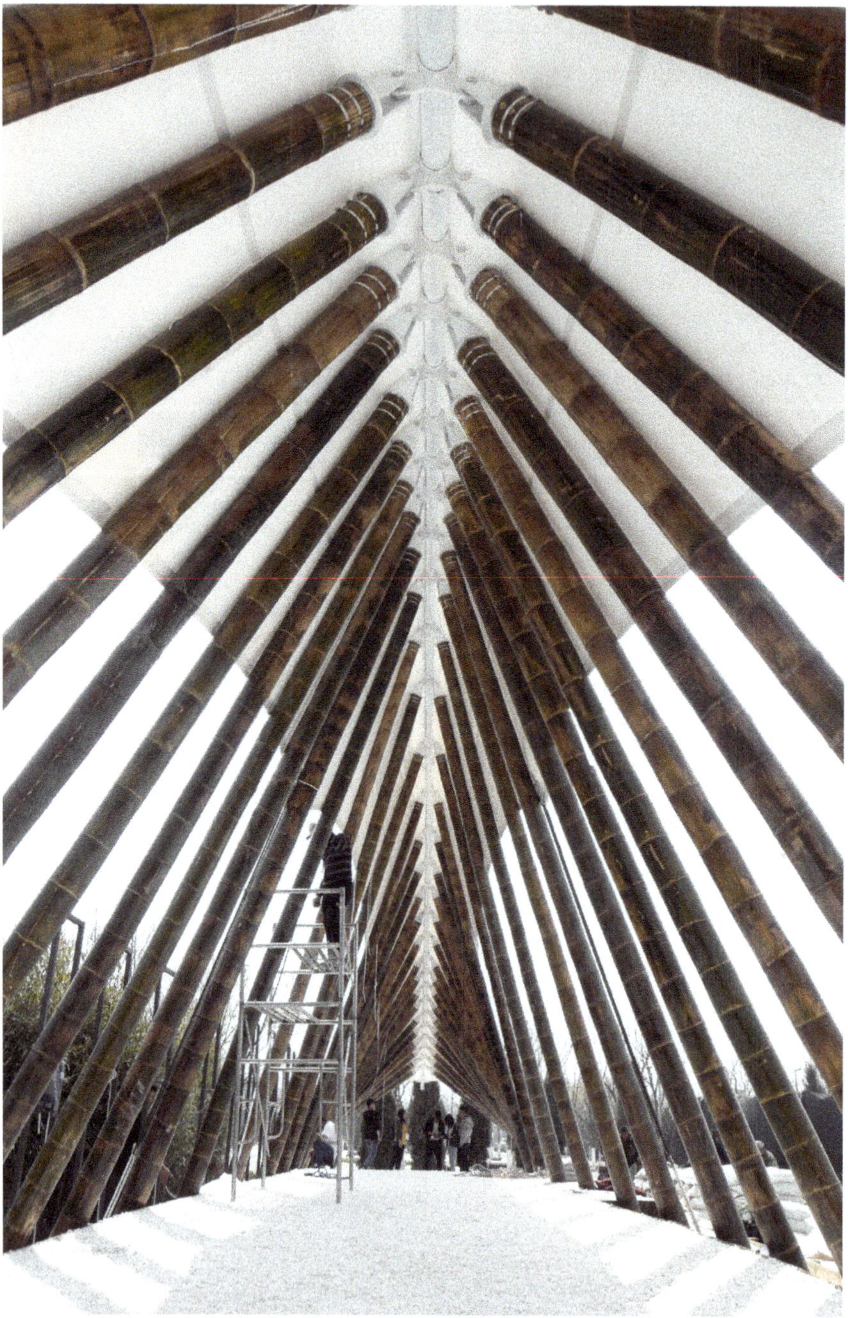

Fig. 6.189 Gallery

Bamboo Canopy and Pavilions, Impression Liu Sanjie

Location: Yangshuo, Guilin, Guangxi, China

Completed: 2020

Architects: Hanxiao Liu, Henry D'Ath, Lexian Hu et al.

Located in the Impression Liu Sanjie, Yangshuo, the project has a magnificent landscape, with endless green plants surrounding large hills and boulders. Architects take natural elements as the core of the concept in the architectural design, especially bamboo. There are two bamboo systems. The first system is an experience area for Impression Liu Sanjie before admission and also a zone for sales of cultural and creative products and leisure. Named "Endless Bamboo Lanterns", it is a space for hand-woven bamboo lanterns, scattered around the visitors' admission area. Another system, the "Handmade Bamboo Art Gallery", creates a staggered hand-woven "bamboo net" among the bamboo clumps, providing a walking area which can avoid light rain. The resonance and echo among the two systems and the surrounding natural environment endow each with stronger vitality.

Application of bamboo products: In the shape of an inverted landscape, the "Handmade Bamboo Art Gallery" staggers and undulates over 140 m from the entrance. The surface of the bamboo shed is made with irregular hand-woven methods similar to bamboo lamps, creating a dreamlike experience of space. While drifting deeper along the waves of the main body, the waiting experience becomes illusory and immersive, extending visitors' perception to the entire natural environment.

Photo source: Figs. 6.190, 6.191 and 6.192 are provided by llLab.

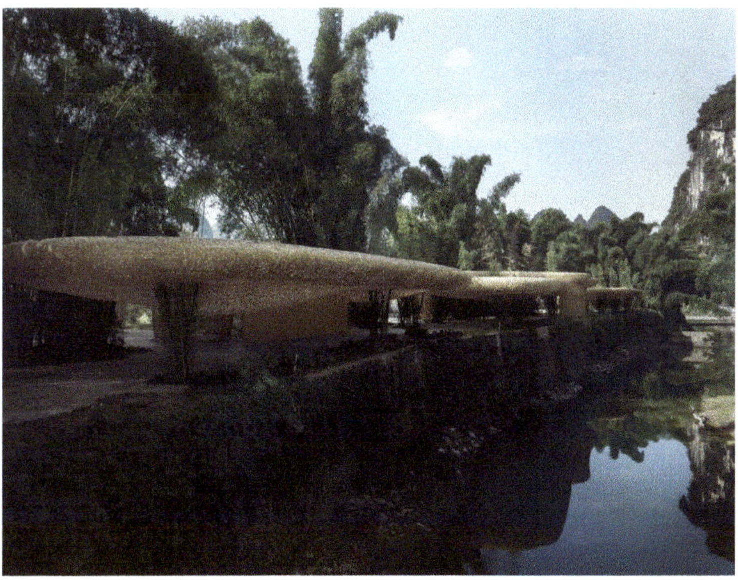

Fig. 6.190 Bamboo canopy, entrance to the open performance stage (*photo by Arch-Exist*)

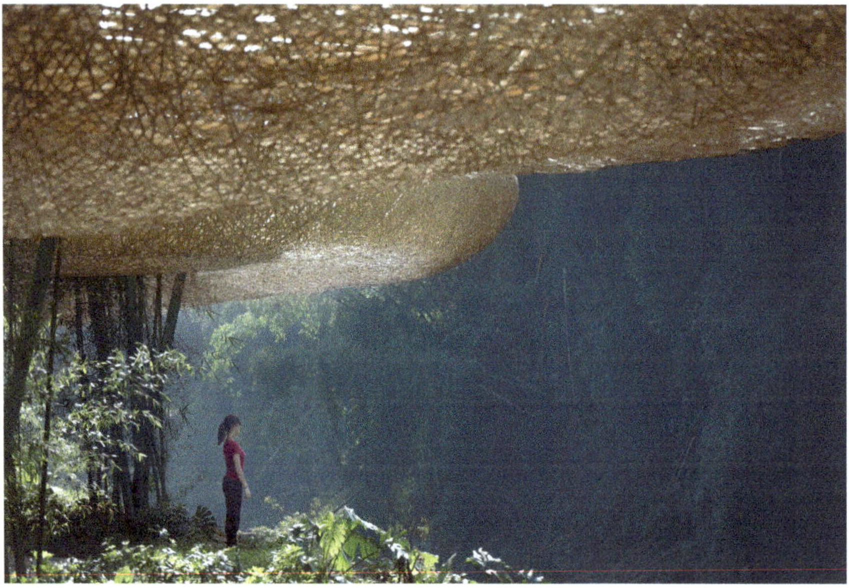

Fig. 6.191 Floating woven bamboo cloud (*photo by Arch-Exist*)

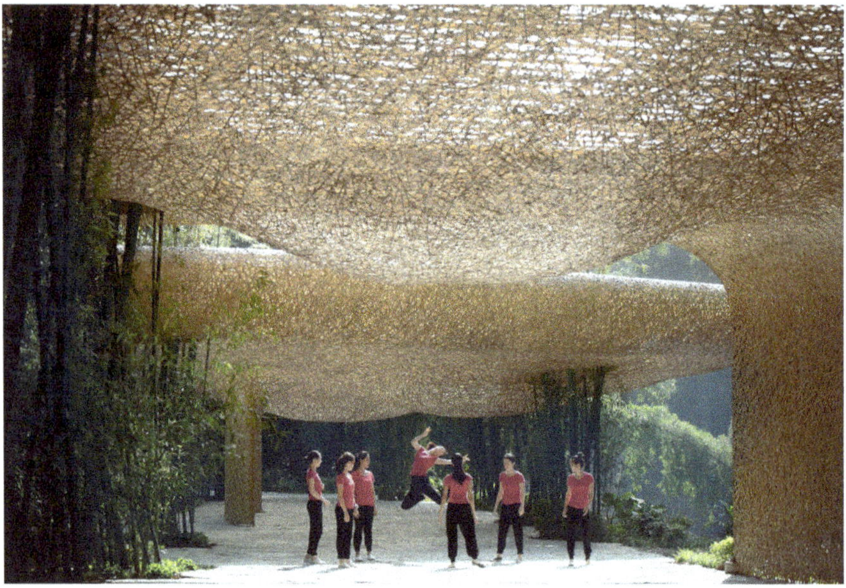

Fig. 6.192 Open theatrical rehearsal stage for the dance school students (*photo by Arch-Exist*)

Peaceful Time, Lingering Charm

Location: Kun Ming, Yunnan, China

Completed: 2017

Architect: Zhu Hongbo

The project explores local characteristics in the way of contemporary gardening, and integrates architecture, interior and landscape design. The design takes "bamboo" as the main material to build three bamboo mountains. Surrounded by bamboo mountains, water is bubbling through. Trees were planted and stones arranged along the edge, and bamboo was used to build a mountain-enclosed courtyard. Stones and trees break through the bamboo hills and emerge through the bamboo corridor, as if in a valley. Among mountains and rivers, visitors can enjoy tea breaks in bamboo pavilions. Surrounded creeks in the valleys, one can look up to the sky between bamboo mountains.

Application of bamboo products: Bamboo scrimber used as landscape materials.

Photo source: Figs. 6.193, 6.194, 6.195 and 6.196 are provided by Planting Design.

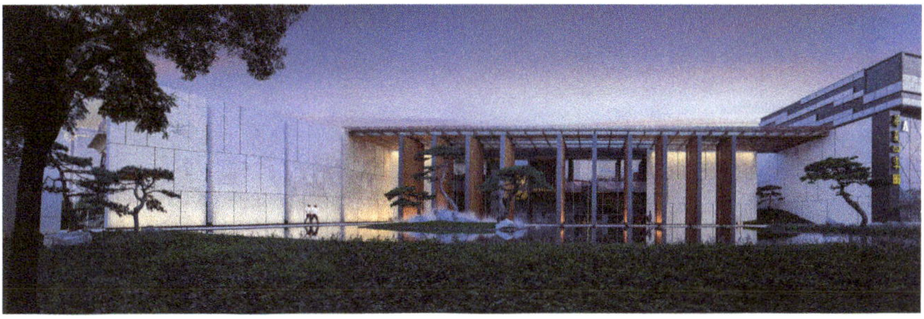

Fig. 6.193 Entrance of the front hall

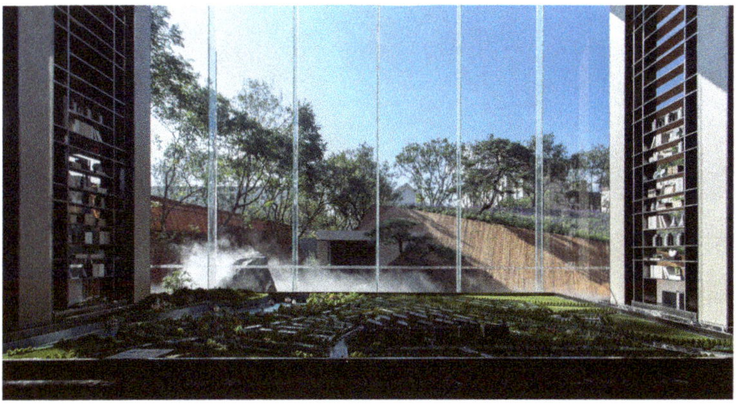

Fig. 6.194 Backyard view from the middle hall

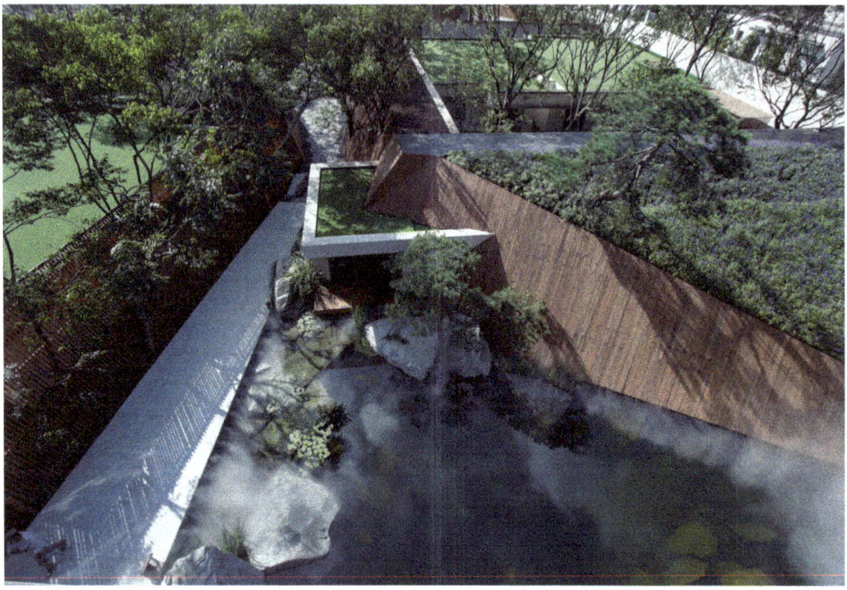

Fig. 6.195 Aerial view of the backyard

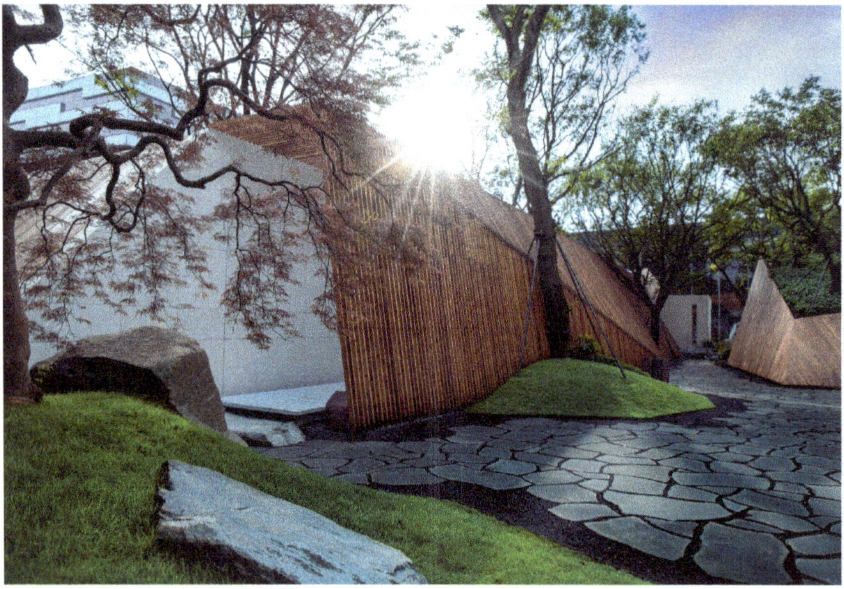

Fig. 6.196 Bamboo mountains

Flexible Landscape

Location: Shanghai, China

Completed: 2015

Architect: Wang Yan

Located in Jing'an Temple Square, this artistic and practical installation came from the public art theme event "Gravitational Field" sponsored by the Group of Architects (goa). The installation was jointly completed by goa and architects of College of Architecture and Urban Planning of Tongji University. The Jing'an Temple Square has sunk 7 m compared to the ground of Nanjing West Road. The installation here is understood as a group of open landscapes, just like the landscape bonsai in the city, which intervene in the whole site with a soft and lightweight gesture. The two arc-shaped uplifts are scattered, like mountains in the basin, cleverly responding to the open-air stage and bustling subway exits. The arc mountain provides shade space for the abundant public activities in the square. The spray of water mist is hidden in the cracks of the poles, which can make the square cooler in summer. At the same time, the cloudy scenery strengthens the visual impression of the urban landscape bonsai.

Application of bamboo products: Because the stone floor of the square is not allowed to be damaged, it was impossible for the whole device to have a foundation or be fixed to the ground. Considering the strong destructive force of typhoon, fifty 20 m long streamlined bamboo scrimber poles were adopted to support the installation.

Photo source: Figs. 6.197, 6.198 and 6.199 are provided by Group of Architects (goa).

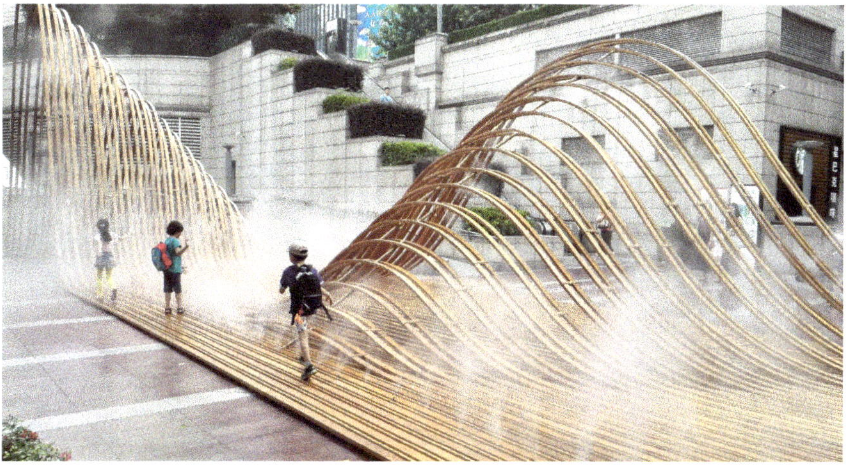

Fig. 6.197 Children playing in the flexible landscape (*photo by Lv Hengzhong*)

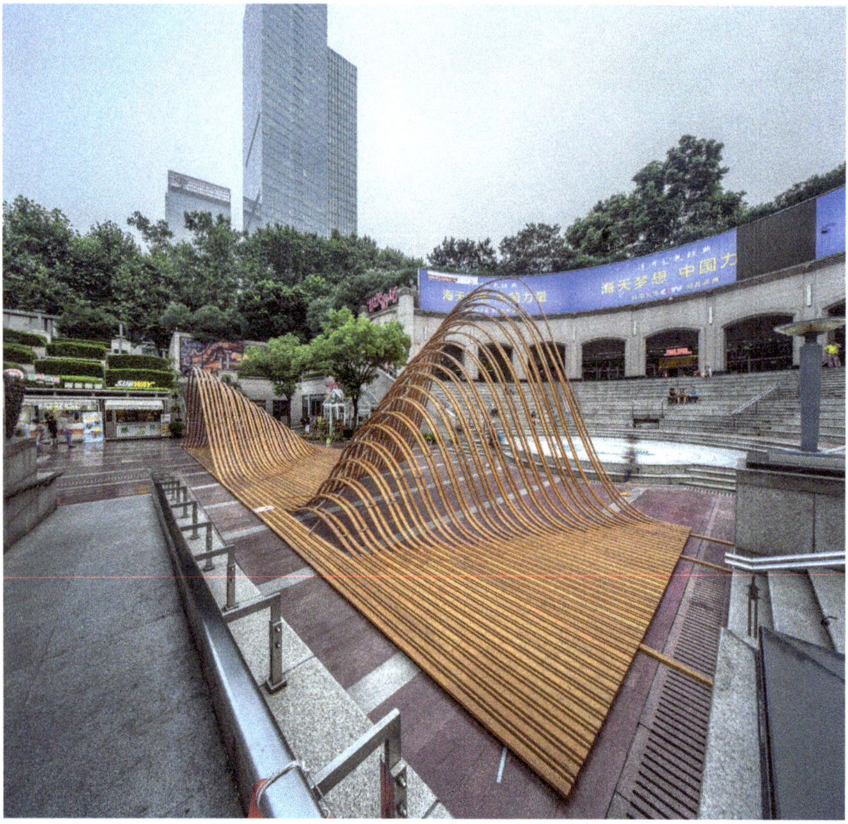

Fig. 6.198 Overall view of flexible landscape (*photo by Lv Hengzhong*)

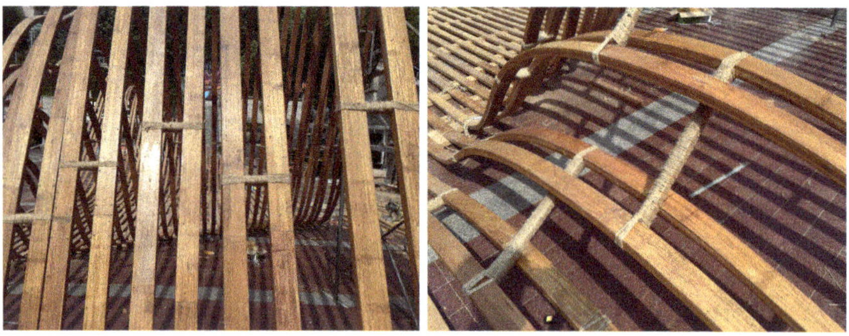

Fig. 6.199 Detail of structure (*photo by Shen Bo*)

Barcelona Re.Set—Identity Pavilion

Location: Barcelona, Spain

Completed: 2014

Architects: Liu Xiaodu, Mengyan, Bernat Riera et al.

In 2014, Barcelona commemorated the 300th anniversary of the city's fall in the 1701–1714 War of the Spanish Succession. Initiated and planned by Fundació Enric Miralles, seven leading contemporary international architect teams were selected to design commemorative public installations for seven historic public spaces in Barcelona in collaboration with internationally renowned architectural institutes. The ancient Plaza Nova Square was a city gate in the past. The remnants of the Roman arch in the corner show the traces of the ancient Roman waterway. "Identity Pavilion" is placed in the site of the ancient Roman canal, its form transitioning from a Roman arch to a Catalonian vault as it traverses the square.

Application of bamboo products: The structure is 6 m high and 30 m long although uses only 10 mm thick bamboo scrimber to form a self-supporting large-span structure. Crosswise bamboo scrimber sheets were prefabricated in factories in China and constructed on site by a team of graduate students from La Salle University in Barcelona and the University of Southern California in the United States under the guidance of the Urbanus Architecture & Design Inc. (URBNUS). This style of building with such a single component is a collective construction, which is in line with the essence of the traditional activity of the famous Catalonian "Castell" and expresses the will of people united together.

Photo source: Figs. 6.200, 6.201 and 6.202 are provided by URBNUS.

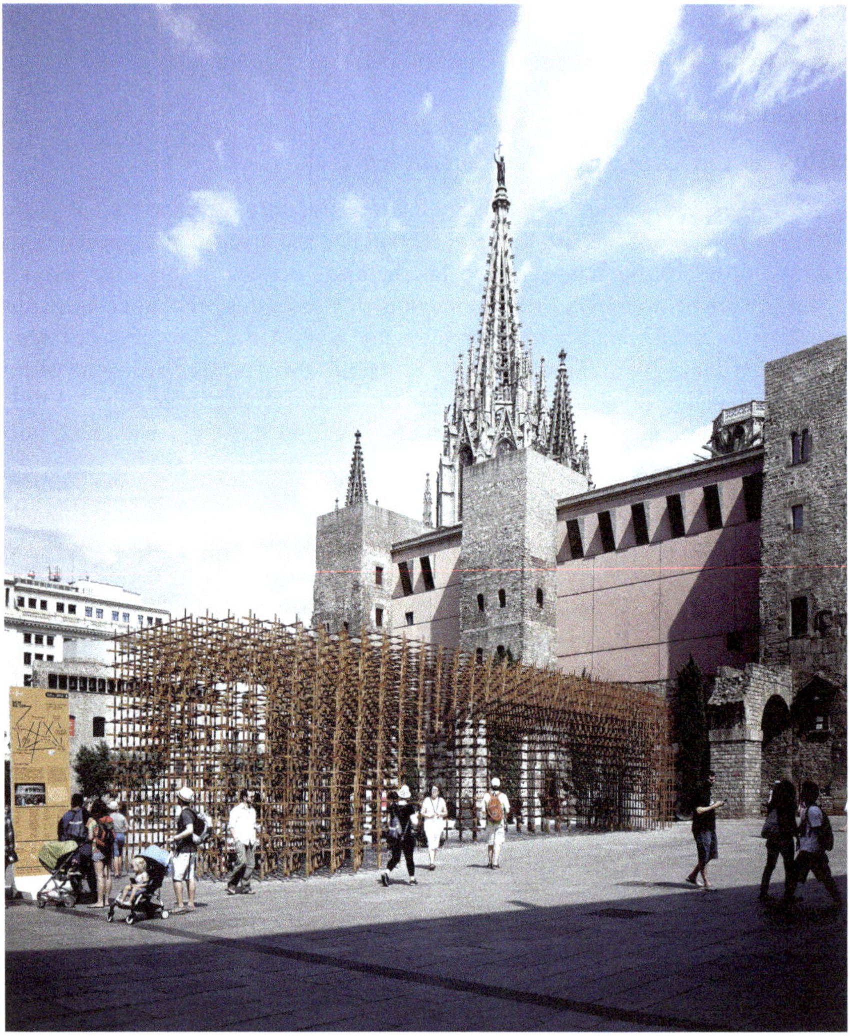

Fig. 6.200 Barcelona Re.Set of URBNUS

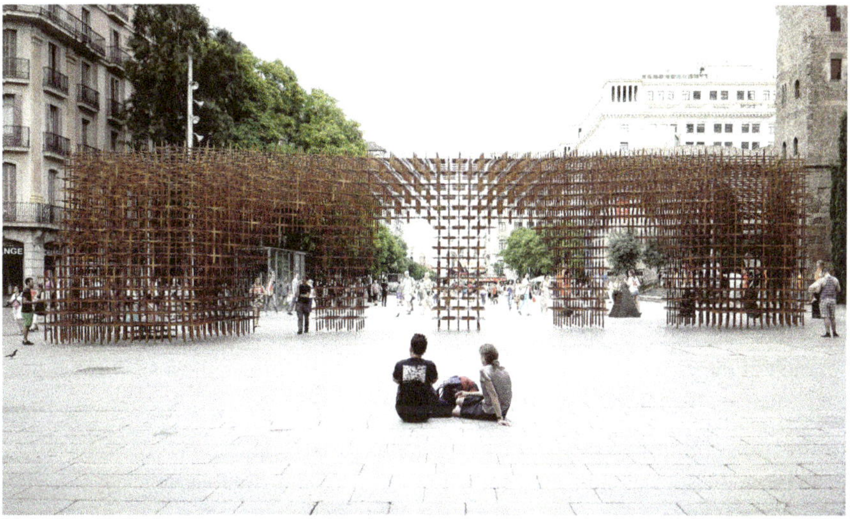

Fig. 6.201 Interaction between local residents and Barcelona Re.Set of URBNUS

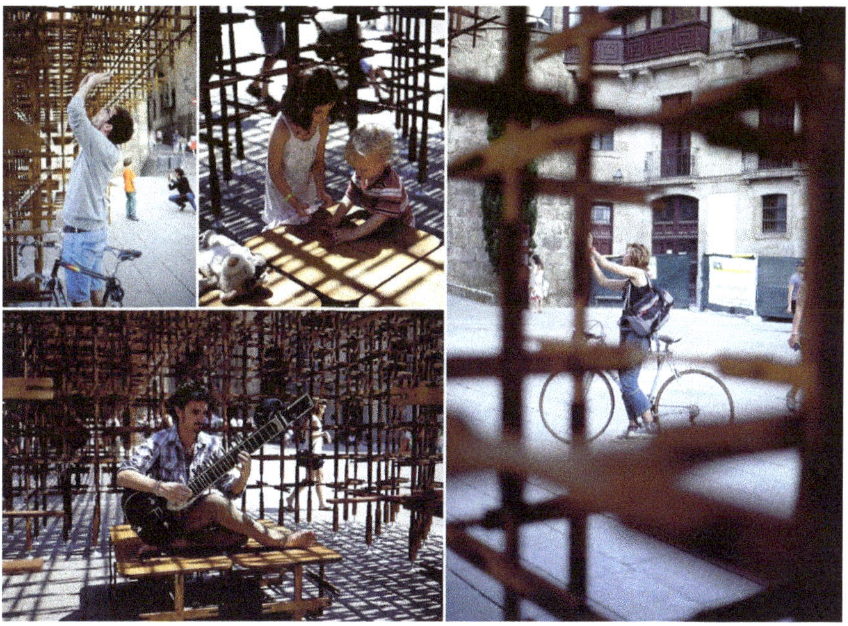

Fig. 6.202 Barcelona Re.Set of URBNUS

Outdoor Bamboo Landscape of Shanghai Qinsen Group at 2019 Beijing Expo

Location: Beijing, China

Completed: 2019

Architect: Qi Yanwei, Feng Sheng, Chen Hanyi et al.

The outdoor special-shaped bamboo structure landscape of Shanghai Qinsen Group carries the application of glued laminated bamboo in outdoor landscape to extremes. The whole design respects the beauty of nature and aims to use natural materials. Its design inspiration comes from the natural curl and undulating three-dimensional space structure of wood shavings. Combining with the environment and the current situation of the site, the designer takes the overall layout to create a sense of natural beauty and spatial hierarchy.

Application of bamboo products: In order to meet the requirements of architectural space form, the special-shaped hyperboloid space structure with double-layer curved mesh was adopted in the structural design of engineered bamboo, which combines the characteristics of strong tensile strength and weak shear strength. The soft and winding engineered bamboo structure shows a rich level of transparent beauty under the light. The design with modern sense perfectly presents the rigidity and softness of bamboo, simple and pure.

Photo source: Figs. 6.203, 6.204 and 6.205 are provided by Shanghai Qinsen Landscape Co., Ltd.

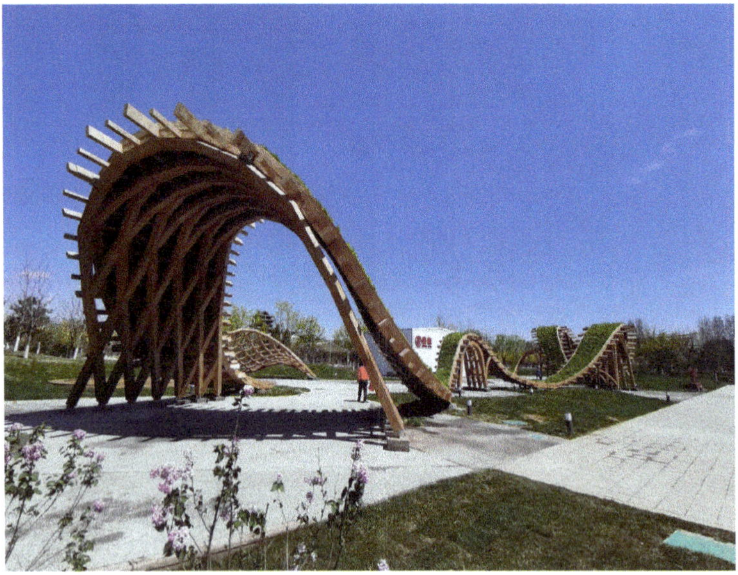

Fig. 6.203 Wood shavings

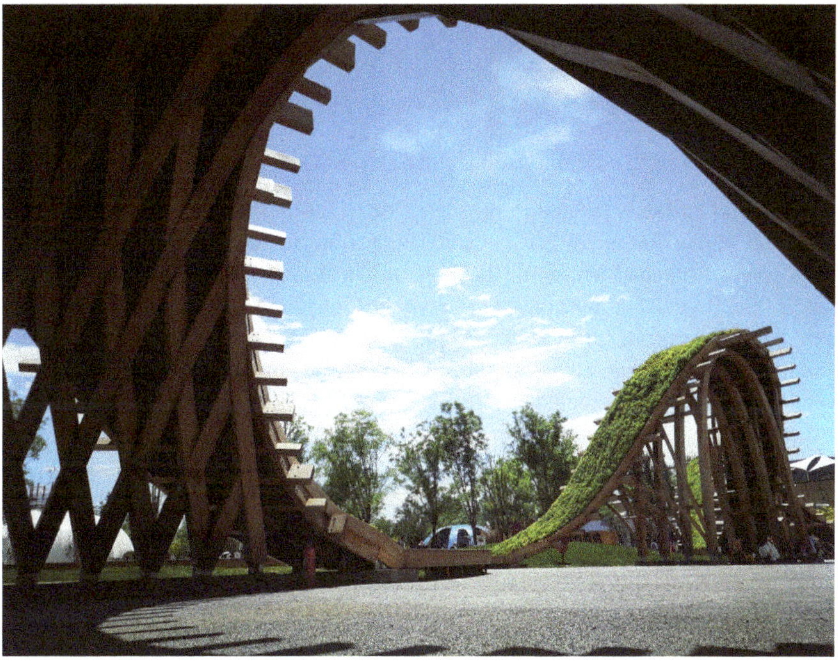

Fig. 6.204 Soft curve

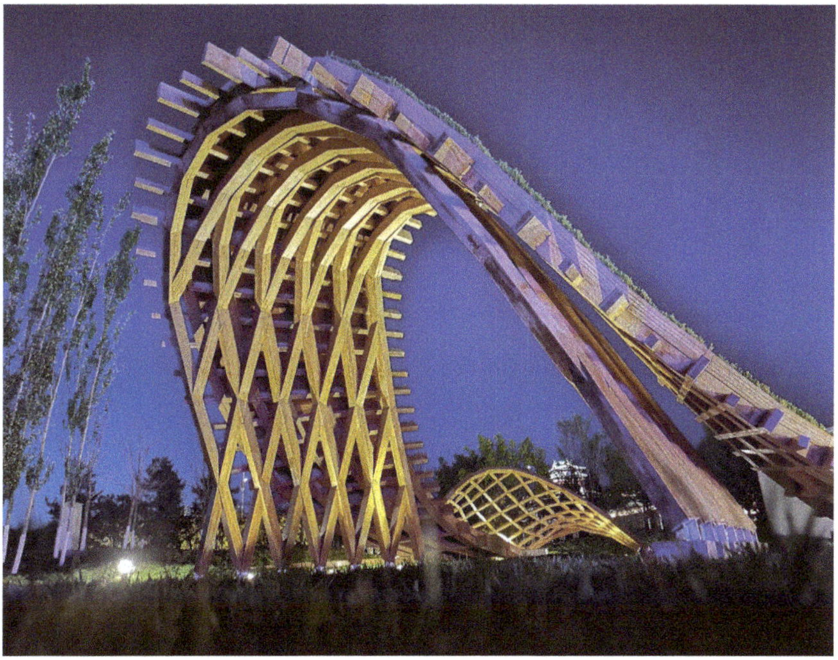

Fig. 6.205 Mortise and tenon structure

Baiguo Garden of Beijing Landscape Bureau at 2019 Beijing Expo

Location: Beijing, China

Completed: 2019

Architect: Beijing Landscape Architecture Design Co., Ltd.

The Baiguo Garden is located in the west side of the 2019 Beijing Expo. It is built along the Vanke Botanical Museum, forming a horticultural exhibition area with the adjacent Botanical Museum, Baishu Garden, Baicao Garden and Horticultural Towns. The Baiguo Garden consists of a series of interesting orchard landscapes. The main bamboo scrimber structures—Apple Gate, Music Fruit Gallery, Float Garden Gallery, Rainbow Fruit Bar, Big and Small Bamboo Houses and Grape Shelves—were prefabricated and erected on site. Each landscape is interesting. The main scene at the entrance, Apple Gate, for example, is an open space surrounded by a micro-terrain, composed of a series of bamboo galleries and waterscapes. The arc-shaped bamboo gallery resembles half an apple which is obliquely planted into the ground. A circular shallow pool is set in the depression where the stem grows. The reflection in the pool expands the bamboo gallery into a circular shape.

Application of bamboo products: Bamboo scrimber as a structural material for landscape.

Photo source: Figs. 6.206, 6.207 and 6.208 are provided by Hongyazhuyuan Science and Technology Company.

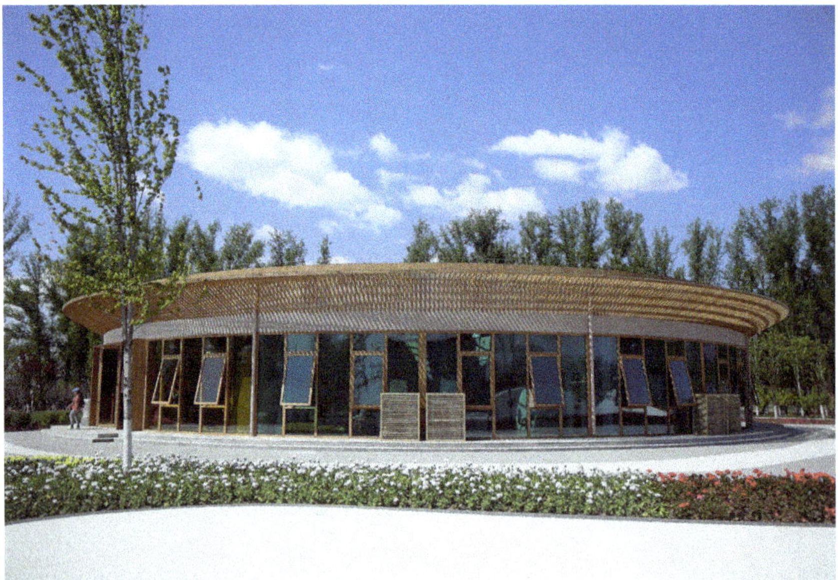

Fig. 6.206 Rainbow fruit bar

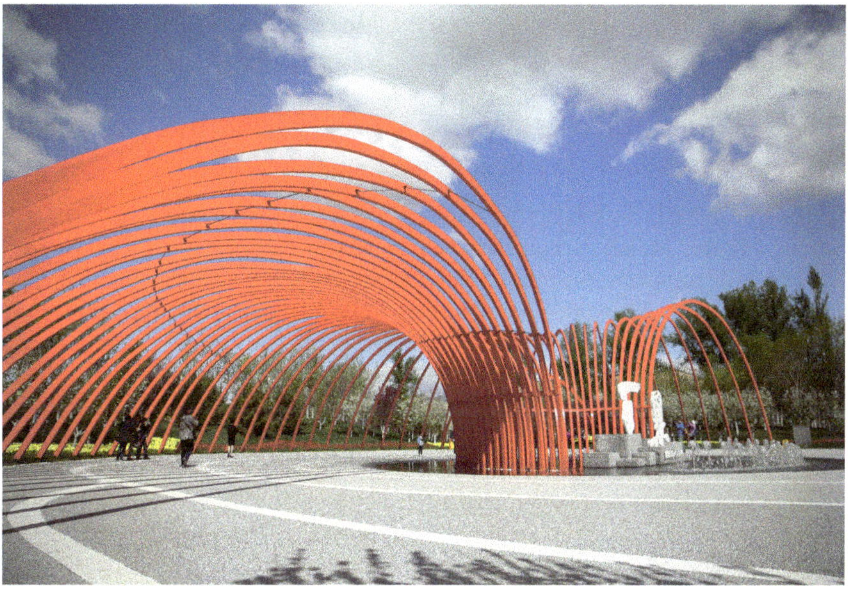

Fig. 6.207 Apple gate

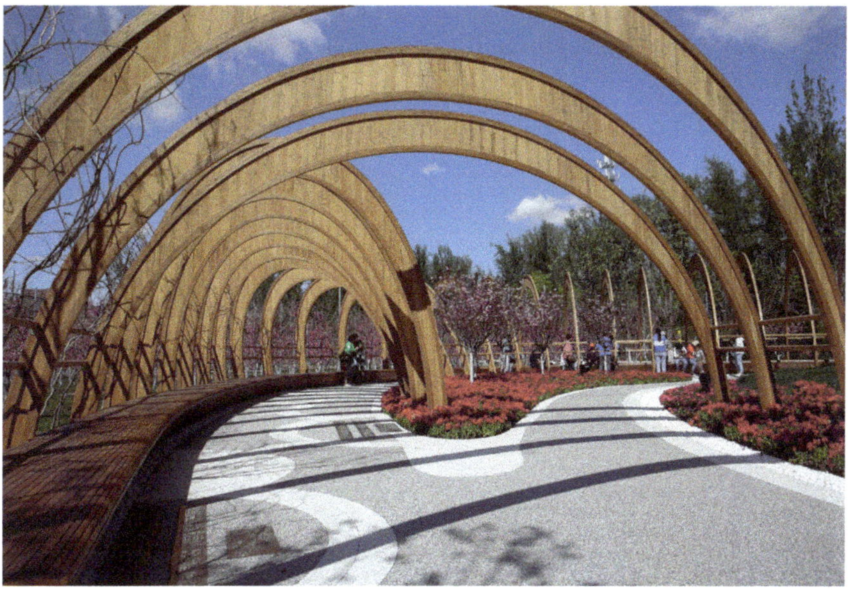

Fig. 6.208 Float garden gallery

China House Vision—Life in the Future: Living Garden

Location: Beijing, China

Completed: 2018

Architects: Ma Yansong, Dang Qun, Yosuke Hayano et al.

"China House Vision—Life in the Future: Living Garden" was planned by International Graphic Master Kenya Hara and sponsored by Great Wall Club (GWC). Ten groups of architects cooperated with industry to build the ideal "future home". "Living Garden" is the expression of the emotional relationship between human and nature in the living environment by MAD Architects and Hanergy Group. It is a group of works with open imagination between architecture and landscape. From a distance, the main "roof" of the building, composed of more than 200 pieces of Hanergy film roof components, is like a cloud floating in mid-air. Each board can be adjusted with the sunlight to achieve efficient power generation. The electricity produced is sufficient for the average daily consumption of three families. Sunlight also penetrates through the transparent film into the space. The gap between the ground and the "roof" is the interior and exterior space of the home. Living space and landscape alternate with each other. Accompanied by the flowing air, warm sunshine and breeze, people's emotions are nourished in the transformation of the environment.

Application of bamboo products: Bamboo scrimber used as a structural material for landscape.

Photo source: Figs. 6.209, 6.210 and 6.211 are provided by MAD Architects.

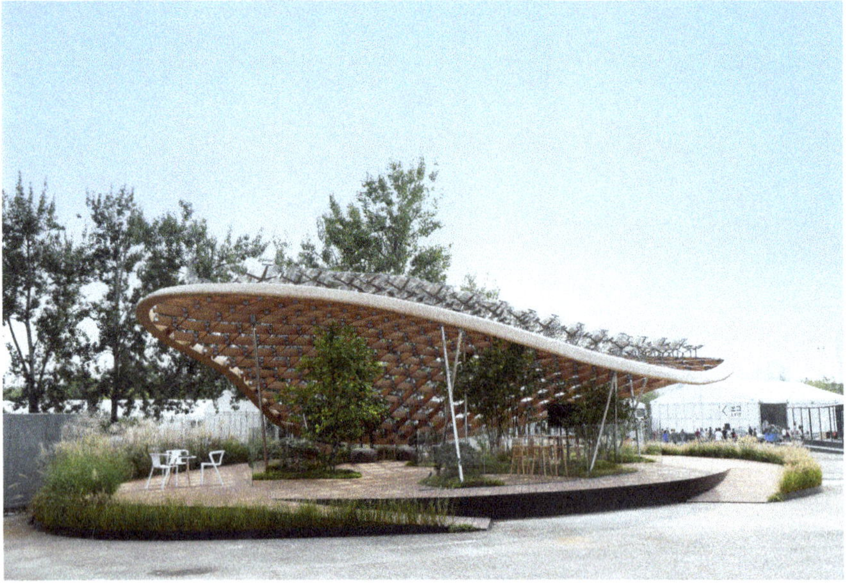

Fig. 6.209 Living garden (*photo by Tian Fangfang*)

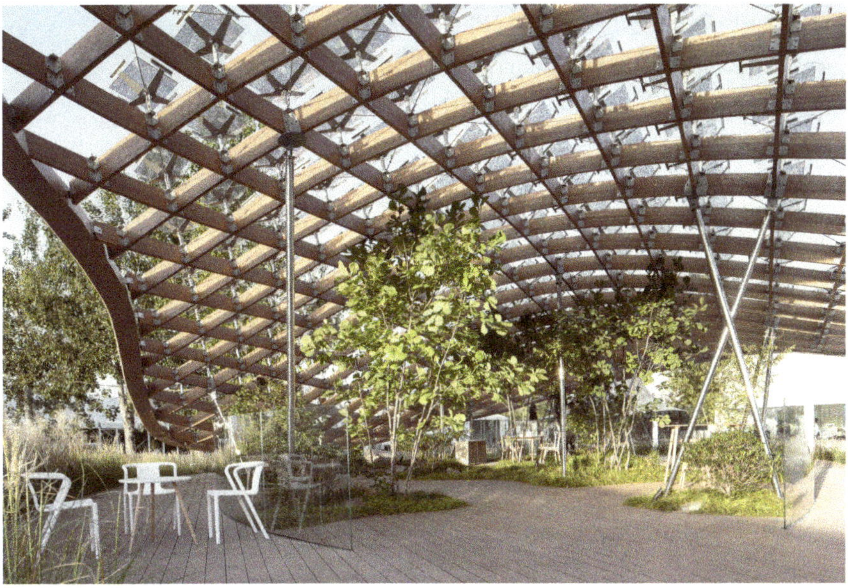

Fig. 6.210 Under the living garden (*photo by Tian Fangfang*)

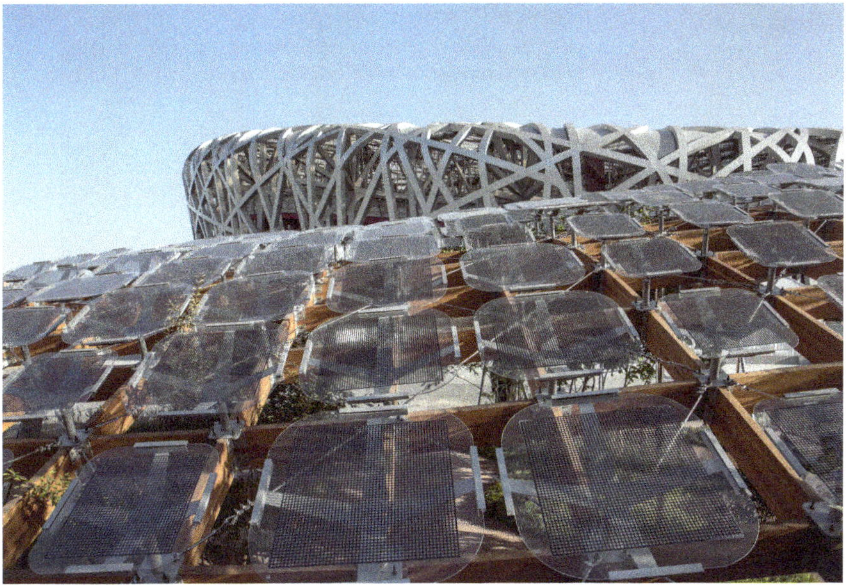

Fig. 6.211 Hanergy film roof (*photo by Tian Fangfang*)

Nanjing Youth Olympic Games Sports Park

Location: Nanjing, Jiangsu, China

Completed: 2015

Architect: Xu Zhiwei

Nanjing Youth Olympic Games Sports Park is located in the Jianye District of Nanjing, Jiangsu, China. The Park covers an area of 500,000 m^2 with a total building area of 97,000 m^2 [2].

Application of bamboo products: In order to present the environmental protection concept of the Olympic Games, the design adopts durable outdoor bamboo materials, covering an area of 30,000 m^2. The park is open to the public free of charge.

Photo source: Figs. 6.212 and 6.213 are provided by Hangzhou Dasso Technology Co., Ltd.

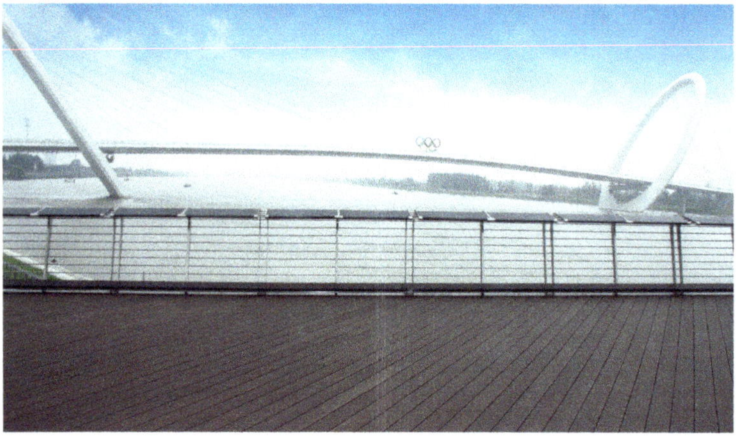

Fig. 6.212 Nanjing Youth Olympic Games Sports Park

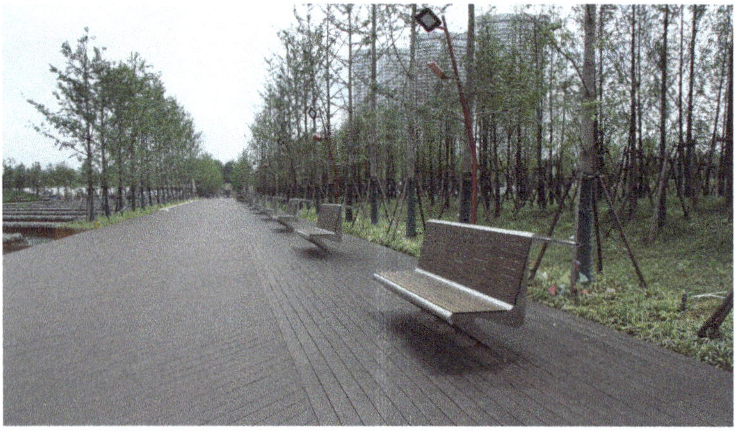

Fig. 6.213 Bamboo benches

Hong Kong-Zhuhai-Macao Bridge Landscape Island

Location: Zhu Hai, Guangzhou, China

Completed: 2018

Architect: Meng Fanchao

The Hong Kong-Zhuhai-Macao Bridge spans the "Lingding Ocean", near Hong Kong in the east, next to Zhuhai in Guangdong Province, and Macao in the west. Its total length is about 55 km. This project, built jointly by Guangdong, Hong Kong and Macao, is also the longest sea-crossing bridge in the world [2].

Application of bamboo products: A large number of outdoor bamboo walkways have been laid on the landscape island of the Hong Kong-Zhuhai-Macao Bridge. The high humidity and salinity of the sea water environment is a highly corrosive environment; yet the durable bamboo materials used achieve up to grade 1 corrosion resistance.

Photo source: Figs. 6.214, 6.215 and 6.216 are provided by Hangzhou Dasso Technology Co., Ltd.

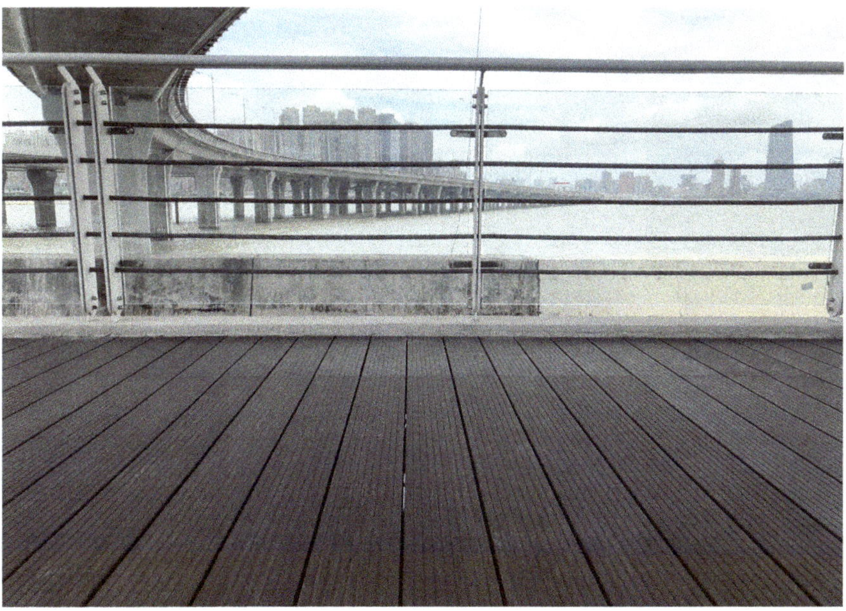

Fig. 6.214 Bamboo decking

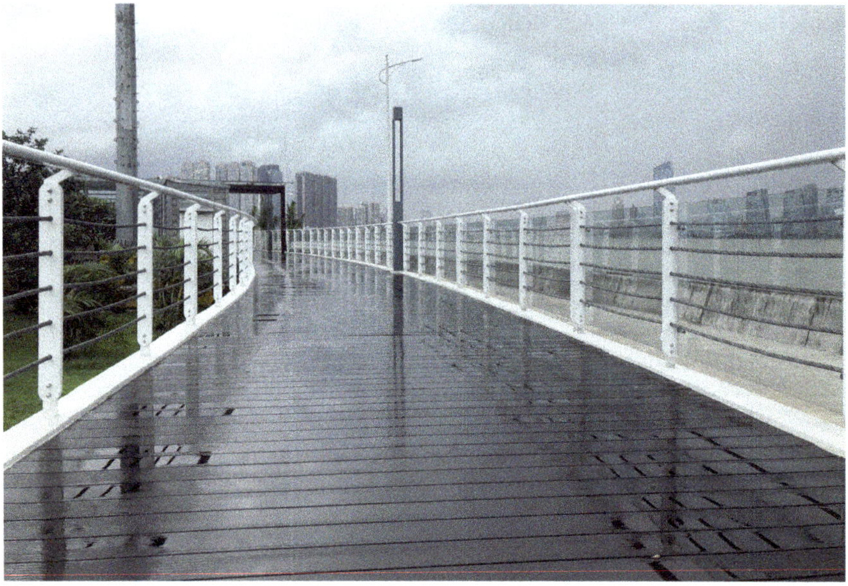

Fig. 6.215 Corridor with bamboo decking

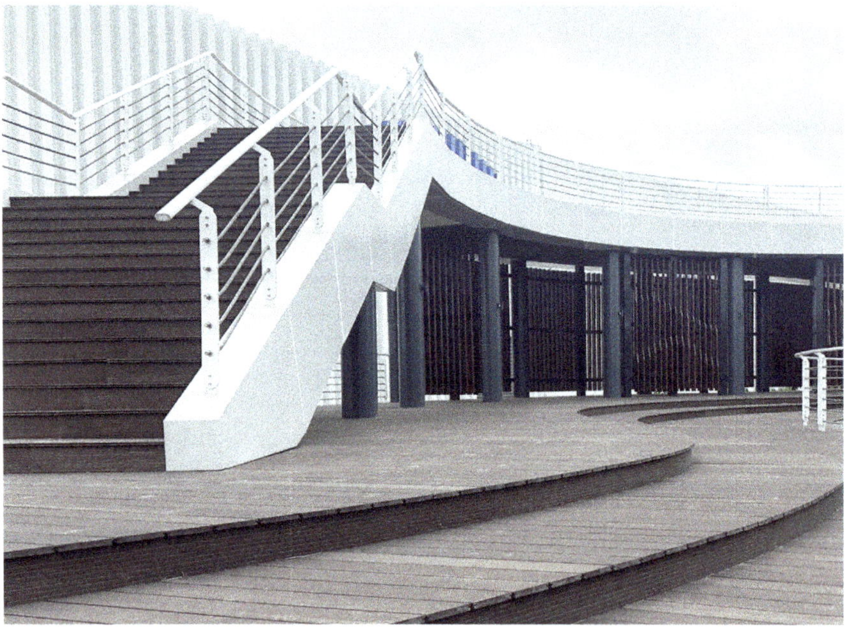

Fig. 6.216 Bamboo stairs and flooring

Xiong'an New District Citizen Service Center

Location: Xiong'an, Hebei, China

Completed: 2018

Architect: Huang Wenlong

Xiong'an New Citizen Service Center is the administrative organization of Xiongan New District in Hebei Province, China. It is located in Xiaobaita and Mazhuang Village, east of Rongcheng, Xiongan New District. Its total construction area is 960,000 m^2, and the total planned area is 2424 ha; the total investment is about 800 million yuan (124 million USD) [3].

Application of bamboo products: Outdoor floor decking, landscape tree circumference and leisure seats.

Photo source: Figs. 6.217 and 6.218 are provided by Hunan Taohuajiang Bamboo Science and Technology Co., Ltd.

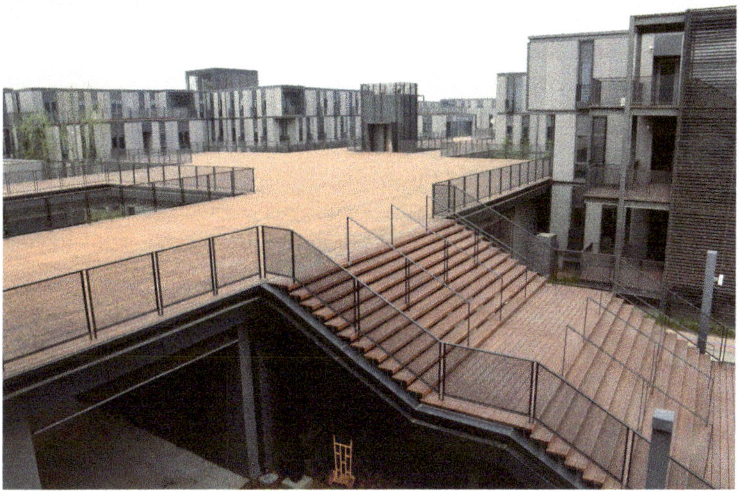

Fig. 6.217 Bamboo stairs and flooring

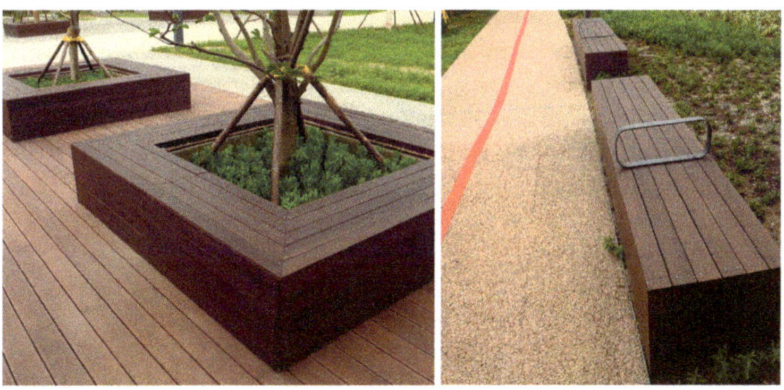

Fig. 6.218 Bamboo benches

6.4 Rural Construction

International Bamboo Architecture Biennial

Location: Longquan, Zhejiang, China

Completed: 2016

Curators: Ge Qiantao: Double Helix Bridge; George Kunihiro (USA): Traditional Ceramic Workshop

Architects: Wise Architecture (South Korea): Bamboo · Children's Handmade Workshop; Anna Heringer (Germany): Design Hotels and Youth Hostels; Kengo Kuma (Japan): Contemporary Celadon Ceramic Museum; Mauricio Cardenas Laverde (Italy): Eco-Energy Efficient Experimental House; Li Xiaodong: Creative and Innovation Center; Keisuke Maeda (Japan): Invited Ceramist Workshop; Vo Trong Nghia (Vietnam): Reception Center; Madhura Prematilleke (Sri Lanka): Public Ceramics Workshop (Unbuilt); Simón Veléz (Colombia): Boutique Hotel; Yang Xu (1967–2015): Art Hotel· Living in the Water, Art Hotel· Living in the Flowers.

The first International Bamboo Architecture Biennial opened in Baoxi Township, Longquan City, Zhejiang Province in 2013. It was co-curated by Chinese artist Ge Qiantao and American architect George Kunihiro. Twelve well-known international and domestic architects were invited to design 16 bamboo buildings in different styles; this reflected the theme of the Biennial Exhibition "Spirit of Place, Rural Construction". The architects, from nine countries, gathered in Baoxi to conduct an investigation and practice at the "local" level. Indigenous people with "low technology" were invited to participate in the construction of the project, so as to awaken cultural memory, reflect technology and lifestyle of the countryside, and use diverse and different architectural styles to blend into the cultural context of the village. Although architects held different opinions of "local" culture, and their architectural design expressions were quite different, their spiritual connotations were identical. They reflected "local" by an ideal concept: Chinese culture by international expression to transcend the concept of time and space and explore the infinite possibility of innovating with bamboo architecture. Their works transcended their experience of specific cultural symbols. At the same time, they worked with villagers to build a spiritual and interactive new rural space. On September 28, 2016, the International Bamboo Architecture Biennial was officially opened. This unique and ongoing Bamboo Architecture Biennial has now become the most attractive artistic landscape in the Chinese countryside. People feel the creativity of bamboo architecture and art as well as the enlightenment brought by the life cycle of natural architecture. The "local" principle of the Biennial has been adopted in China; its influence has become the cultural choice of "rural construction" across the country.

Application of bamboo products: Full-culm and engineered bamboo as decorative and structural materials.

Photo source: Figs. 6.219, 6.220, 6.221, 6.222, 6.223, 6.224, 6.225, 6.226, 6.227, 6.228, 6.229, 6.230, 6.231, 6.232, 6.233, 6.234, 6.235, 6.236, 6.237, 6.238, 6.239, 6.240 and 6.241 are provided by Qiantao Studio.

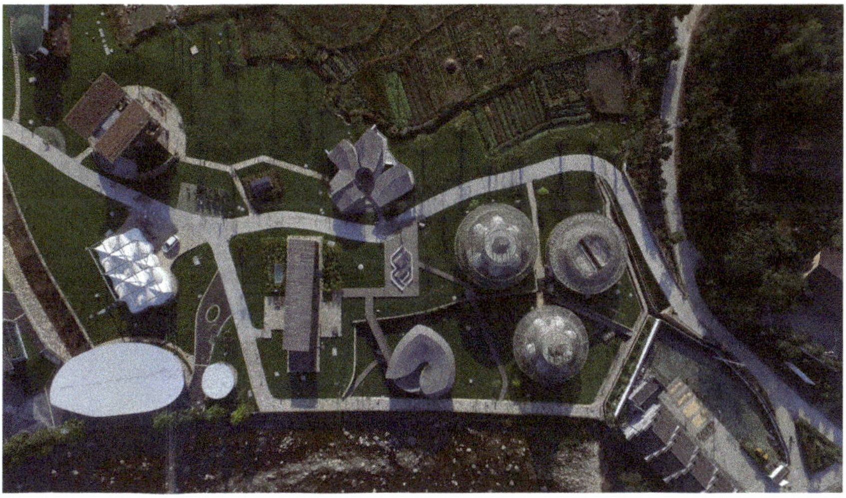

Fig. 6.219 Aerial view of International Bamboo Architecture Biennial

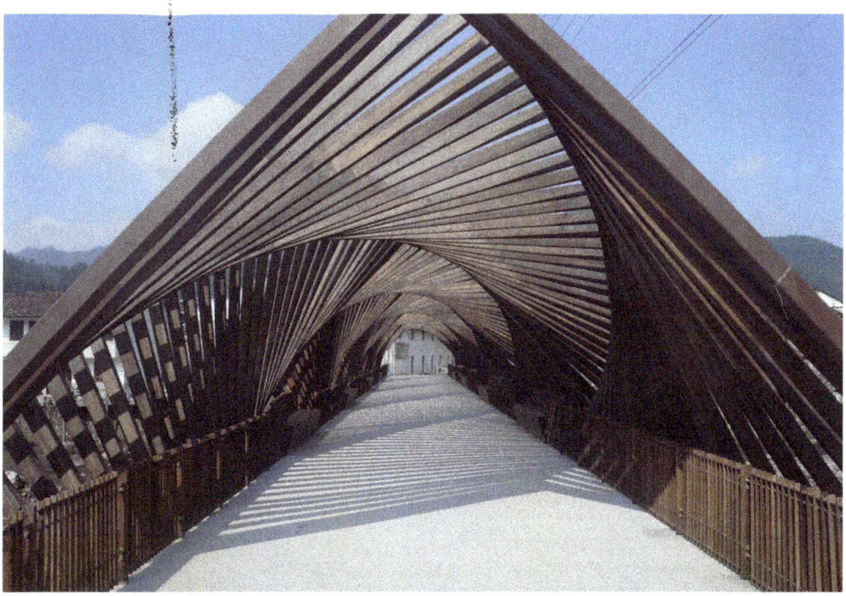

Fig. 6.220 Double Helix Bridge

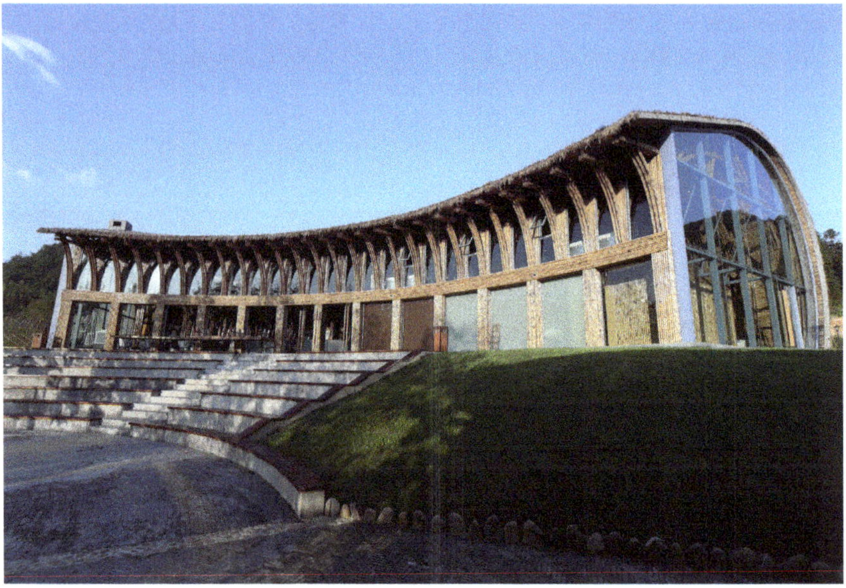

Fig. 6.221 Reception Centre

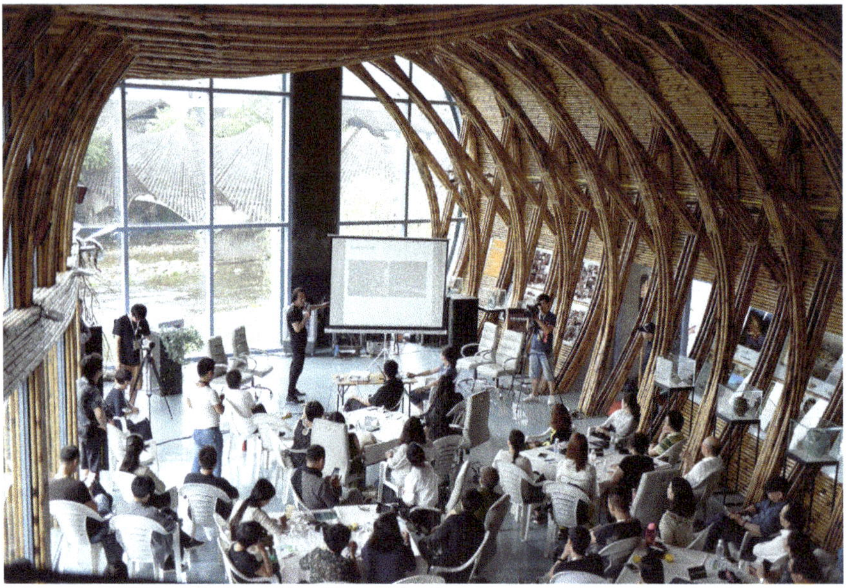

Fig. 6.222 Launch Ceremony of International Bamboo Architecture Biennial in the Reception Centre

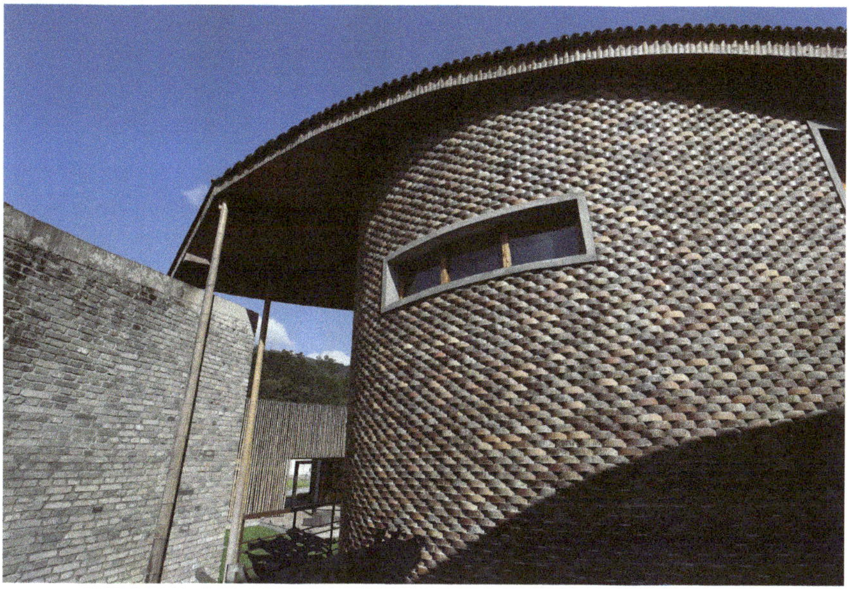

Fig. 6.223 Art Hotel·Living in the Water

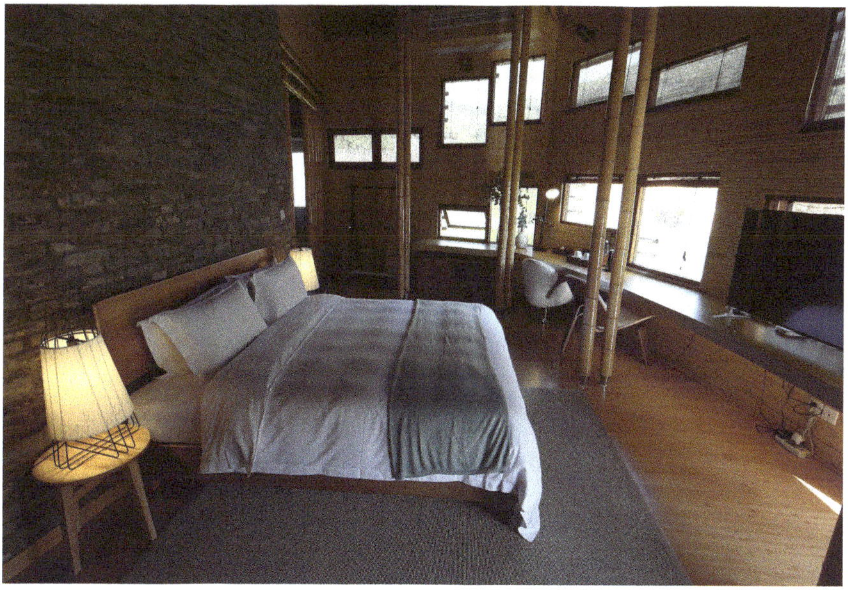

Fig. 6.224 Art Hotel-Living in the Water

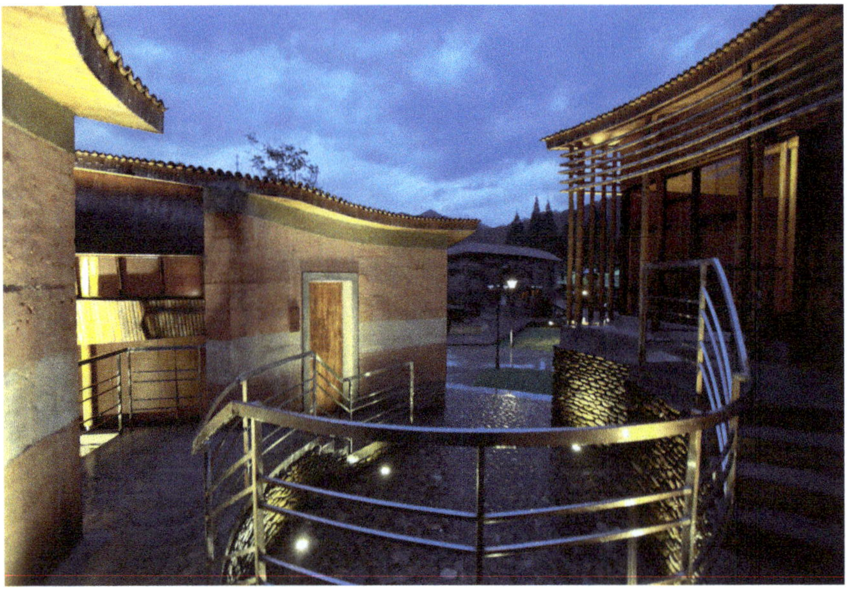

Fig. 6.225 Art Hotel·Living in the Flowers

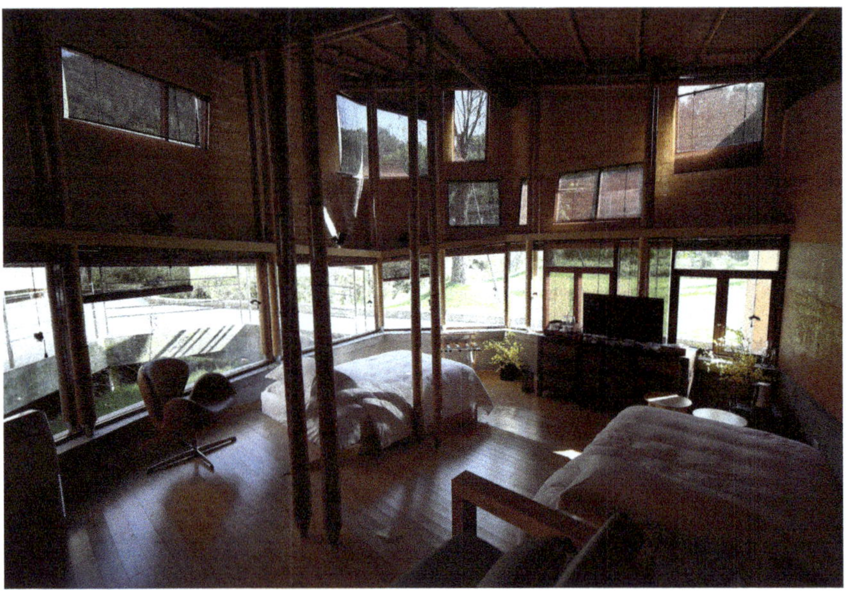

Fig. 6.226 Art Hotel·Living in the Flowers

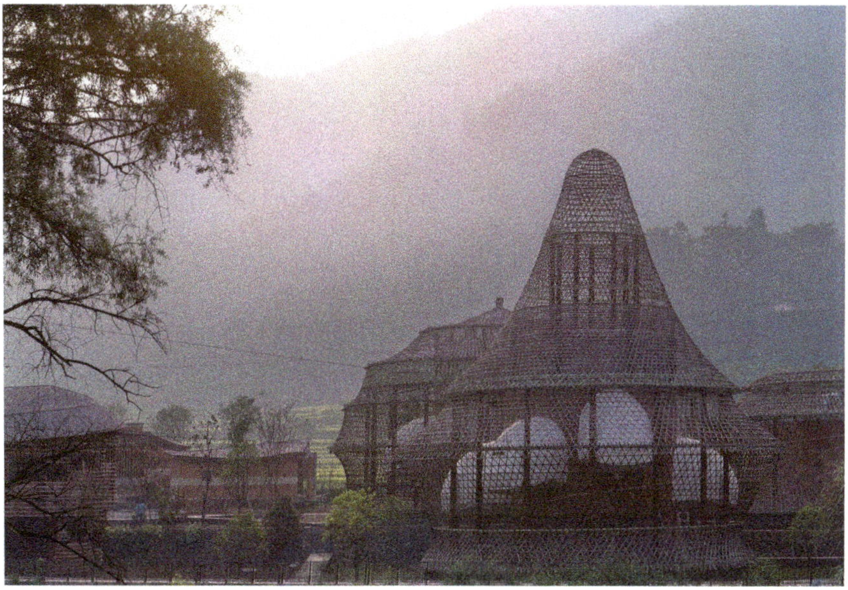

Fig. 6.227 Design Hotels and Youth Hostels

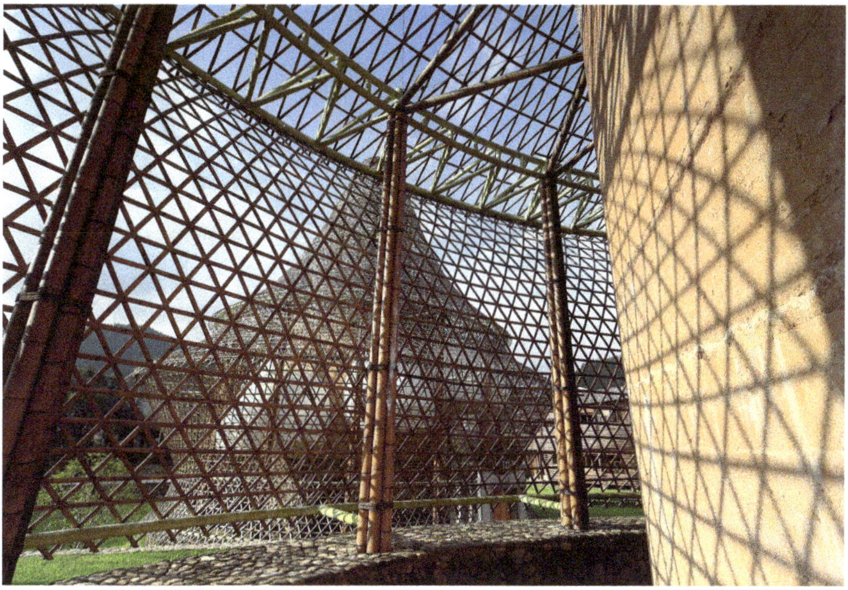

Fig. 6.228 Design Hotels and Youth Hostels

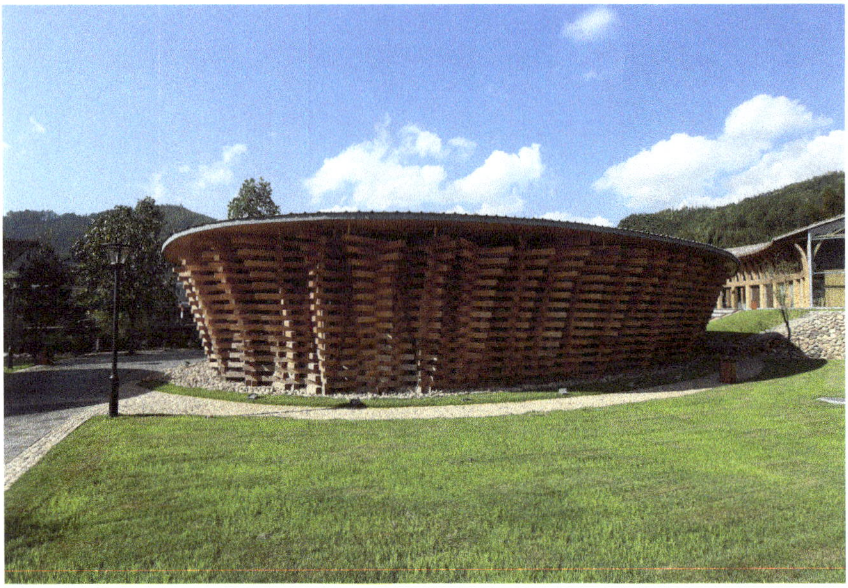

Fig. 6.229 Contemporary Celadon Ceramic Museum

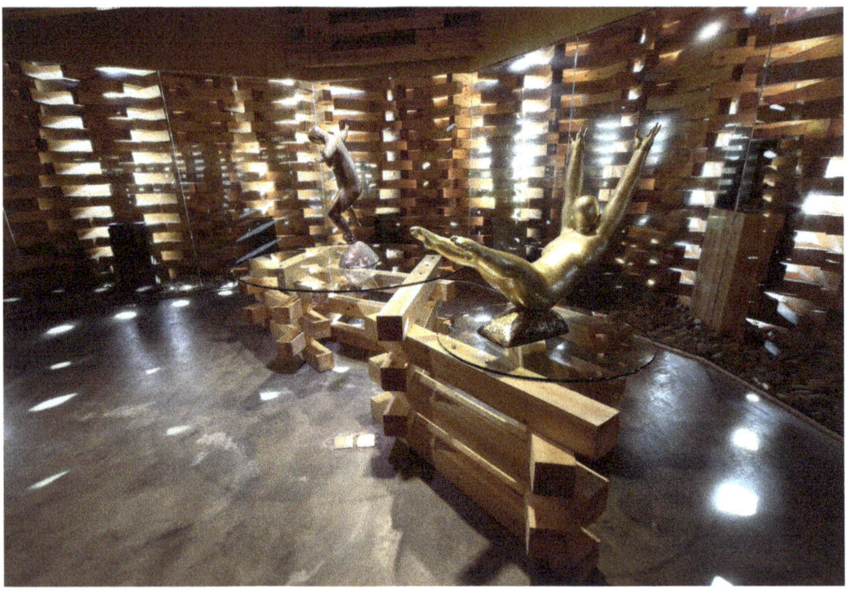

Fig. 6.230 Exhibition in the Contemporary Celadon Ceramic Museum

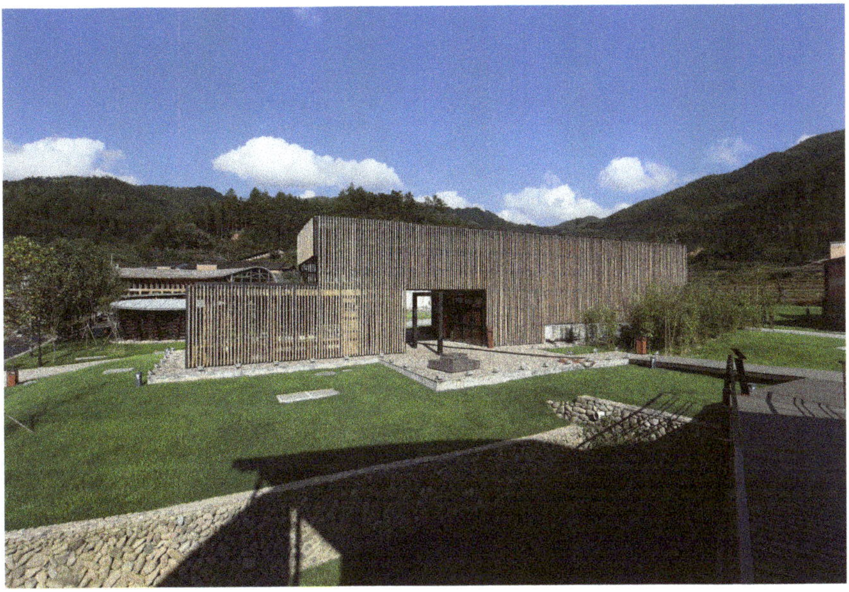

Fig. 6.231 Creative and Innovation Centre

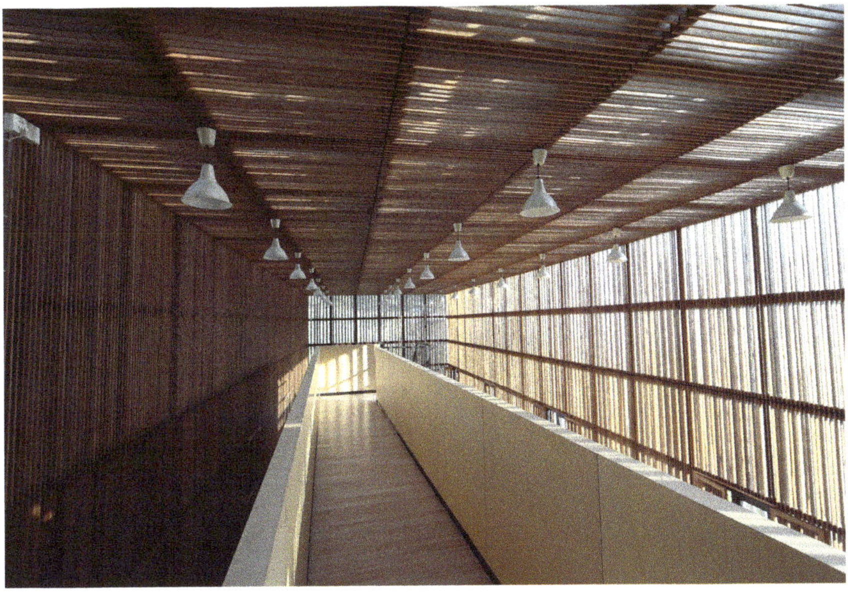

Fig. 6.232 Interior Space of Creative and Innovation Centre

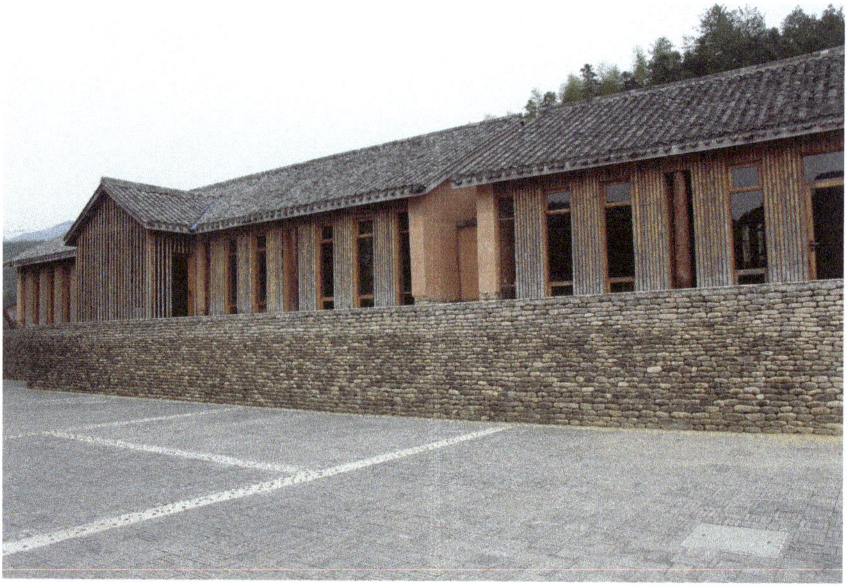

Fig. 6.233 Traditional Ceramic Workshop

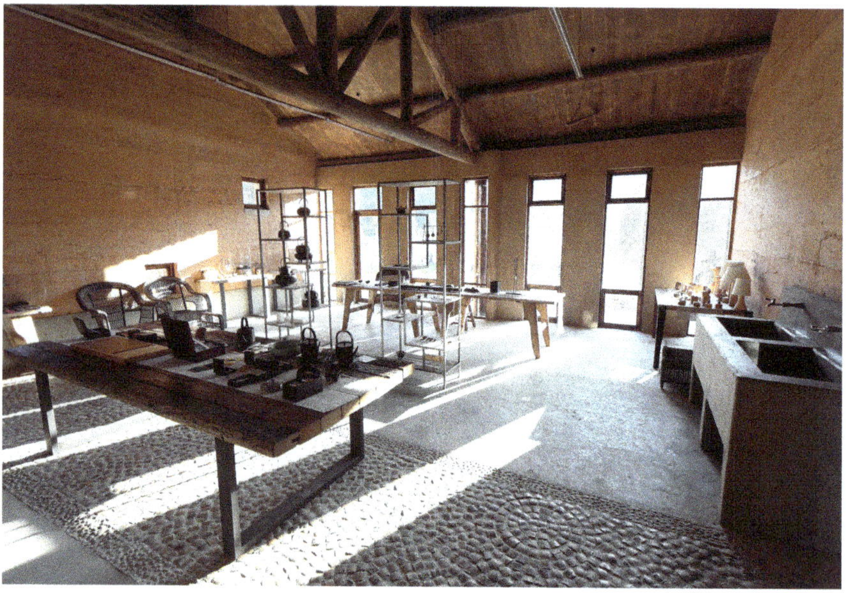

Fig. 6.234 Traditional Ceramic Workshop

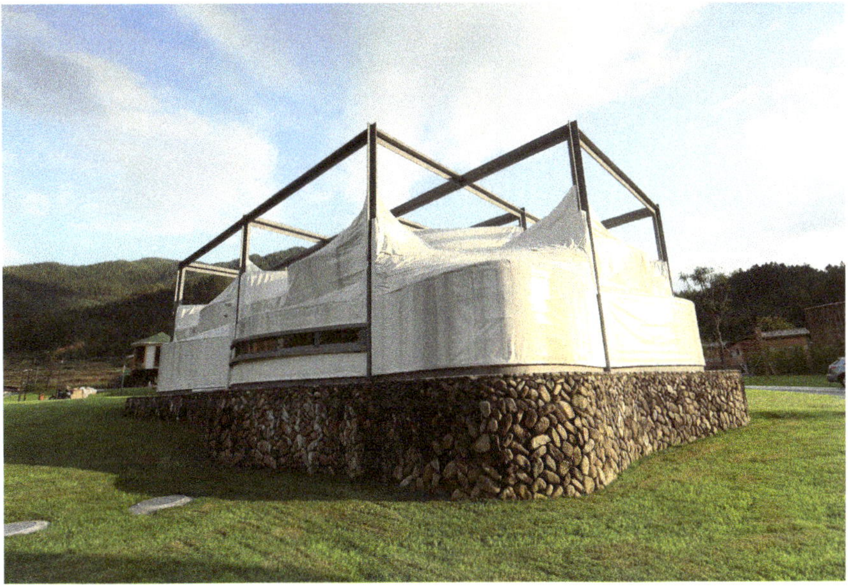

Fig. 6.235 Bamboo Children's Handmade Workshop

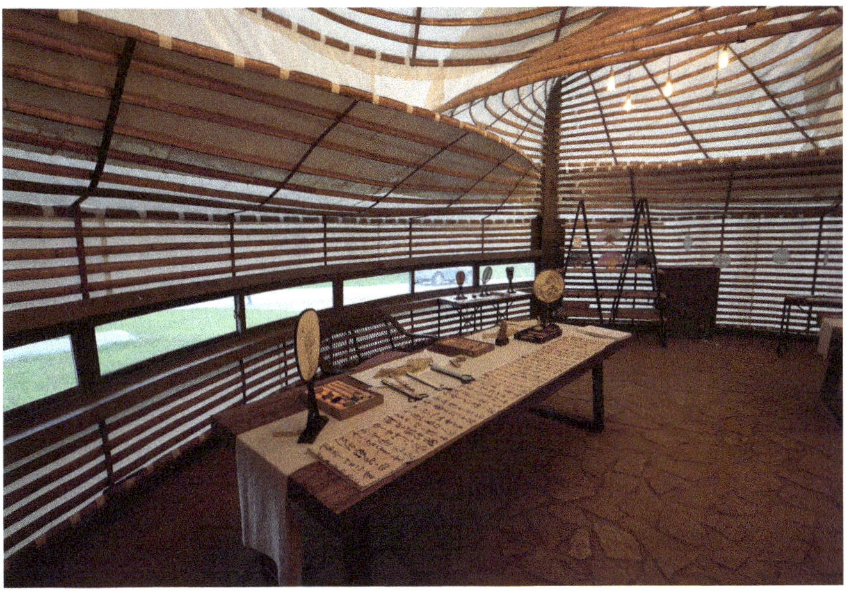

Fig. 6.236 Bamboo Children's Handmade Workshop

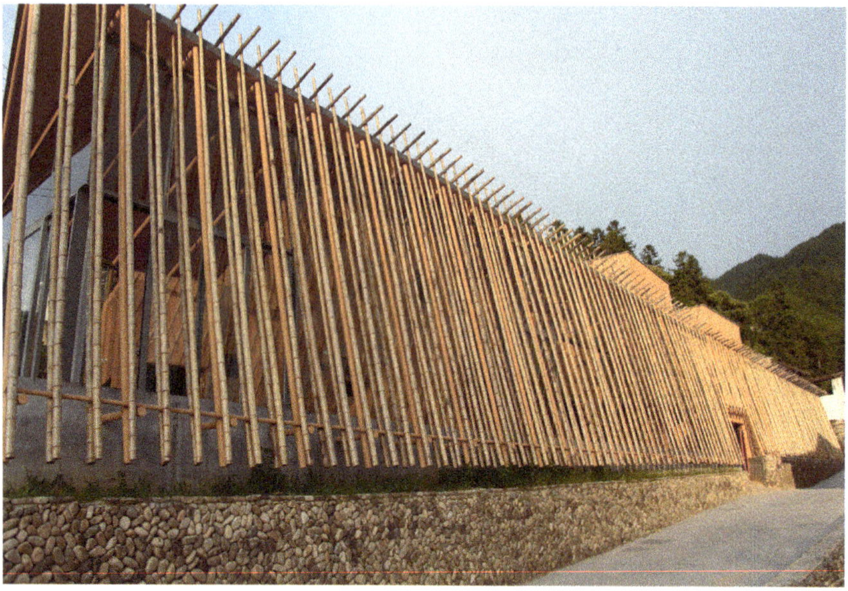

Fig. 6.237 Invited Ceramist Workshop

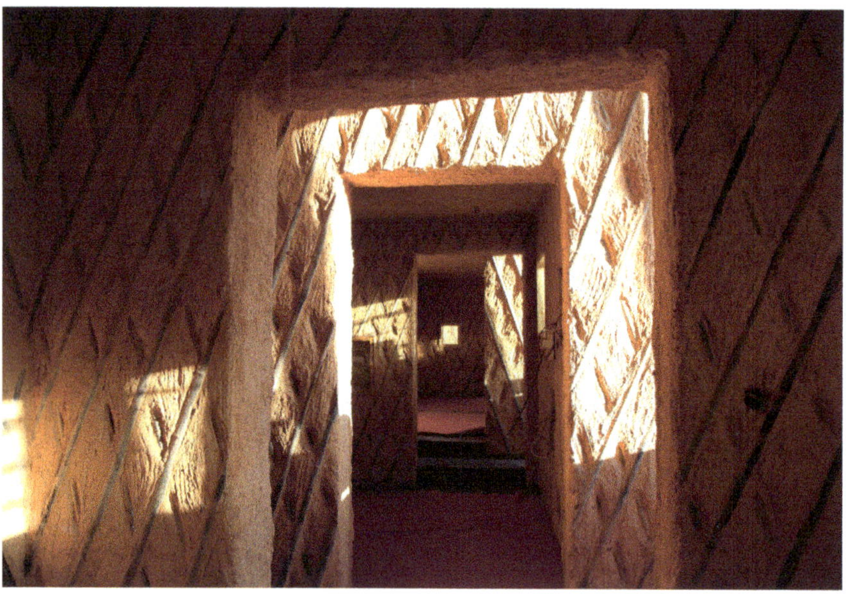

Fig. 6.238 Invited Ceramist Workshop

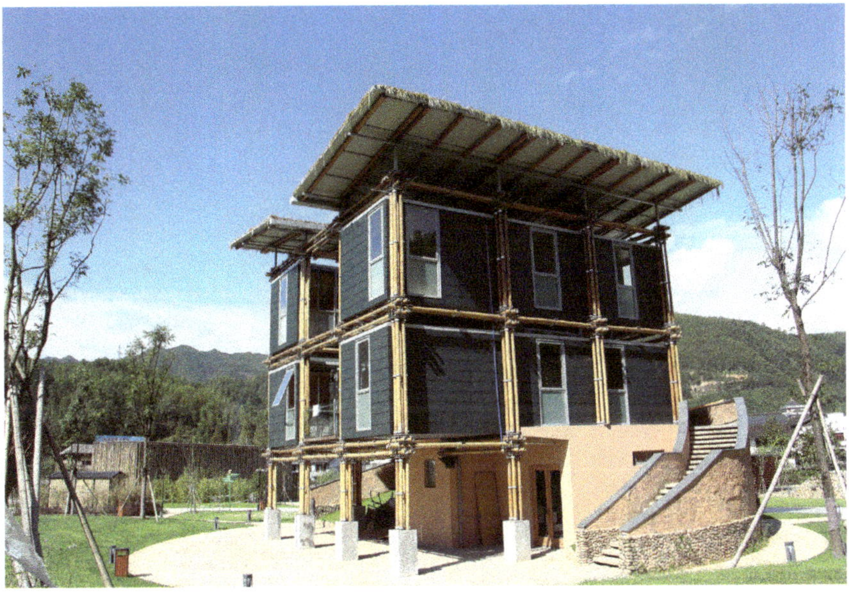

Fig. 6.239 Eco-energy Efficient Experimental House

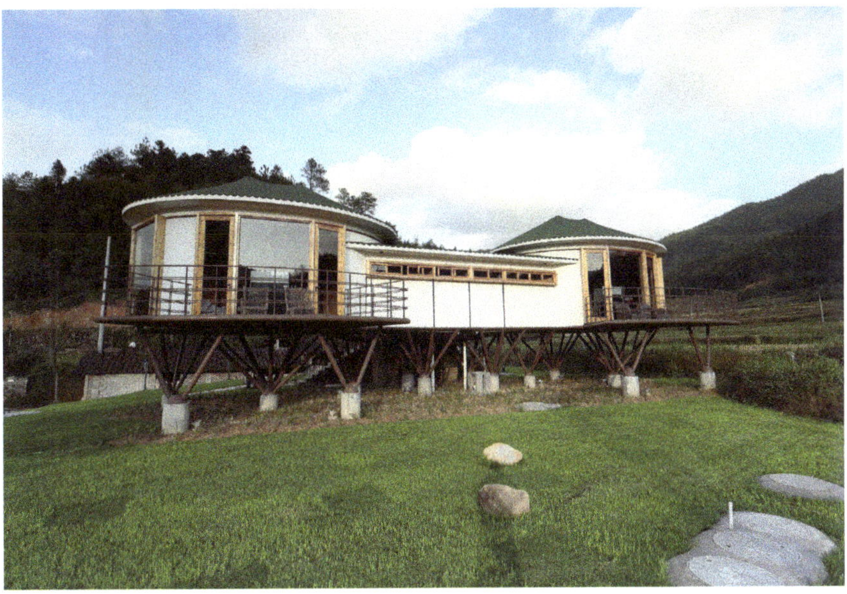

Fig. 6.240 Boutique Hotel

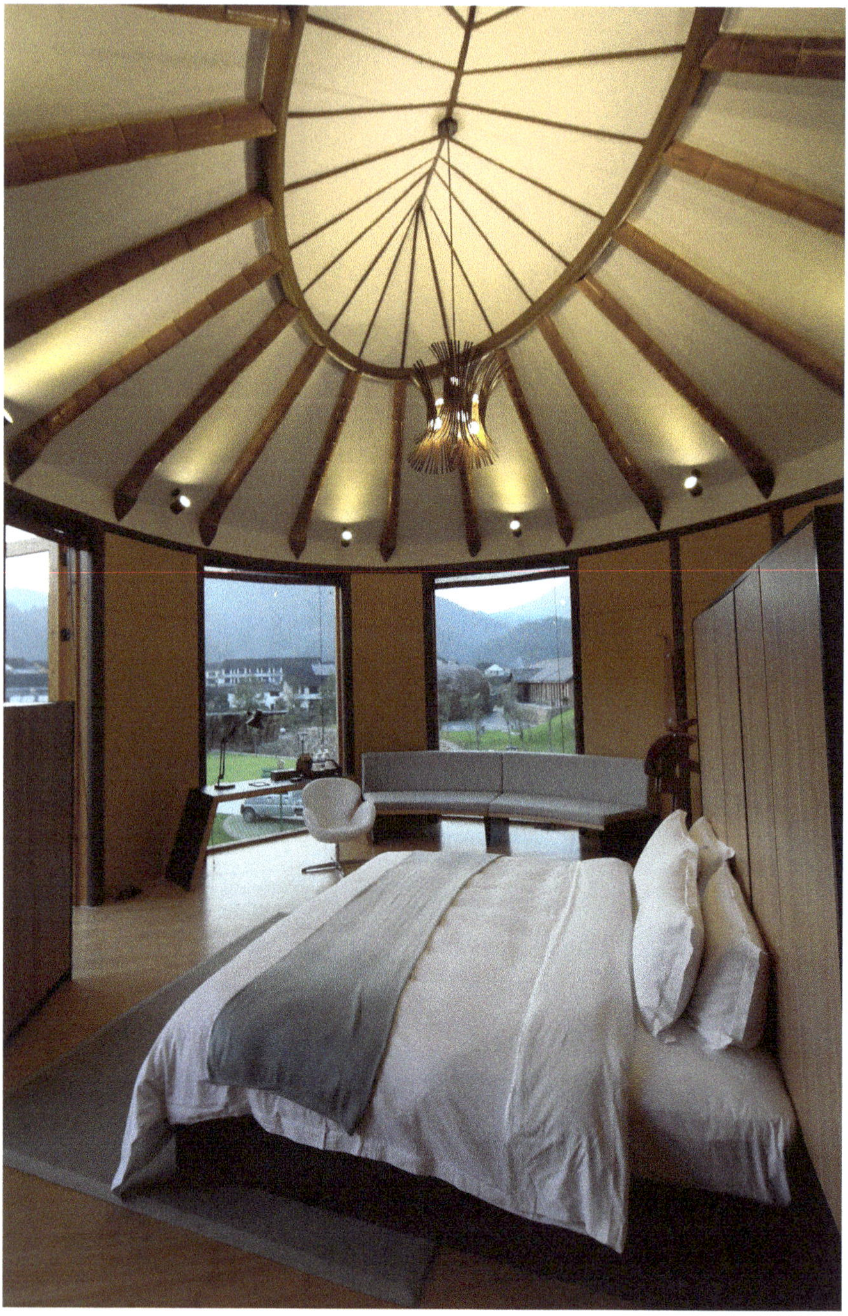

Fig. 6.241 Boutique Hotel

6.5 Transport Facilities

Yixin Bridge in Dujiacun Village (2018)

Bamboo Structural Parking Lot in Shuangyi Village (2018)

Bamboo Bus Platform in Yiyang, Hunan Province (2018)

Safety and Life Protection Project of X730 Highway in Xianju County Road, Taizhou, Zhejiang, China (2017)

Yongxiu Wucheng Water Road Plank Road (2020)

Yixin Bridge in Dujiacun Village

Location: Chongqing, China

Completed: 2018

Designer: Shao Changzhuan

Yixin Bridge Project is a research-oriented public welfare project jointly sponsored by the Chinese University of Hong Kong and Tsinghua University. The left-behind elderly and children in Dujiacun village of Chongqing, despite potential hazards, have been crossing the river for a long time. The design team took full consideration of the environment, economy and society, to explore a sustainable path for the development of rural transport facilities in China. The span of Yixin Bridge is 21 m, the width is 3 m, and the deck width is 2 m. The bridge is constructed with more than 700 bamboo culms. The section of Yixin Bridge is in the shape of "合", which embodies the philosophy of "harmony between man and nature" in traditional Chinese culture.

Application of bamboo products: Treated Moso (*P. edulis*) bamboo was selected as the bridge structural material. The modern full-culm bamboo bridge technology used in this project has great potential for small rural transportation facilities and urban pedestrian bridges.

Photo source: Figs. 6.242, 6.243 and 6.244 are provided by Tsinghua University.

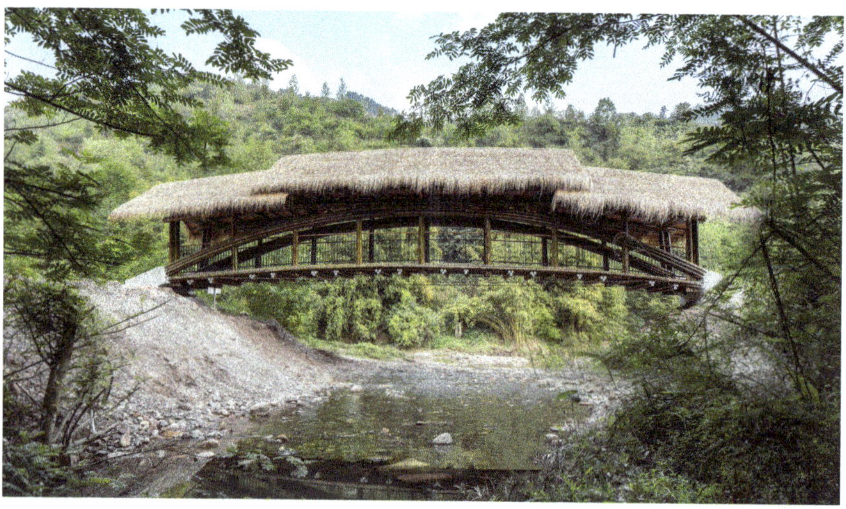

Fig. 6.242 Yixin Bridge

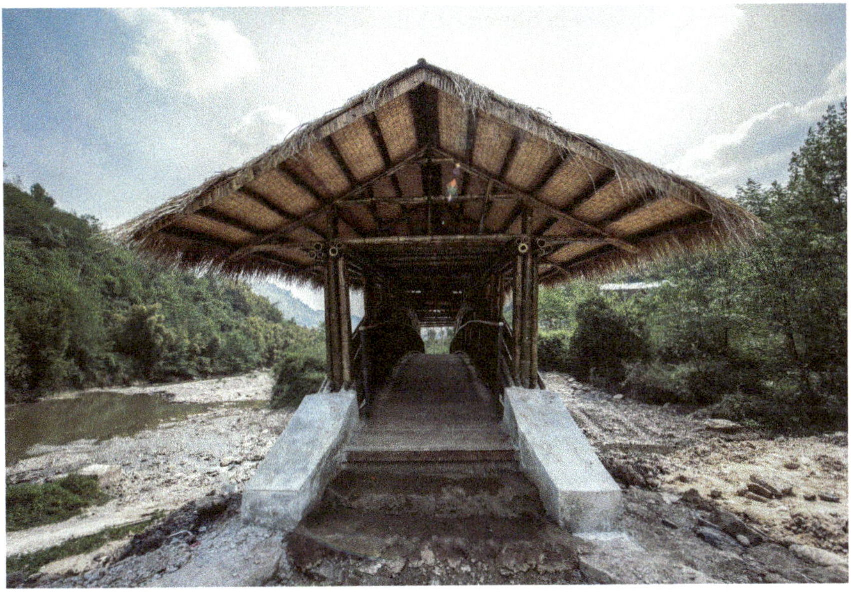

Fig. 6.243 Front view of Yixin Bridge

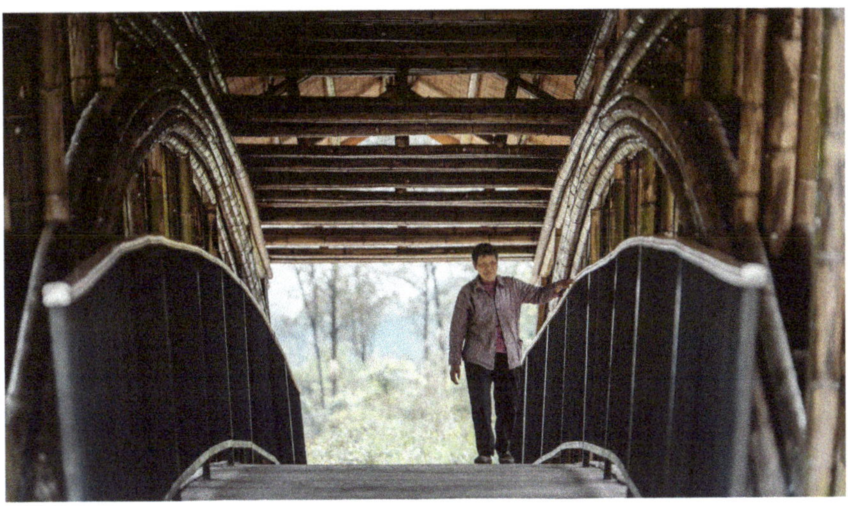

Fig. 6.244 On the bridge

Bamboo Parking Shelters in Shuangyi Village

Location: Anji, Zhejiang, China

Completed: 2018

Designer: Zeng Weiren

With a thousand-year history, Shuangyi Village, is located in Anji, the famous bamboo county in China. The designer participated in the beautiful countryside reconstruction project, aiming at developing the countryside with ecological balance based on the cultural characteristics of the bamboo county. The ingeniously designed bamboo parking shelter for small vehicles, has become a beautiful landscape in rural construction, while providing a more convenient life for local villagers.

Application of bamboo products: The main components of the structure were adopted full-culm bamboo with a diameter of more than 10 cm. These were processed to ensure good durability. Metal connections are used for joints, and components can be quickly installed and replaced. Under normal maintenance, the service life of the bamboo parking shelter can be more than 30 years.

Photo source: Figs. 6.245, 6.246 and 6.247 are provided by Zhejiang Jiangnan Bamboo Research and Design Center.

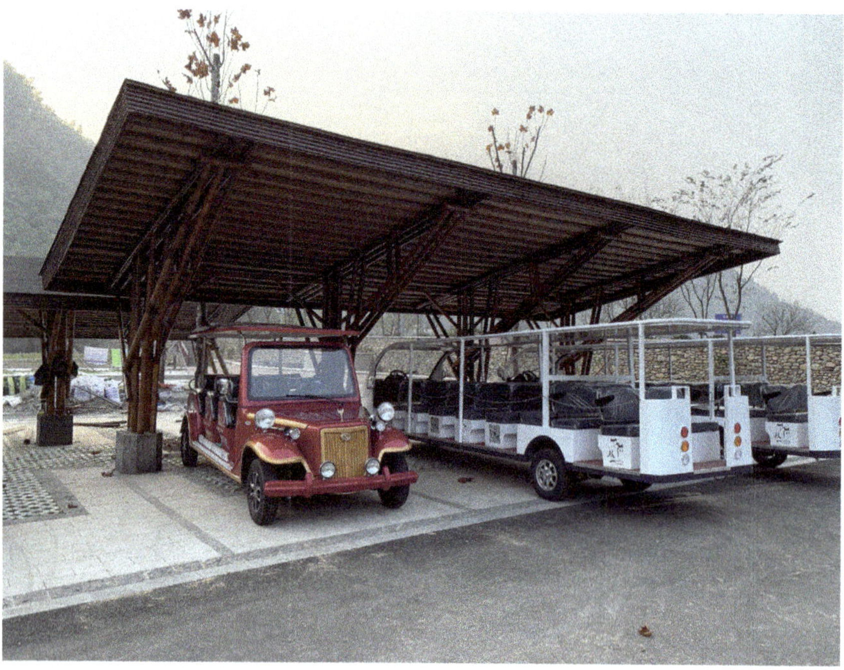

Fig. 6.245 Bamboo Parking Shelter

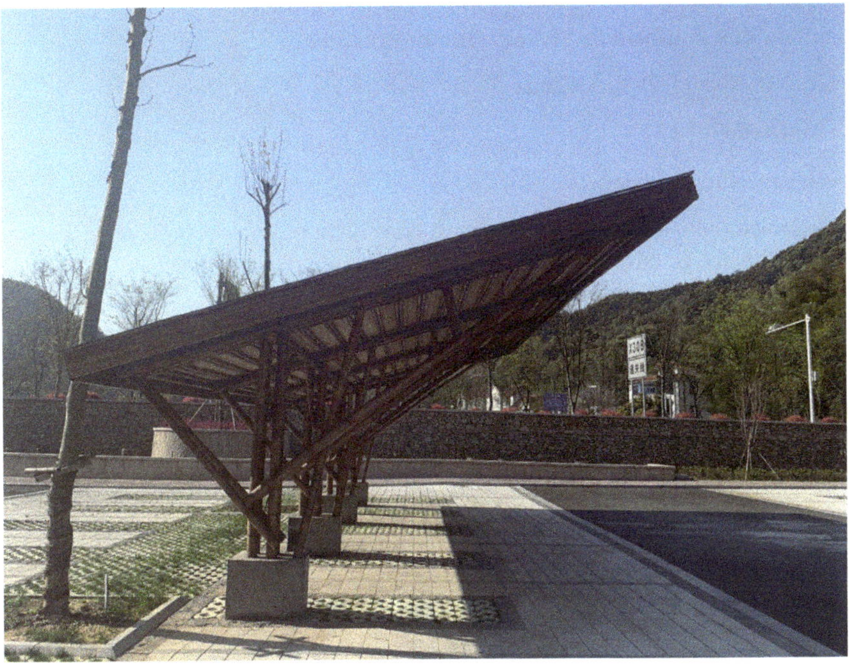

Fig. 6.246 Bamboo Parking Shelters

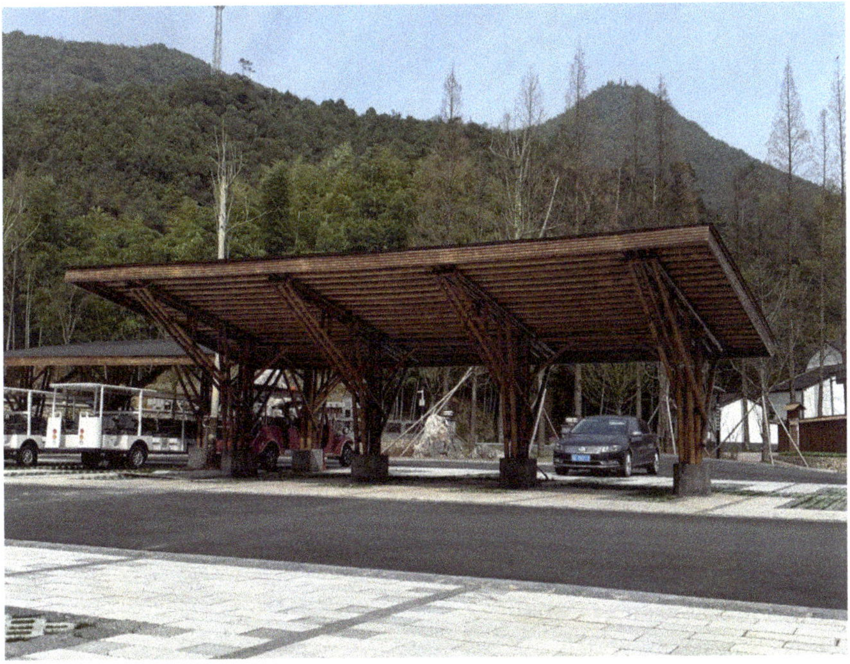

Fig. 6.247 Bamboo Parking Shelters

Bamboo Bus Platform in Yiyang, Hunan Province

Location: Yiyang, Hunan, China

Completed: 2018

Designer: Xue Zhicheng

This project belongs to one of the rural construction projects in Yiyang, Hunan, China. The design of the bamboo bus platform is simple and classical, becoming a beautiful spot in the rural environment.

Application of bamboo products: The main structure of the bus platform is steel, and all the exterior of the bus platform was decorated with durable outdoor bamboo scrimber grilles and wallboards.

Photo source: Figs. 6.248 and 6.249 are provided by Hunan Taohuajiang Bamboo Science and Technology Co., Ltd.

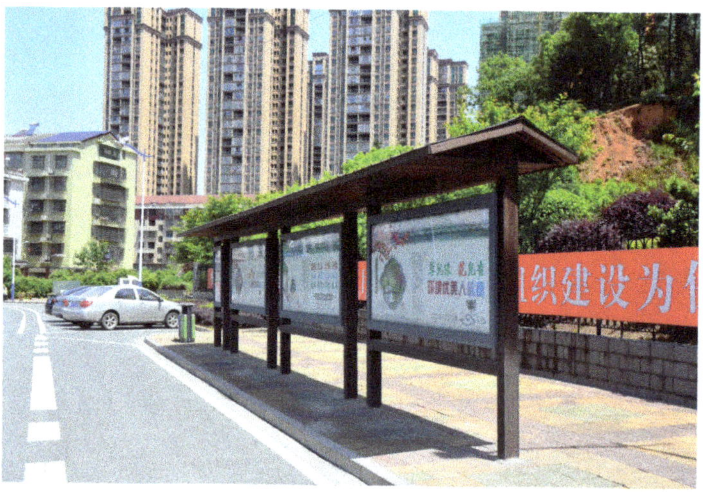

Fig. 6.248 Bamboo bus platform

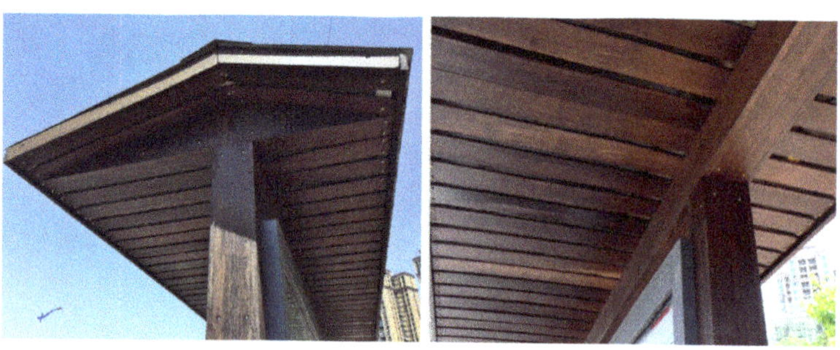

Fig. 6.249 Detail of structure

Safety and Life Protection Project of X730 Highway in Xianju County Road, Taizhou, Zhejiang Province

Location: Taizhou, Zhejiang, China

Completed: 2017

Designer: Taizhou Traffic Design Institute

County Road X730, located in Taizhou, Zhejiang Province, is a tourist highway leading to the immortal dwelling in a deemed 5A scenic spot. Its highway features exquisite scenery, beautiful mountains and rivers along its path. The highway guardrail made of bamboo scrimber is concise and smooth in line and elegant in appearance. The bamboo scrimber guardrail not only blends naturally into the surrounding scenery, it adds a pastoral feel to the whole tourist highway and becomes a beautiful scenic link connecting the city and scenic spots.

Application of bamboo products: Bamboo scrimber and steel were combined by cold pressing and gluing to produce this durable, corrosion resistant guardrail. The guardrail meets the safety and performance requirements for a Grade B (Grade II) highway. Due to the good durability, the service life of the bamboo scrimber rail can meet the 15-year design standard stipulated in the *Design Specifications for Highway Safety Facilities* (JTG 081–2017).

Photo source: Figs. 6.250, 6.251 and 6.252 are provided by Hangzhou Fangqiao Transportation Facilities Co., Ltd.

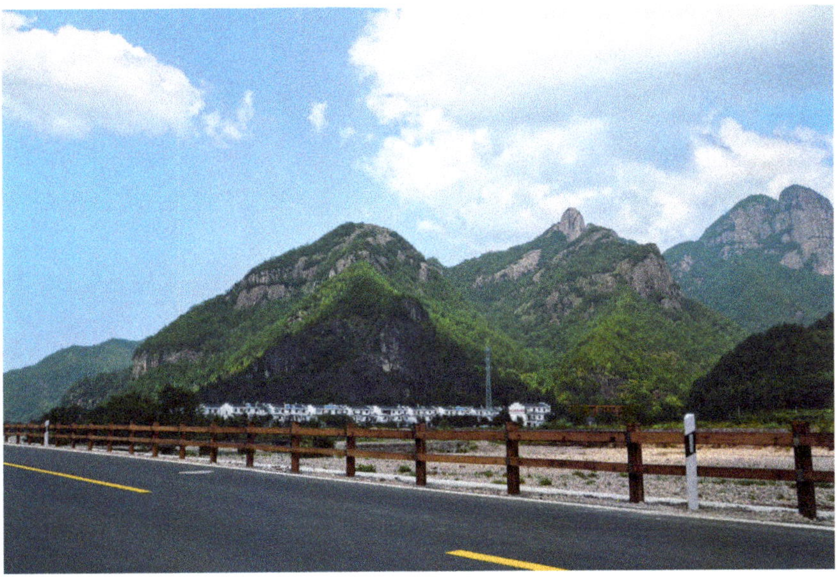

Fig. 6.250 X730 Highway with bamboo guardrail

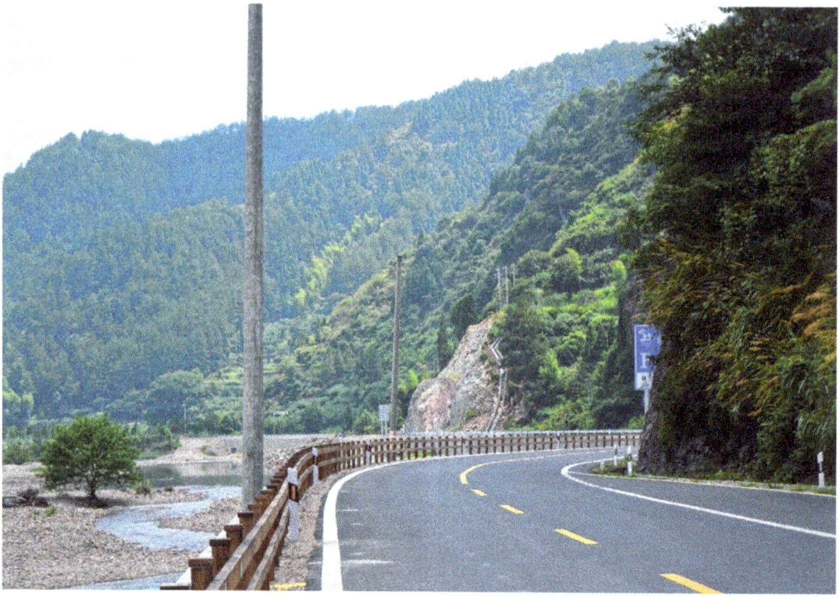

Fig. 6.251 X730 Highway with bamboo guardrail

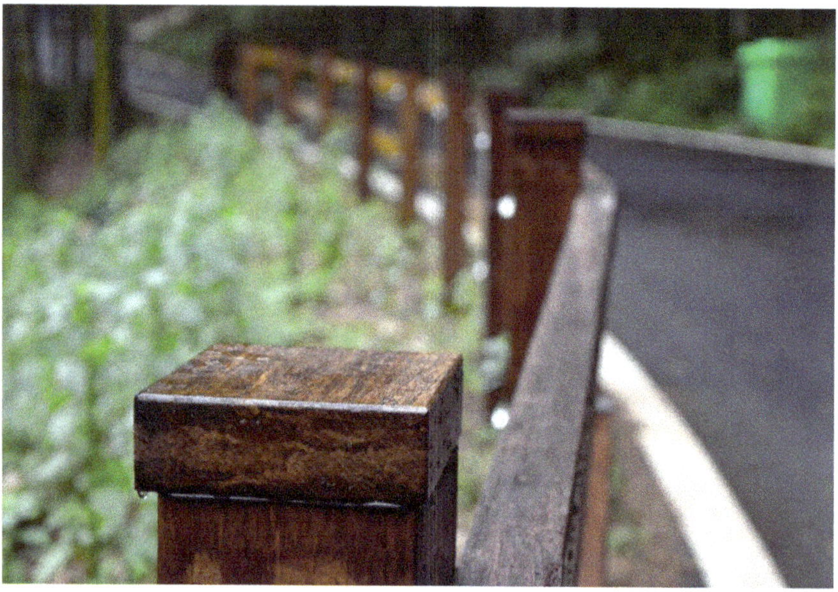

Fig. 6.252 Detail of bamboo guardrail (*photo by Qin Hongguang*)

Yongxiu Wucheng Water Plank Road

Location: Jiujiang, Jiangxi Province, China

Completed: 2020

Designer: Shenzhen Urban Transport Planning Center Inc., Ltd.

Wucheng Town, Yongxiu County, Jiujiang City, Jiangxi Province, China is one of four famous towns in Jiangxi Province. It is where the World Class A Migratory Bird Reserve and the core area of the Jiangxi Poyang Lake National Nature Reserve are located. The town takes up an area of 368 km^2, including about 26,667 ha of water surface, 12,800 ha of marsh grass, and 42 large and small lakes. Every summer, when the rainy season comes, water from Poyang Lake rises and flows over the Da Lake section of the Yongwu Highway, located on the banks of the Da Lake in the Poyang Lake Migratory Bird Reserve. The road surface is submerged by the lake, which then resembles a dragon in the water. Such peculiar landscape is known as the "most beautiful water road". The plank road, which is the best "companion" of the highway, paves 5000 m and has a width of 3 m. The total length of the plank road railing is more than 10,000 m, with the total investment of more than 40 million RMB ($6.2 million USD).

Application of bamboo products: In order to meet the demands of being submerged for more than one month per year, the plank road uses more than 66,000 high-resistance outdoor bamboo scrimber with a density of 1200 kg/m^3. The processing and construction took only 70 days. The 5000 m long boardwalk is supported on about 900 concrete bents connected by over 50,000 m of steel truss beams.

Photo source: Figs. 6.253 and 6.255 are provided by Liu Yuanqing. Fig. 6.254 is provided by Hangzhou Dasso Technology Co., Ltd.

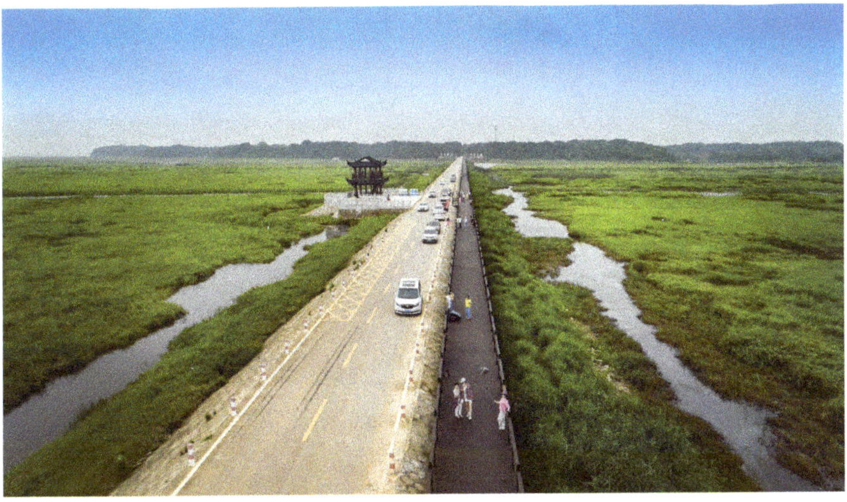

Fig. 6.253 Aerial view of Yongxiu Wucheng Water Plank Road (*photo by Liu Yuanqing*)

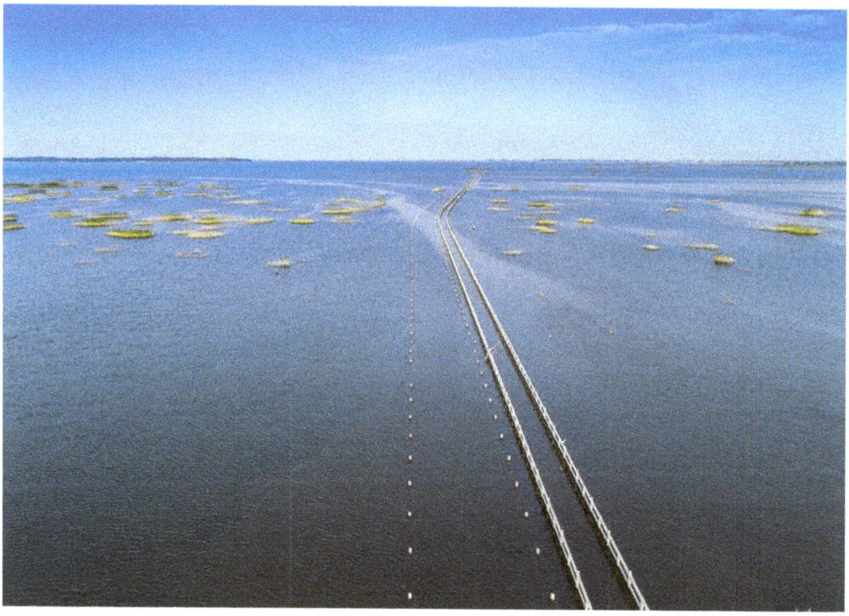

Fig. 6.254 Yongxiu Wucheng Water Plank Road submerged by the lake

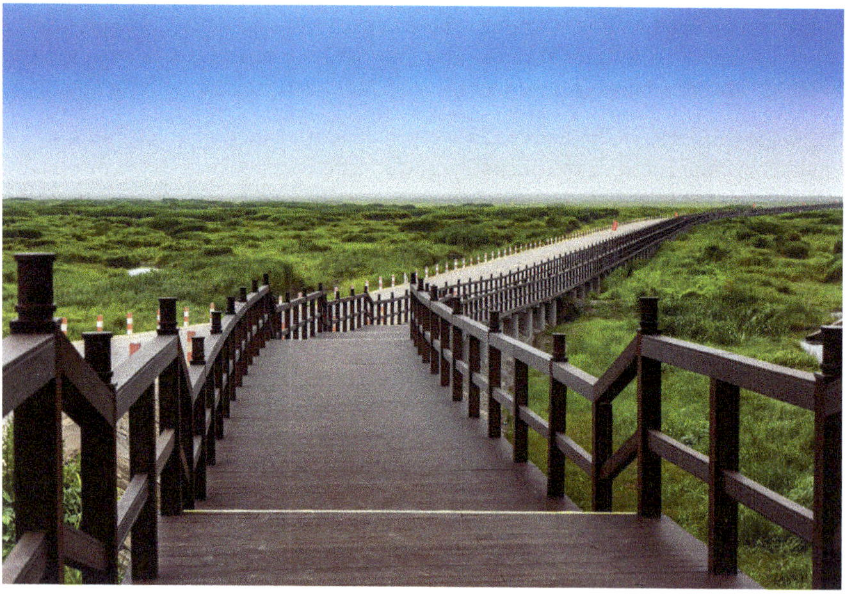

Fig. 6.255 Details of Yongxiu Wucheng Water Plank Road (*photo by Liu Yuanqing*)

6.6 Water Pipelines and Urban Municipal Tunnels

Bamboo Filament-Wound Water Pipeline (2017)

Bamboo Filament-Wound Urban Municipal Tunnels (2018).

Bamboo Filament-Wound Water Pipeline

Location: Linyi, Shandong, China

Completed: 2017

The bamboo filament-wound water pipeline is a water conveyance pipeline which is made of bamboo as its primary structural material. It is designed and fabricated through an optimization of multi-layer structure of anticorrosion, waterproof and winding processes. The available diameter ranges from 0.6 to 2.2 m, ring stiffness reaches more than 1000 N. Operating pressure is less than 1600 kPa and service temperature is less than 110 °C.

Application of bamboo products: Bamboo filament-wound water pipeline has the advantages of light weight, high strength, low temperature tolerance and humidity resistance. It can be widely used in many fields including water transportation, irrigation, urban drainage, power plant circulating water and so on.

Photo source: Figs. 6.256 and 6.257 are provided by Zhejiang Xinzhou Bamboo-Based Composites Technology Co., Ltd.

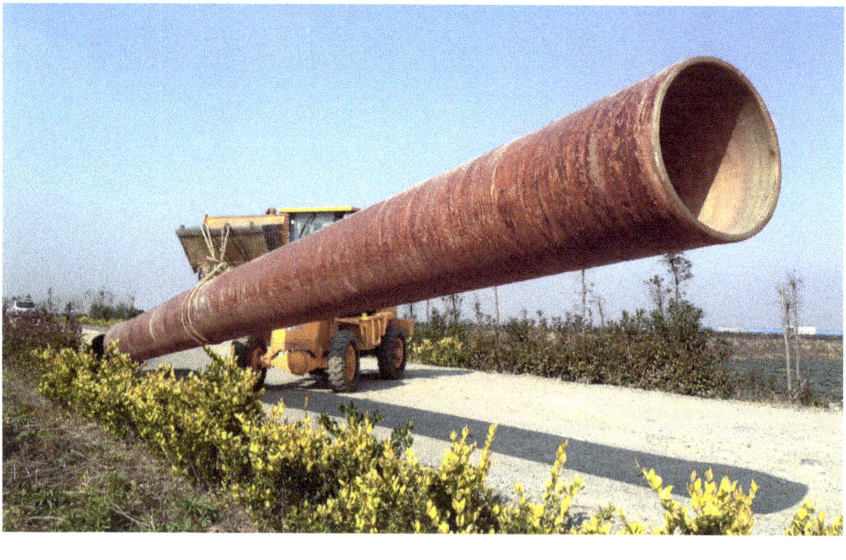

Fig. 6.256 Bamboo filament-wound water pipeline

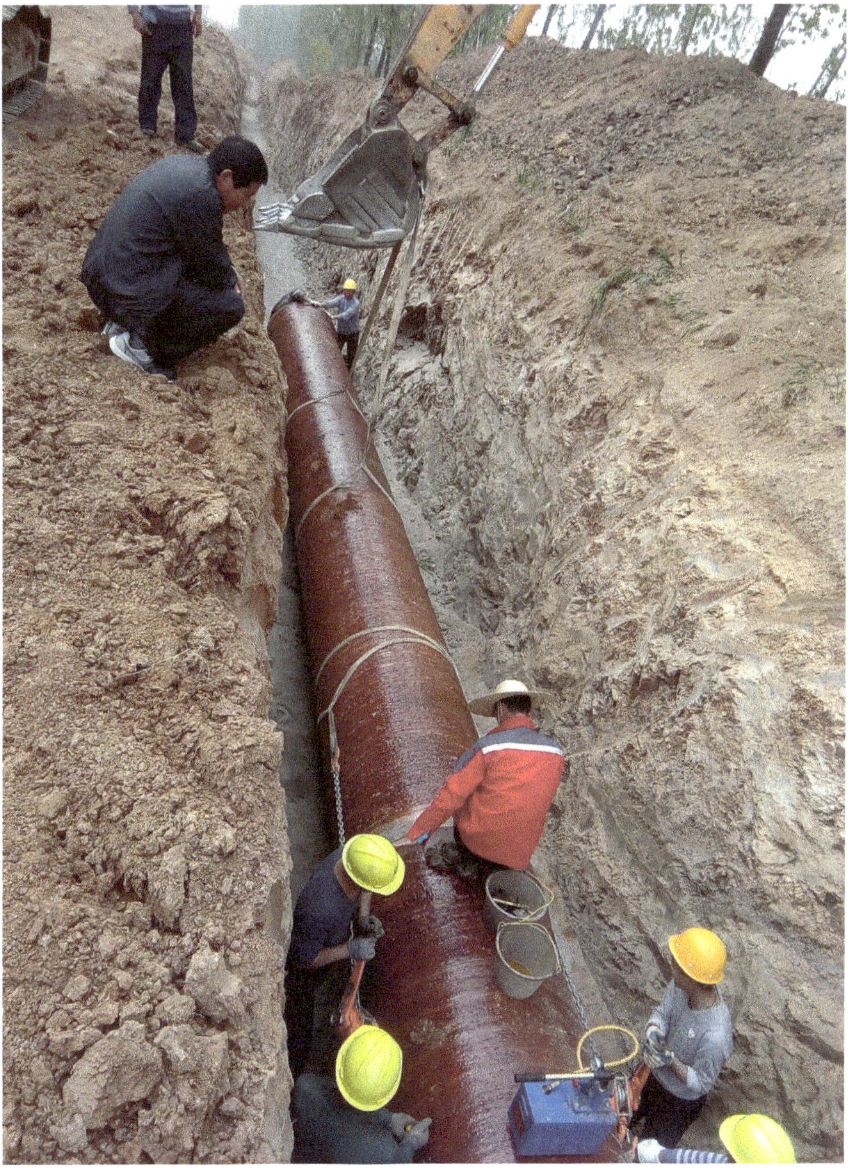

Fig. 6.257 Bamboo filament-wound water pipeline

Bamboo Filament-Wound Urban Municipal Tunnels

Location: Hohhot, Inner Mongolia, China

Completed: 2018

Located in Yuanheng Graphite Industrial Park, Hohhot, Inner Mongolia, China, this is the world's first bio-based urban municipal tunnel. Different from reinforced concrete urban municipal tunnels that are widely used all over the world, the biological-based tunnel is made of bamboo filament-wound composite materials. In terms of material properties, its compressive strength is equal to that of C30 concrete, and it meets the requirements of technical specifications for urban municipal tunnels. Each section of tunnel is 3 m in diameter, 12 m in length and weighs 10 tons—approximately one tenth the weight of a comparable concrete tunnel. Installation can reach 120 m per day using only two 25-ton cranes; this greatly reduces the difficulty of installation and shortens the construction period. Compared with conventional concrete tunnels, the overall project cost is reduced by 10%-30%. At the same time, a structural health monitoring system was used to monitor tunnel stress and deformation in real time. For conditions of the same burial depth and overburden conditions, the seismic performance, settlement resistance, thermal insulation and frost-resistance performance of bamboo filament-wound urban municipal tunnels are superior to those of concrete tunnels.

Photo source: Figs. 6.258, 6.259 and 6.260 are provide by Zhejiang Xinzhou Bamboo-based Composites Technology Co., Ltd.

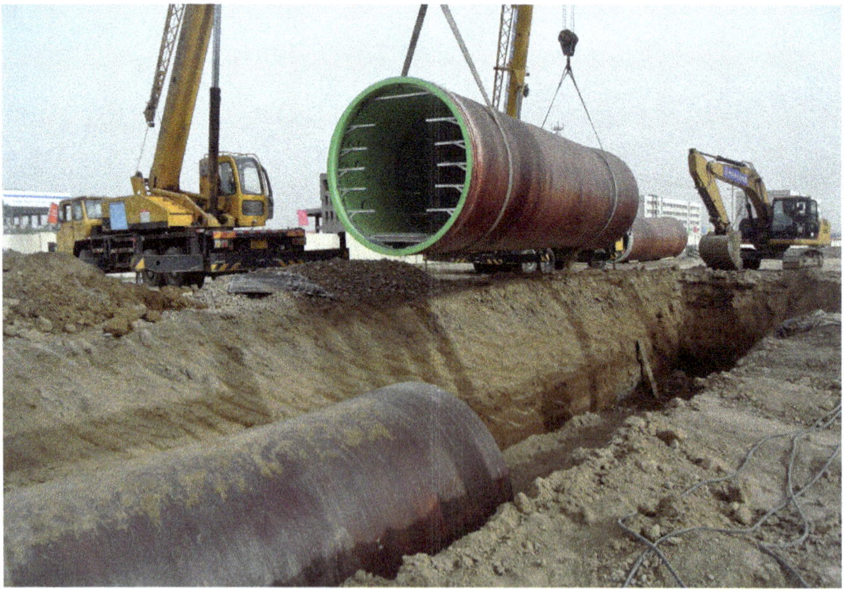

Fig. 6.258 Bamboo filament-wound urban municipal tunnel

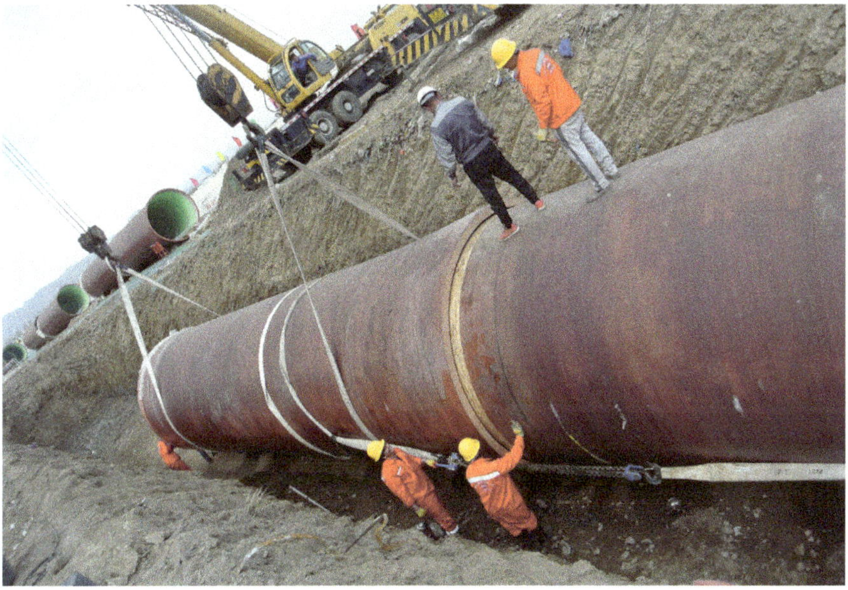

Fig. 6.259 Bamboo filament-wound urban municipal tunnel

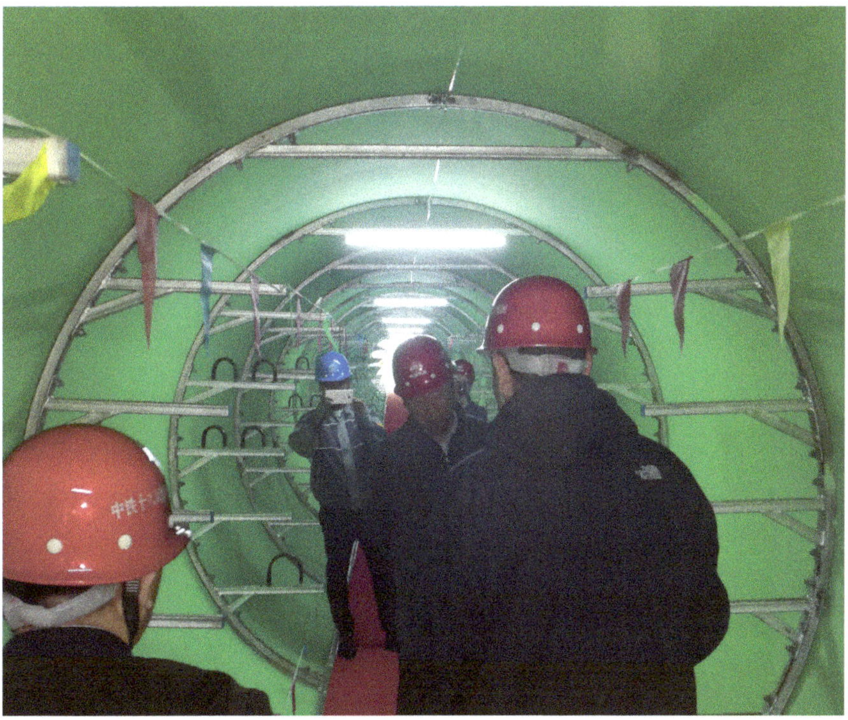

Fig. 6.260 In the tunnel

References

1. https://baike.baidu.com/item/公望美术馆/20398857. Retrieved in April, 2019.
2. https://baike.baidu.com/item/南京青奥文化体育公园/20133005?fr=aladdin. Retrieved in April, 2019.
3. https://baike.baidu.com/item/雄安市民服务中心/22378021?fr=aladdin. Retrieved in April, 2019.

Chapter 7
Opportunities and Challenges for the Modern Bamboo Construction Industry in China

Although China has the advantage of significant bamboo resources and a cultural foundation for the development of the bamboo construction industry, the construction technology and theoretical basis of modern bamboo construction has not been fully established. Building on the previous chapters, this chapter will use the SWOT (strengths—weaknesses—opportunities—threats) method to analyse and summarise (Table 7.1) the opportunities and challenges faced by China's bamboo construction industry. The SWOT analysis method, is based on situational analysis of the internal and external competitive environment and conditions. SWOT lists the main internal strengths, weaknesses, external opportunities and threats closely related to the research object through investigation and analysis. Subsequently, using the concepts of system analysis, various factors are matched and analysed, from which a series of conclusions are drawn. By using this method, a comprehensive, systematic and accurate study of the scenario in which the research object is located is made. Corresponding development strategies, plans and countermeasures are formulated according to the research results.

7.1 Advantages for Competition

7.1.1 Resources and Material Advantages

As described in Chap. 1, China has the greatest bamboo resources in the world, with almost 7 million ha of bamboo forest. Almost 5 million ha are *Phyllostachys edulis* (moso) bamboo, a large diameter species well suited to use in both full-culm and engineered bamboo construction products. In general, the properties of moso, and moso-derived engineered products are comparable to those of conventional softwood timber.

K. W. Liu et al., *Contemporary Bamboo Architecture in China*,
https://doi.org/10.1007/978-981-16-8309-1_7

Table 7.1 SWOT analysis of opportunities and challenges for bamboo construction industry in China

Strengths	Weaknesses
1. China has abundant bamboo species for construction. These are widely distributed across the country. It is suitable for developing a bamboo construction industry in areas rich in bamboo resources	1. More basic research is still needed
2. Bamboo is structural materials with excellent mechanical properties and good seismic performance	2. The production and processing of bamboo construction materials are labour-intensive. There is a lack of automation and continuous processing equipment
3. Bamboo grows fast and can be harvested in 3–5 years. Once cultivated, managed harvesting can realise sustainable development	3. Lack of necessary engineering standards
4. Processing performance of bamboo in China is good. China is a world leader in bamboo processing technology	4. Because of the limited scale of industrial production, there is no obvious competitive advantage in terms of price
5. The natural form of bamboo is unique, and its industrial products also maintain the natural texture; this is popular and favoured by architects	5. The durability and fire resistance of bamboo still restrict its development; these topics require further study
6. Bamboo is a green building material worthy of wide promotion considering global sustainable development goals and dwindling forest areas	6. Research productivity and patent development technology is inadequate
7. China has a long history of bamboo culture and has the basic conditions to promote bamboo architecture	7. Lack of relevant professional practitioners and training mechanisms
Opportunities	*Threats*
1. Bamboo construction has development potential under the promotion of national macro-policies in China. Since 2010, China has issued relevant policies such as "green building materials", "green buildings", "prefabricated buildings"	1. China's labour costs are rising, and it is difficult to harvest bamboo in mountainous areas
2. At present, China is vigorously developing characteristic towns. Bamboo can be selected as a construction material in areas rich in bamboo resources to develop villages and towns with bamboo buildings and landscapes. Tourism can drive the development of related industries	2. At present, the majority of the public's understanding of bamboo architecture is limited by the impression of traditional bamboo houses; it will take a long time to change public awareness
3. INBAR is headquartered in China and can actively promote green bamboo construction products and industries by taking advantage of its location	3. Policies supporting and promoting bamboo construction by relevant departments in China are insufficient

(continued)

Table 7.1 (continued)

Strengths	Weaknesses
	4. At present, almost all bamboo production and processing enterprises in China are small and medium-sized. There is a reliance on overseas markets, increasing risk

Sustainably harvested, high-yield moso bamboo forests can produce about 1200 mature culms per hectare per year. Reasonably, this represents a potential of harvesting about 3 billion mature culms annually in China or a volume of material on the order of 100 million cubic meters. Current timber production in China is on the order of 365 million cubic meters whereas domestic consumption is on the order of 450 million cubic meters with exports accounting for a relatively modest 12 million cubic meters. This leaves China importing approximately 100 million cubic meters of timber annually (2017 data reported by FAO [1]). A mature bamboo industry can therefore make a considerable impact in reducing this market gap.

Bamboo becomes mature in 3–5 years, much faster than softwood. This, along with the fact that China has the greatest rate of afforestation in the world, approximately 1.5% annually over the last 40 years [1] indicates that the potential for bamboo production growth will outpace that of timber in years to come.

7.1.2 Industrial and Technological Advantages

As described in Chap. 5, the Chinese bamboo industry leads the world in research and development innovation. China accounts for almost 8000 patents related to bamboo (about 50% of the world's total) and publishes more than 100 national and profession standards (described in Chap. 4). China is a dominant source of technological innovation in engineered bamboo processing and leads the world in bamboo exports (Chap. 1). Development of China's bamboo industry occurs at all stages, making an unbroken chain from bamboo seedling expansion, resource cultivation, processing and utilization, innovative development, through to domestic and foreign trade. The industry has successfully developed more than 10,000 kinds of bamboo products and exports these throughout the world [2]. Nonetheless, the proportion of the export market represented by construction products remains modest: about 14% of Chinese bamboo exports, mostly flooring products [3]. Thus there remains considerable room for innovation and growth. The potential for engineered bamboo construction products such as bamboo plywood, glued laminated bamboo, strip laminated board, scrimber, flattening bamboo, bamboo-polymer composites, bamboo bundle and veneer laminated composites, wound composites, etc., is greatest. Although the present market for such products is nascent, innovation, coupled with advocacy in the construction industry should lead to rapid growth.

7.1.3 *The Advantage of Eco-culture Value*

From an ecological point of view, bamboo is a natural renewable material which, among other advantages, can absorb more carbon dioxide than other timber species. Developing new bamboo construction materials and products supports the objectives of sustainable or "green" development. Energy consumption associated with processing and use of bamboo is approximately 1/8th that of concrete, 1/3rd that of timber and 1/50th that of steel for a building of a given area [4]. Additionally, the complex graded structure of vascular bundles and thin-walled tissues in bamboo leads to good thermal insulation performance. For example, the thermal conductivity of bamboo-based panels is 0.14–0.18 W/(m K), much lower than that of reinforced concrete and clay bricks [5].

From the cultural point of view, China has a long history of bamboo culture. China is the first country to use bamboo as a construction material. Bamboo architecture has been closely related to people's lives since ancient times and its development and resurgence in China has a deep-rooted cultural foundation. Nonetheless, the challenge of creating contemporary expression of this traditional natural material and thereby building a context of contemporary bamboo architecture requires the participation of more architects and designers.

7.2 Disadvantages for Competition

7.2.1 *Inadequate Basic Research*

Compared with conventional construction materials such as steel, timber and concrete, the basic research on bamboo materials, structural components and construction systems is relatively nascent and is, as yet, insufficient to see bamboo construction into the main stream. China has not yet established a standard system for design, construction and quality acceptance for bamboo construction. For full-culm bamboo structures, there are many studies of unidirectional mechanical properties of bamboo, but test methods to establish mechanical properties under compound forces have not been established [6]. Research on classification or grading of full-culm bamboo and joint design for full-culm bamboo is scarce. For engineered bamboo, tests from the timber industry are often adopted which rely on "small clear specimens". The microstructure of bamboo and the fabrication of engineered bamboo make the use of such test, in some instances, inappropriate. Calibration of resulting mechanical properties with full scales tests is required and remains rare. Finally, research on the long-term performance, durability and fire resistance of both full-culm and engineered bamboo is in its infancy, and require greater research investment.

7.2.2 *Inadequate Penetration of Automation in Bamboo Processing Industries*

Compared with the industrial utilization of wood, the automation, continuity and scale of bamboo processing and utilization in China are insufficient [7]. Currently, the production of glued laminated bamboo and bamboo scrimber are labour-intensive, using a relatively small degree of automation. This results from the relatively weaker technological foundation: mechanical equipment for processing bamboo is often adapted from another process, and suffers from poor efficiency of process and precision. As a result, processed and engineered products may be of reduced quality and consistency, affecting market competitiveness. For example, due to the lack of precision machine tools and processing technology, large glued laminated bamboo beams need to be transported to Japan for processing and then shipped back to China for installation [8], resulting in high costs and poor efficiency. Finally, most bamboo processing enterprises are small or medium-sized, with fewer resources for investment in innovation and research.

Many bamboo products have a problem with consistency and, as a result, manufacturers are often focused at very narrow applications, such as flooring, rather than products having more universal utility, such as products comparable to dimensional lumber [9]. The development and utilization of bamboo in the construction field has high added value, but remains in its infancy, not yet coalescing around large-scale production and applications. There is a need for large-scale machinery and equipment suitable for the production and processing of bamboo construction materials. Innovation should be directed toward products having broad utility and predictable consistency (as is the case in other sectors of the construction industry), rather than niche applications.

7.2.3 *Lack of Standardisation*

As described in Chap. 4, global standardisation of bamboo products and construction is only just beginning. Although Chinese standards exist, these are, as yet, insufficient to instill a broad acceptance of bamboo in the construction industry. Standardardisation is the basis for incentivising the industry and is essential for broad acceptance of bamboo construction materials, methods and adoption in projects [10].

Standardisation requires advocacy and, ironically, advocacy requires a basis of standardisation [10]. In China, several forestry industry standards and association standards are being developed to provide the necessary technical foundation for the development of China's bamboo construction industry. These efforts are promulgated by INBAR and their advocacy at the International Organisation of Standards Committee ISO165. As standards are developed (see Chap. 4), INBAR also advocates for adoption in the international community through its member nations. As a result, Chinese expertise is supporting the development of both national and international bamboo standardisation and therefore industry. The final step, with association and industry standards in hand, these must be advocated for recognition and/or inclusion in national building codes.

7.2.4 *Lack of Capacity-Building for Professionals*

Like timber, bamboo professionals require training in forestry sciences, engineering and construction. Development of viable engineered bamboo structures requires knowledge all along the production chain. At present, only a few institutions in China have established optional or short-term courses for bamboo architecture. Similarly, few research institutes or universities are engaged in bamboo-related engineering or architectural research. From the perspective of industry, there is a lack of architects and engineers qualified to design and build bamboo structures. There is a similar lack of professionals able to develop the processing side of the industry. Strengthening the capacity building of professionals is critical. Students should have access to bamboo as part of their education; they should have exposure to not only the materials and engineering aspects of the use of bamboo but also the social aspects of the modern use of this traditional material.

7.3 Opportunities

7.3.1 *Support by National Policies*

The construction, steel and automobile industries are known as the three pillar industries of China's national economy. In recent years, China has attached great importance to the healthy development of the construction industry, especially the development and application of green building materials and green buildings. Many encouraging policies and guiding documents have been produced. These indicate the future development needs of the construction industry in the broadest sense focusing on sustainability and green development, reducing energy footprint, and promoting prefabrication and industrialization. Bamboo meets these needs but requires a concerted effort and advocacy from across the industry.

7.3.2 *Towns with Local Characteristics and Revitalization of Rural Areas*

In addition to policies aimed at the construction industry, China is vigorously developing towns with local characteristics and implementing the "rural revitalization strategy". Seizing the historic opportunity of building characteristic towns and beautiful countryside, choosing bamboo as a construction material in areas rich in bamboo, developing characteristic bamboo building villages and towns, and promoting the development of related industries such as tourism are paths worth exploring. Chapter 6 reports multiple examples of construction with local characteristics and the use of bamboo in rehabilitating the built infrastructure.

The first International Bamboo Architecture Biennial introduced in Chap. 6 is an excellent case. For the Biennial, we can see that bamboo buildings can integrate ecology, culture, architecture, art and interest in the construction of Characteristic Towns and beautiful countryside, and has a broad market prospect. Practitioners can combine innovative marketing methods to develop tourism and other business services, creating innovative links between the bamboo industry and regional culture. Examples may include bamboo buildings (shelter), bamboo shoot cuisine (food), bamboo bicycle manufacture (travel) and bamboo fabric production (housewares and handicrafts). Consumers are thereby introduced to new, high-value bamboo products and services creating and integration of manufacture, culture, tourism and ecology in one material.

7.4 Threats

7.4.1 *The Rising Cost of Labour*

With the development of the economy, the cost of labor is gradually increasing in China. Labour costs for bamboo harvesting and processing have risen quickly, threatening the industry with contraction. Because bamboo grows in mountainous areas, technologies for mechanical harvesting and large-scale management are not yet mature. At present, the harvesting of bamboo mostly accomplished by hand. In the initial processing steps of engineered bamboo—cutting, slicing, drying, transportation, stripping, decomposition and weaving—also remain labour intensive. The process of fabricating bamboo panels—including dipping (applying glue), paving, hot pressing, forming—is also largely manual. Much of the bamboo processing chain is characterized by high labour intensity, noise and dust, and generally poor working environments. Bamboo processing enterprises in Jiangsu, Zhejiang, Fujian and other regions face difficulty recruiting workers: many young people are unwilling engage in such production, and wages are increasing year by year.

At the production stage, the industry must improve the level of mechanization and automation in the processing of bamboo. Simultaneously, developing prefabricated bamboo buildings will enhance standardization, consistency and offer opportunities for intelligent processing of bamboo for construction. Both approaches shift the nature of work for front-line workers: gradually changing from labour-intensive physical oriented work to knowledge- and skill-based work. Automated production lines, CNC machine tools and intelligent robots are all a part of a thriving bamboo construction industry.

7.4.2 Low Public Awareness of Bamboo Architecture

Bamboo architecture is not well recognized by the public. On the one hand, for historical reasons, for most Chinese people, bamboo is a cheap and low-quality material, its durability is poor, and its physical and mechanical properties do not meet the requirements of permanent buildings. As illustrated in Chap. 6 and throughout this book, modern bamboo building materials and modern bamboo buildings fully meet the requirements of modern architecture. As described previously, the lack of appropriate building standards is a barrier to broad adoption of bamboo as a construction material—this affects and limits the market and therefore public perception. At present, the cost of bamboo building materials is typically higher than that of other building materials. As a result, it is difficult for the public to have a direct perception or experience of modern bamboo architecture. Through advocacy, expanding market scale and raising public awareness we actively promote bamboo architecture and promote the development of the bamboo construction industry.

7.4.3 Risk in Bamboo Enterprises

At present, the vast majority of bamboo production and processing enterprises in China are small and medium-sized. Bamboo processing enterprises often engage limited technology, have limited capacity, and lack the capital to innovate or grow. For many, their production relies heavily on overseas markets, increasing exposure to risk. As a natural material that is often highly processed, the bamboo supply and production chain is complex and diverse. The use of bamboo products further requires architectural, engineering and construction phases. All entail risk. To some extent the entire production chain is not seamless and cooperation at all phases is often fractured. A seamless integration of the entire production and utilisation chain is necessary to mitigate the risk inherent in the industry. Improved training at all levels is required. Finally, the development of a "brand" and brand alliances with other, similar enterprises can promote the development of the whole industry.

7.5 Summary

China's bamboo construction industry faces many challenges, but there are also many opportunities. In view of the unique role played by bamboo construction in alleviating poverty and improving people's livelihood, developing a green economy and coping with climate change, the healthy development of China's bamboo construction industry can make a unique contribution to deepening South-South cooperation, promoting the construction of "one belt and one road", implementing the 2030 Agenda of Sustainable Development, and building a clean and beautiful world. Only by recognizing the competitive advantages and disadvantages of bamboo architecture, innovating bamboo processing technologies and equipment, training professional personnel, and improving the industrial chain, can China's bamboo construction industry develop and prosper.

References

1. Food and Agriculture Organization of the United Nations (FAO). (2020). 2020 Global Forest Resources Assessment (FRA). http://www.fao.org/forest-resources-assessment/en/.
2. Zhou, C. (1998). *Bamboo cultivation.* China Forestry Publishing House (in Chinese).
3. INBAR. (2019). *Trade Overview 2018: Bamboo and Rattan Commodities in China.* https://resource.inbar.int/upload/file/1577946385.pdf.
4. Xiao, Y., & Li, J. (2015). The state of the art of modern bamboo structures. *Industrial Construction, 45*(4), 1–6 (in Chinese).
5. Chen, G., Xiao, Y., Shan, B., et al. (2011). Design and engineering application of modern bamboo single house. *Industrial Construction, 41*(4), 66–70, 74 (in Chinese).
6. Akinbade, Y., & Harries, K. A. (2021). Is the rule of mixture appropriate for assessing bamboo material properties? *Construction and Building Materials, 267*, 120955.
7. Chen, F. (2014). *Technology and theory of bamboo bundle laminated veneer lumber with continuous plate process.* Beijing: Chinese Academy of Forestry (in Chinese).
8. Ren, H., Zhou, H., Fei, B., et al. (2006). Opportunity and challenge on development of modern wood frame house in China. *Wood Processing Machinery, 1*, 28–31 (in Chinese).
9. Wang, G., Jiang, Z., Chen, F., et al. (2014). Manufacture situation and problem analysis on large size bamboo engineering material in China. *China Forest Products Industry, 41*(1), 48–52 (in Chinese).
10. Harries, K. A., Ben Alon, L., & Sharma, B. (2019). Chapter 4: Codes and standards development for nonconventional and vernacular materials in nonconventional and vernacular construction materials. In Harries, K. A. & Sharma, B. (Eds.) *Characterisation, properties and applications* (2nd ed.). Woodhead (Elsevier) Publishing Series in Civil and Structural Engineering.

GPSR Compliance
The European Union's (EU) General Product Safety Regulation (GPSR) is a set of rules that requires consumer products to be safe and our obligations to ensure this.

If you have any concerns about our products, you can contact us on

ProductSafety@springernature.com

In case Publisher is established outside the EU, the EU authorized representative is:

Springer Nature Customer Service Center GmbH
Europaplatz 3
69115 Heidelberg, Germany

www.ingramcontent.com/pod-product-compliance
Ingram Content Group UK Ltd.
Pitfield, Milton Keynes, MK11 3LW, UK
UKHW022213051026
15624UKWH00001B/143
9789811683084